Generations at Risk

D0702201

Generations at Risk

Reproductive Health and the Environment

Ted Schettler, M.D., M.P.H., Gina Solomon, M.D., M.P.H., Maria Valenti, and Annette Huddle, M.E.S.

The MIT Press
Cambridge, Massachusetts
London, England

First MIT Press paperback edition, 2000
© 1999 Massachusetts Institute of Technology

All rights reserved. No part of this book may be reproduced in any form by any electronic or mechanical means (including photocopying, recording, or information storage and retrieval) without permission in writing from the publisher.

Illustrations by Lori Messenger.

This book was set in Bembo by Achorn Graphic Services, Inc., and was printed and bound in the United States of America.

Library of Congress Cataloging-in-Publication Data

Generations at risk : reproductive health and the environment / Ted Schettler . . . [et al.].
    p.   cm.
    Includes bibliographical references and index.
    ISBN 0-262-19413-9 (hardcover : alk. paper), 0-262-69247-3 (pb)
    1. Reproductive toxicology. I. Schettler, Ted.
RA1224.2.G46   1999
615.9′02′09744—dc21   98-32432
                                                CIP

In every deliberation we must consider the impact on the seventh generation.
—Six Nations Iroquois Confederacy

Contents

Acknowledgments

We thank the following people who contributed written material to some portion of this book: Paul Burns, Terry Greene, Sarah Handeyside, and John Loretz.

We gratefully acknowledge the following people who reviewed early drafts of this book and the report on which it is based, noting that their review does not constitute an endorsement of the recommendations or findings of the book: Frank Bove, Paul Burns, Theo Colborn, Maxine Garbo, Carolyn Hartmann, Howard Hu, Jonathan Kaplan, Michael McCally, Lawrie Mott, Paul Orum, Laurie Sheridan, and Ellen Silbergeld. Their thoughtful comments have helped to make this book more concise and richer in content. We are also grateful for the insightful comments and suggestions of five additional reviewers who remain anonymous.

The Board of Greater Boston Physicians for Social Responsibility (GBPSR) provided early support for the work that has become this book, and for that we thank its members. A number of people participated in discussions with us during the writing of the original report and this book that provided greater insight into and information about issues, and we appreciate their input: Ruth Hennig, Cynthia Lopez, Edward McNichol, Jr., Peggy Middaugh, Cynthia Palmer, Rachel Pohl, Shanna Swan, and Gary Timm.

We are most grateful to the Jessie B. Cox Charitable Trust and the John Merck Fund for major funding for the original report on which this book is based and to the Cox Trust for continued funding for GBPSR's Reproductive Health and Environmental Exposures Program; to the W. Alton Jones Foundation for helping to promote awareness of the book and its contents; and to the Pew Charitable Trusts and the C. S. Mott Foundation for early support of the original report.

Introduction

In the early 1950s, when the post–World War II industrial boom was not yet tempered by concerns about environmental or health threats, reports began to emerge from a village in Japan about a strange neurological disease. Particularly affecting infants and children, the condition bore a striking resemblance to cerebral palsy. As the story unfolded, it gradually became clear that fetal development was being poisoned by a chemical toxicant. Investigators discovered that mercury had been released into Minamata Bay for years from a vinyl chloride plant. This toxic metal had been transformed into organic mercury in the sediments of the bay and accumulated to dangerous levels in the fish eaten by the villagers. Organic mercury interferes with the function and development of the brain. This local epidemic was an early signal that human reproduction and infant development may be vulnerable to the effects of chemical toxicants.

Meanwhile lead, another toxic metal, was being added to gasoline in the United States. Emitted from tailpipes throughout the country, lead was polluting the air, accumulating in water and soil, and insidiously making its way into the blood, bones, and brains of every person in this country. The epidemic caused by the subtle effects of this toxicant on the fetal and infant brain was not recognized for another three decades.

The generations since the Minamata Bay tragedy have been affected by numerous other epidemics of reproductive or developmental dysfunction due to chemical exposures. In the 1960s, mercury-contaminated grain resulted in an outbreak in Iraq that resembled the one seen in Minamata. In Japan and Taiwan, cooking oil contaminated with poly-chlorinated biphenyls (PCBs) resulted in major health effects in new-borns and children.

In the 1970s, a group of pesticide manufacturing workers in the United States was rendered sterile from a chemical known as DBCP. Young women whose mothers had taken diethylstilbestrol (DES) during pregnancy were developing vaginal cancers and other reproductive problems. Meanwhile, returning Vietnam veterans were wondering what effects exposure to Agent Orange might have on their fertility and in their children, and citizens in places such as Love Canal, New York, and Woburn, Massachusetts, were creating a new grass-roots environ-mental movement, due in part to reproductive and childhood develop-mental problems they believed stemmed from chemical exposures in their community.

These epidemics continue to occur, yet it is the day-to-day encoun-ters with individuals concerned about chemical exposures and their own reproductive dysfunction that have prompted us to write this book. Some anecdotes come to mind:

- A young woman came to see one of us in consultation, concerned because she was pregnant and exposed to solvents at work. Her em-ployer refused to allow a job transfer, and the patient was subsequently overexposed as a result of a chemical spill. Her fetus was stillborn a few weeks later.
- A colleague asked one of us about her brother. He and his wife had never been able to have children, and he wondered whether his job might have had something to do with this infertility. The company where he worked is a major emitter of glycol ethers, chemicals known to cause testicular toxicity in animals.
- A pregnant patient was seen in clinic with a blood lead level revealing lead poisoning. An environmental investigation had failed to uncover any obvious source. She was considering having an abortion out of fear that lead would poison her fetus. The medical staff was frustrated at the inability to identify the source of the exposure and distressed over the decision confronting this patient.
- A woman recounted that she had operated a home business many years ago that involved handling large amounts of solvents in her kitchen.

She had had a miscarriage at the time, but it was years before she realized that there may have been a connection with her chemical exposures.

- A pregnant colleague contacted one of us; she was anxious because there had been a plane crash at an airport a few miles from her home, and her home had been enveloped in choking smoke for hours. She had heard that dangerous chemicals could be emitted from burning plastics and other materials and wanted to know if there was a risk to her fetus.

Anecdotal experiences help to shed light on the kinds of situations some people are encountering and the kinds of questions some people are asking. In most of the situations described above, it is almost impossible to attribute a specific outcome to a specific exposure, just as it is almost impossible to ascertain an individual risk from an exposure. Yet we do know that people are being exposed to substances that may affect reproduction and that they are experiencing reproductive health problems. At least one in twelve couples in the United States is dealing with infertility, 35 to 50 percent of all pregnancies result in spontaneous abortion, and major structural birth defects currently complicate 2 to 3 percent of all live births.[1-3] There is evidence that some of these problems are increasing: sperm counts are declining in many regions, certain birth defects are increasing in incidence, and childhood cancers, whose origin may be during fetal life, are also on the rise.[4,5]

Reproductive and developmental health problems are often unexplained by known risk factors, among them, genetics, alcohol consumption, smoking and drug use, and exposures to nonchemical factors such as radiation, infection, heat, and stress. This book focuses on the associations between environmental and occupational chemical exposures and reproductive and developmental health problems. In some cases, the evidence of a cause-and-effect relationship between exposure to certain substances and reproductive or developmental disorders is compelling. In others, the science is less clear, but a developing body of evidence is suggestive. Often crucial missing information makes it impossible to draw definitive conclusions. These data gaps may result from incomplete toxicity testing, inadequate or inconclusive epidemiological studies, personal or study biases, funding priorities, or even misapplication of research methods.

Although it has been known for decades, even centuries, that various harmful health effects may result from environmental exposures, cancer has often dominated the public agenda. There is also ample reason for concern about other health outcomes, which may be subtle, delayed, difficult to diagnose, and not easy to link causally to specific exposures. Reproductive and developmental effects are of concern because of important consequences for couples trying to conceive, and because exposure to certain substances during critical periods of fetal or infant development may have lifelong and even intergenerational effects.

Health care personnel are often ill equipped to recognize, much less treat, illnesses with environmental causes, since medical education pays inadequate attention to the relationship between human health and the environment. A comprehensive awareness of that relationship requires an understanding that illness is not just an individual condition but also a public health concern. Practicing medicine with an expanded public health perspective offers additional insights and opportunities to serve the health needs of both individuals and society. This approach does not shy away from using appropriate political action as a tool for protecting human health.

Just as many people assume that their physicians are able to diagnose and treat environmentally related conditions, they also assume that government agencies are protecting them from environmental hazards. Many people believe that if chemicals are sold in consumer products, or are present in the air, food, or water, then these chemicals have been adequately tested and found to be safe. In fact, government oversight of consumer products and other industrial chemicals is generally poor. The burden often falls on a regulatory agency to prove an exposure unsafe rather than on the industry to show that the chemical they are distributing to the environment is safe. The result is human exposure to untested or poorly tested materials for economic and political reasons.

Our analysis reveals that scientific, social, economic, and political systems often unintentionally cooperate in this human experiment. We have also identified extensive gaps in scientific data about the health effects of nearly all suspected reproductive toxicants; alarming instances of institutional suppression of information about health effects, resulting in tragic human consequences such as sterility; archaic scientific and policy models that are no longer adequate to address the magnitude of the problem; and regulatory fragmentation, failures, and weaknesses that have not been protective of public health.

The body of information we have to make informed decisions is characterized as much by what we do not know as by what we do. If we are to protect the generations to come, we must act boldly to correct past mistakes, replace dysfunctional models with new ones fit to the task, and reexamine our social and public health priorities.

How to Use This Book

This book is the result of a collaboration between public health professionals, physicians, environmental educators, and policy advocates. It is designed to bring scientific information to the public in a readable form while also serving as a resource to the medical, public health, and activist communities, policymakers, and industry. We are seeking a common language and scientific foundation that will allow activists, medical professionals, and regulators to talk to one another rather than continuing the saga of miscommunication and fragmentation that characterizes the history of public health protection efforts.

This book grew out of a Massachusetts report published by Greater Boston Physicians for Social Responsibility and the Massachusetts Public Interest Research Group Education Fund. The current version retains the original goals of raising awareness about known and suspected threats from reproductive toxicants in our environment and bringing a broader public health perspective to clinical medicine, scientific research, and regulatory activity.

Several levels of detail are offered throughout this book. When appropriate, we provide an analysis of what the information means, what is missing, and why. This book does not offer absolute answers. Taken as a whole, it summarizes the current state of the science and provides information with which to make informed decisions. It is designed to help readers assess the health risks of certain chemicals and to provide tools and resources for health-protective action. Organized in a modular format, it can be read from cover to cover or used as a reference on specific issues.

Part I outlines concepts in reproductive physiology and toxicology, and points of vulnerability in human reproduction and development.

These chapters provide an important foundation for some readers but will be unnecessary review for others. Next is a discussion of the role and limitations of science in decision making, including discussions and critiques of toxicology, epidemiology, and risk assessment.

Part II consists of reviews of metals, solvents, and pesticides. It is impossible to address the reproductive toxicity of all substances to which humans are potentially exposed. With over seventy-five thousand chemicals on the Environmental Protection Agency's chemical inventory, the task is enormous. In many cases, the health effects of these chemicals are unstudied and unknown. Consequently we have focused on classes of substances to which people are regularly exposed. The section on endocrine disruptors, which spans numerous classes of chemicals, outlines our current understanding of an important emerging area of concern.

The risk of an adverse health effect depends on more than the presence of a hazardous substance. People must also be exposed to the substance. All too often accurate assessments of human exposure are unavailable, making the likelihood of harm impossible to estimate. In chapter 7, on human exposure, we trace the path of the chemicals previously discussed, starting from data on production and sales, through releases to the environment, levels in food, water, and air, and studies on chemicals in exhaled breath, blood, urine, fat, and breast milk.

Part III includes a guide to the regulatory system—its history, major agencies, and laws. Identified weaknesses of the system lead to a failure to protect public health adequately. Finally, there are two practical guides: one designed for members of the public who are concerned about potential exposures at work, at home, or in their community (chapter 9) and the other for health care workers. These chapters provide detailed guidance for those seeking to address specific issues.

In this book, we have identified a pattern of continuing exposure to some known, highly likely, or suspected chemical reproductive and developmental toxicants. The potential consequences of these exposures are largely unknown to the general public, occupationally exposed workers, and health care providers. One of our goals is to shed some light on this important topic for those who wish to make more informed decisions. Beyond that, we consider this material an example of the need for a precautionary, public health–oriented perspective for analyzing and controlling human exposures to environmental toxicants.

Understanding and Using the Science: Reproductive Physiology and Toxicology

I

Reproductive and Developmental Physiology

The reproductive system is a complex, interconnected set of organs, tissues, and hormones that together make possible the birth of new generations. Not only the structure but also the function of the reproductive system may be harmed by exposure to environmental agents. Reproductive toxicants may directly damage the ovaries, testes, or other critical organs and cause abnormalities in sperm, ovulation, or hormone levels. The genetic makeup of an individual (genotype) only partially determines the course of fetal and childhood development. Ultimately, the outward expression of genetic makeup (phenotype) is a result of interactions between the genotype and the environment. Embryonic or fetal development may be altered by developmental toxicants, resulting in problems that range from birth defects to neurobehavioral disorders to cancer.

We have learned about normal and abnormal reproduction and development from observations in humans and other mammalian species with remarkably similar processes. From the earliest moments of its development, the reproductive system is exquisitely fine-tuned, and at times it is quite vulnerable. Similarly, normal fetal development depends on an orchestrated cascade of events characterized by periods of vulnerability to disruption. Communications among organs or tissues, essential for normal reproduction and development, are mediated by chemical messengers, including hormones.

The following brief summary of normal human reproduction and development, and their vulnerability to hazardous exposures, sets the foundation for the chapters that follow. Readers interested in more detailed information can consult a general textbook of reproductive physiology, endocrinology, or toxicology.[1-4]

Hormones

Coordinated function of the organs of the reproductive system is essential for reproductive success. Hormones, which are chemical messengers circulating in the bloodstream, cycle and establish feedback loops that integrate function. Feedback loops can be positive or negative. A thermostat regulating room temperature is an example of a negative feedback loop. As the temperature rises, the thermostat shuts the furnace off, allowing the room to cool to a predetermined level, at which time the thermostat signals the furnace to reignite. Negative feedback loops in hormone systems tend to maintain fairly constant levels of the circulating chemical. A positive feedback loop differs in that a rising level of a hormone results in even more being produced. Although they are less common than negative feedback loops, those that are positive are essential to some reproductive processes—for example, the events immediately preceding ovulation.

Hormones exert their effects by binding to specific molecular receptors located on the surface of or inside cells. Their ability to influence the biochemical inner workings of a cell depends on attachment to these receptors. When the hormone attaches to its specific receptor, much like a key fits into a lock, the linkage causes changes in the shape of the receptor, triggering a cascade of biochemical events. This sequence may result from a small amount of hormone attaching to few receptors.

Peptide hormones, which are proteins composed of a string of amino acids and include luteinizing hormone (LH) and follicle-stimulating hormone (FSH) from the pituitary, attach to receptors on the cell surface (figure 1.1). Steroid hormones, like the sex hormones testosterone, estrogen, and progesterone, are derived from cholesterol. They pass through the cell membrane, attach to their specific receptors in the cell, and are transported to the cell nucleus (figure 1.2). The hormone-receptor complex interacts directly with DNA in the nucleus, triggering genes to produce their programmed chemicals (gene products).

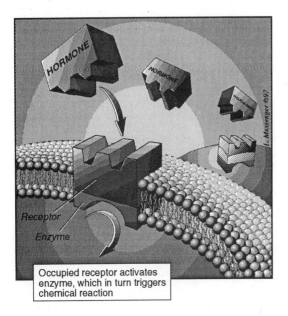

Occupied receptor activates enzyme, which in turn triggers chemical reaction

Figure 1.1 The lock-and-key model of hormone-receptor interaction necessary for a hormone to trigger biochemical activity in a cell.

Hypothalamic–Pituitary–Gonadal System

The hypothalamus, a portion of the base of the brain, produces its own hormones, which heavily influence pituitary output. In women, a pulsatile release of gonadotropin-releasing hormone (GnRH) controls production of FSH and LH from the pituitary, lying just beneath the brain. These hormones stimulate the ovaries to produce estrogen and to ovulate and the testes to produce testosterone and sperm.

In order to keep the system balanced and in check, estrogen and testosterone circulate back to the pituitary and hypothalamus in feedback loops, fine-tuning pituitary hormone levels (figure 1.3). Normal functioning of a feedback loop may be disrupted at any point by chemicals, drugs, malnutrition, or other factors that cause a change in hormone production.

In men, the loop maintains testosterone and pituitary hormones at fairly constant levels (figure 1.4). But in women, at a critical level of estrogen, the feedback loop becomes positive, and a surge of pituitary hormones stimulates the ovary to release an egg, which may then be

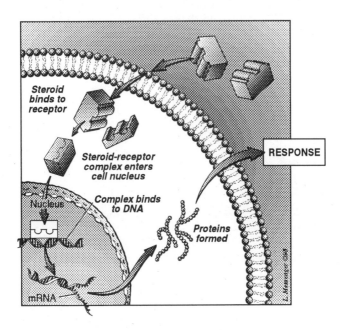

Figure 1.2   Steroid hormones attach to an intracellular receptor. The hormone receptor complex then attaches to the DNA, triggering gene transcription and a biochemical response.

fertilized (figure 1.5). The ovarian follicle from which the egg is released, now known as a corpus luteum, continues to produce both estrogen and another hormone, progesterone. The two once again suppress the pituitary. If the egg is not fertilized, the corpus luteum degenerates, and the uterine lining is shed during menstruation. The pituitary once again begins to stimulate the ovary to produce estrogen in the next menstrual cycle.

Prolactin is another pituitary hormone with a variety of actions. Its primary and best-known role is maintenance of milk production by the breast during lactation. Instead of requiring stimulation, prolactin secretion is spontaneous and is regulated by the inhibiting effects of dopamine, a chemical reaching the pituitary from the hypothalamus. In this case, reduced levels of dopamine result in increased prolactin production. Although all the functions of prolactin are not well understood, it is also involved in testosterone production by the Leydig cells in the testes of male mammals. Elevated levels of prolactin may be associated with di-

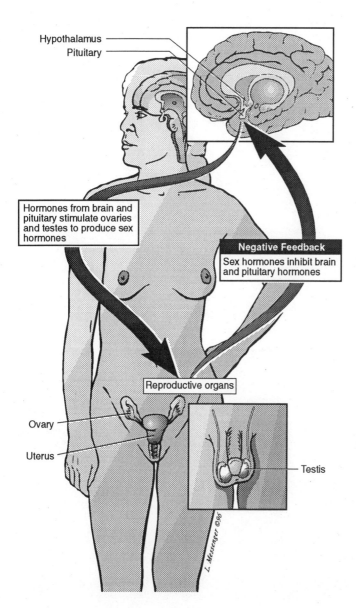

Figure 1.3   A negative feedback loop in hypothalamic-pituitary-gonadal (HPG) hormonal communication tends to keep sex hormones at constant levels. In males, the feedback loop is always negative. In females, it fluctuates between negative and positive.

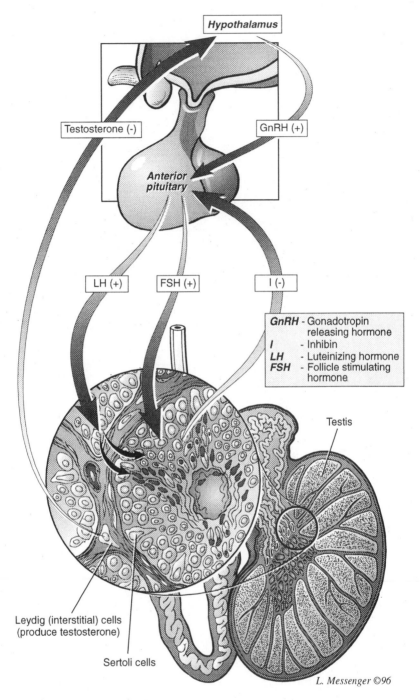

Figure 1.4   Negative feedback loops characterize the male hypothalamic-pituitary-gonadal system. + = stimulate; − = inhibit.

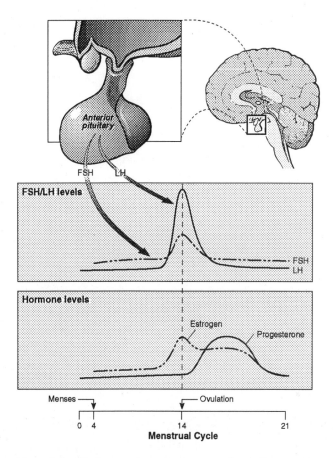

Figure 1.5 Pituitary and ovarian hormone levels fluctuate throughout the female menstrual cycle showing both negative and positive feedback loops. FSH (follicle-stimulating hormone) and LH (luteinizing hormone) are from the pituitary, estrogen and progesterone from the ovaries.

minished FSH and LH secretion, causing disorders of ovulation or infertility.

The Ovaries

The two primary functions of the ovaries are to produce eggs and the hormones estrogen and progesterone. An infant girl is born with all of the immature eggs in her ovaries that she will ever have. Ovaries consist of follicles, each containing an immature egg (the germ cell or oocyte) surrounded by an envelope of cells capable of producing hormones. With the onset of each menstrual cycle, and in response to hormonal stimulation from the pituitary, a group of follicles begins to mature in the ovaries. Eventually one of them releases its egg at the time of ovulation; the others deteriorate.

If the egg is fertilized, dividing cells of the new embryo produce human chorionic gonadotropin (HCG), a hormone that maintains the corpus luteum, enabling continued preparation of the uterus for implantation.

The Testes

The testes serve two important functions: production of sperm (spermatogenesis) and hormones. They consist of several different kinds of cells: (1) the germ cells or immature sperm, (2) those that produce the hormone testosterone (Leydig cells), and (3) those that protect and nourish the developing sperm (Sertoli cells). Sertoli cells help to form a blood-testis barrier that isolates the developing sperm from harmful substances. LH stimulates testosterone production by Leydig cells. FSH enhances the effects of LH on the Leydig cells and also maintains Sertoli cells, which are necessary for sperm production. Sertoli cells also produce inhibin, a hormone that reduces FSH production, forming a negative feedback loop.

Males continue to produce sperm throughout their lifetime, provided the immature germ cells (stem cells) remain healthy. Immature sperm mature in seminiferous tubules in the human in about seventy days.

Normal Fetal Development

General Principles

The newborn infant consists of millions of different types of cells that come from a single cell: the fertilized egg. Usually they arrive at the right place at the right time, functioning normally. This extraordinary series of events is accomplished by a number of mechanisms, including physical, electrical, and chemical signaling among cells and organs as they develop.

Early in development, each cell is flexible and capable of developing in a variety of ways. For example, the heart, kidney, and intestine all develop from the same primitive cell type. At an early stage of embryonic development, each of the primitive cells still has the capacity to develop into each of the three organs. Later in development, as the cells continue to divide and become more specialized, that diverse capacity is lost. The very young embryo therefore is often less susceptible to structural birth defects than to lethal damage from exposure to chemicals or radiation. If a chemical or physical injury does not kill it, it may still be able to develop into a normal infant since the cells still have considerable flexibility, and damage can be repaired as uninjured cells take over the role of those that are injured.

Formation of Organs (Organogenesis)

Not all parts of the body are formed at the same time. Development of the eye and brain begins first, followed weeks later by the palate and genitals. Biochemical enzyme systems, which are important for the metabolism of toxicants and drugs, the reproductive system, and the immune system, continue to mature throughout pregnancy and well after birth. The period of time necessary to complete basic brain nerve connections extends from the first part of pregnancy until several years after birth.

Sexual differentiation of the body and brain begins early. Müllerian inhibiting substance (MIS), another Sertoli cell hormone, promotes male sexual development by suppressing the cascade of events leading to female characteristics.

The very early embryo has the capacity to develop into a child of either sex but will develop as a female unless specifically directed toward maleness by genetically determined hormones. Male sex is determined by the presence of a Y chromosome (genotype). Normal expression of the male genotype (phenotype) depends on adequate MIS and testosterone, which masculinize many different organs and tissues, controlling endocrine function and sexual behavior. Inadequate production of MIS or testosterone may lead to incomplete masculinization of the male fetus.

Sexual differentiation of the brain also takes place during fetal development (figure 1.6). Fetal testes produce testosterone and chemically alter the hormone to dihydrotestosterone (DHT), which continues to masculinize the genitals. In the brain, testosterone is converted by the enzyme aromatase to estrogen, which is largely, but not exclusively, responsible for masculinizing nerve and hypothalamic-pituitary-gonadal connections. Although we are accustomed to thinking of estrogen as a female hormone, in the fetal and childhood brain estrogen is necessary for male-type brain development. Diethylstilbestrol (DES), an estrogen-like compound, given to female rats soon after birth, will masculinize the hypothalamus.[5] Estrogen receptors are present not only in the hypothalamus but also in other portions of the brain, such as the cerebral cortex, responsible for more advanced neurological functions.

Development of the hypothalamus is critical during the fetal period and early years of life. It bathes the pituitary with its hormone-regulating chemicals and during development sets lifelong baseline levels of pituitary hormones.

The adult brain is partially protected from toxic substances by a blood-brain barrier, which keeps many chemicals circulating in the blood from coming into contact with brain tissue. The embryo, however, has no blood-brain barrier. It is not complete until about six months of life in humans (three weeks in rats—an important difference for toxicity testing of chemicals because the brain of rodents is better protected by the blood-brain barrier at a younger age than the human infant). Even then, the hypothalamus has no blood-brain barrier and remains vulnerable throughout life.[6]

This sequence of events suggests that many functions that develop more fully later in life, as an individual matures sexually, are largely determined during fetal life and early childhood—when the brain is developing its lifelong tendency for receptor and hormone levels and when it is less fully protected from toxic exposures by the blood-brain barrier.

How Toxicants Can Affect Reproduction and Development

Any of the organs or processes whose coordinated function is essential for normal reproduction and subsequent development of the fertilized egg is a potential target of toxic exposures. Normal reproductive func-

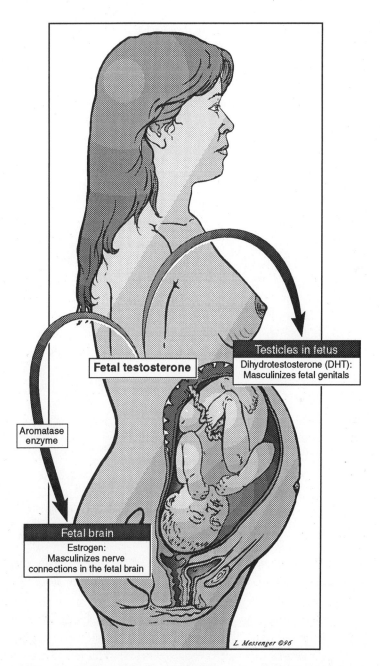

Figure 1.6   Testosterone from the fetal testes is chemically altered in two ways. It is converted by an enzyme to estrogen, necessary, along with testosterone, for masculinizing the male fetal brain. It is also converted by a second enzyme to DHT, required for further male genital development.

tion requires timing, balance, properly set feedback loops, and communication among cells and organs from the time of conception through the reproductive years. Normal fetal, infant, and child development depends on genetic makeup, a healthy environment, and interactions between the two. There are numerous opportunities for disruption.

Reproductive Toxicology

Toxic chemicals may directly damage cell structure or biochemical function. Some trigger the production of enzymes that transform other chemicals into more toxic substances. This mechanism often explains how mixtures of chemicals may be more harmful than individual exposures. Some agents require metabolism or breakdown into a different chemical before causing harm. In those cases, test tube studies that do not provide for metabolism will fail to reveal toxicity.

Chemicals may also exert harmful effects if they are similar to normally occurring compounds. For example, if a portion of a chemical molecule structurally resembles a hormone, it may stimulate or block the hormone's receptor, either triggering a cascade of inappropriate events or blocking events required for normal function. A small amount of a hormone mimic or antagonist may influence a system that functions by amplifying the effects of individual hormone-receptor linkages.

Toxic Effects on the Ovaries

Any of the basic cell types in the ovary may be damaged by toxic exposures, potentially interfering with either of the two basic ovarian functions: hormone synthesis and ovulation. In humans, malfunction of ovulation results in abnormalities of the menstrual cycle or reduced fertility. These, of course, may have causes other than toxic exposures. Since every woman is born with all of the ovarian follicles and eggs that will be available throughout her reproductive life, any toxicity that results in egg or follicular destruction may result in premature menopause. Egg or follicular toxicity is difficult to detect since there are always significant numbers of follicles that degenerate in each menstrual cycle. Even in animal studies, where the ovaries are microscopically examined after exposure of the animal to a toxicant, it is difficult to determine if an excess number of follicles has been lost.

Abnormalities of hormone production may result from damage to cells in the ovarian follicles, leading to abnormal function of the corpus luteum. Recall that the primary role of the corpus luteum is to produce the hormones necessary for implantation of the fertilized egg in the uterus and for early development of the placenta and fetus. Therefore abnormal luteal function may result in very early spontaneous abortion. Examples of substances that interfere with corpus luteum function are estrogens and estrogen-like compounds, which suppress progesterone production, and polycyclic aromatic hydrocarbons, which are found in cigarette smoke and in the products of fuel combustion.

Toxic Effects on the Testes

Chemicals toxic to the testes may decrease sperm counts, damage sperm, cause infertility, or alter hormone production. Each of the various cell types is a potential target. Leydig cell toxicity results in decreased testosterone production. Sertoli cell toxicity indirectly lowers sperm counts since these cells are necessary for the health of immature sperm. A toxicant that reduces the number of Sertoli cells formed during fetal development results in permanently lower sperm counts.[7]

The blood-testis barrier may be disrupted. This barrier results, in part, from tight connections between Sertoli cells; when it is intact, it protects immature sperm from toxic exposures. Its breakdown may result in sperm damage. In sufficient quantity, lead, cobalt, cadmium, and other toxicants are harmful to the blood vessels in the testes, also causing damage to the Leydig cells and seminiferous tubules.[8] As sperm cells mature through progressive stages of development, their susceptibility to toxicity varies.

Much about sperm production and cellular interactions in the testes is not well understood. Toxicity studies usually examine the number and quality of sperm. Both the form (morphology) and activity of sperm are important factors, although toxicologists disagree about which is more predictive of abnormal fertility or pregnancy outcome in humans.

Emerging evidence also indicates that fetal exposures to chemicals with hormone-disrupting properties may predispose an individual to the development of testicular cancer years later.

Toxic Effects on Hypothalamic-Pituitary-Gonadal Connections

Chemicals that are toxic to the hypothalamus or pituitary gland may cause malfunctions resembling those associated with direct toxicity to the ovaries or testes. Examples are toluene and other organic solvents, a wide variety of pharmaceutical drugs, marijuana, and environmental agents that possess hormone-like activity.[9]

Developmental Toxicology

Abnormal fetal development may result from toxic exposures before or after fertilization of the egg. Chemicals that alter the genes on chromosomes in the sperm or egg may cause developmental abnormalities.[10] Alternatively, fetal exposure to chemicals that cross the placenta may directly disrupt development. Fetal death, altered growth, and structural or functional defects are possible outcomes. Visible deformities (e.g., cleft lip or palate) and those detectable on additional physical examination (e.g., congenital heart disease) or at autopsy are structural defects. The causes are often obscure. In humans it is estimated that about 50 percent of major malformations are due to genetic or inheritance abnormalities, 3 to 4 percent to a specific, identifiable toxicant, and over 40 percent to unknown causes.[11]

Functional developmental defects are not necessarily visible but are revealed in the way that various organs or systems work. A normal-appearing brain, pituitary, or thyroid gland may actually function abnormally because of a fetal toxic exposure or genetic abnormality. For example, fetal lead or mercury exposure may interfere with neurological function. In animal studies, small exposures to dioxin during a critical time of fetal life permanently alter certain hormone levels.[12] There is some evidence that childhood cancer may result from fetal exposures (see chapter 5).

Developmental neurotoxicity—one kind of functional damage—is an important subject of study. Some chemicals with only a temporary or weak effect on the brain in adults may have permanent effects on the developing brain, influencing intellectual function, sexual differentiation, and behavior. Fetal exposure to hormones or hormone mimics at levels that have no effect on adults may inappropriately set lifetime base-line levels of hormone production and hormone receptors. Some fetal exposures influence the development and competence of the immune system with lifelong implications for the offspring.

Principles of Abnormal Development

Several principles of teratology (the study of abnormal development) provide a basis for understanding the developmental toxicity of chemicals:

1. Abnormal development may result in fetal death, altered growth, malformation, or functional disorders.
2. The likelihood of abnormal development depends on the genetic makeup of the embryo, its environment, and interaction between the two.
3. The ability of toxicants to cause abnormalities varies with the *stage* of fetal development at the time of exposure. Moreover, the minimal amount of toxicant required to harm the embryo or fetus may change with the stage of development. Some toxicants are lethal to an embryo at one stage of development but cause structural or functional deficits at other stages. Some can cause structural defects in offspring even when the mother is exposed at the time of fertilization or preimplantation.[13] Recent studies demonstrate that there may actually be multiple periods of particular susceptibility to developmental harm in specific organs.[14]
4. The physical-chemical nature of a potential toxicant determines its access to developing tissues. An environmental agent must be able to cross the placenta to affect a developing fetus directly. Most maternally administered substances have the potential to cross the placenta but at varying rates, depending on the chemical nature of the agent.[15] A toxicant may exert its adverse effects only if it is first absorbed, ingested, inhaled, or otherwise internalized and makes its way to a target tissue. If it is not blocked or metabolized into a harmless substance, is present in sufficient quantity, the timing and duration of the exposure are appropriate, and the target tissue is unable to repair the damage, it may then exert an adverse effect.

Identifying and Understanding Adverse Outcomes

Some adverse reproductive outcomes are difficult to identify, and many are caused by multiple factors, making it difficult to attribute a particular outcome to a particular cause or event. For example, early miscarriages are frequently unrecognized, even by the mother, who may believe that she has reduced fertility or is experiencing a slightly delayed menstrual period. Furthermore, miscarriages are common, occurring in up to 35 to 50 percent of all pregnancies.[16] Genetic studies indicate that the most common cause of recognized early pregnancy loss is fetal chromosome abnormalities. Other causes include toxic exposures, maternal genital tract abnormalities, maternal illness, and immunological abnormalities. Therefore, even with extensive testing, determining the cause of a particular miscarriage may be very difficult.

Infertility or reduced fertility may result from maternal, paternal, or couple-dependent factors. Investigating the cause in any couple requires assessment of each of these possibilities, and toxic exposures are only one cause to consider.

Possible harmful developmental effects of toxic exposures include fetal or infant death; prematurity; malformations; retarded growth; neurological, metabolic, and immunologic abnormalities, abnormal sexual differentiation; and cancer. In populations, an altered sex ratio may be a readily discernable signal that a toxic exposure is differentially affecting male and female embryos. Some of these outcomes are rapidly and readily apparent, while others are delayed or subtle. Delayed diagnoses make it difficult to link an adverse developmental outcome to an earlier toxic exposure. Multiple possible causes for each outcome make it difficult to determine the role of a potentially harmful exposure with certainty.

Interpreting Toxicity Studies

Most studies of reproductive toxicants have been conducted in animals, but there is often uncertainty about how well animal tests predict human toxicity. Biochemical and developmental pathways often differ from one species to another, and sometimes in ways not fully understood. The timing of brain growth spurts during fetal and infant life varies considerably, even among mammalian species.[17] Since actively growing and dividing cells are more vulnerable to chemical or environmental injury than resting cells, the timing of windows of vulnerability also varies among species. Translation of animal data to humans must be done with care.

There is no consensus among reproductive toxicologists about the relative importance of various outcomes or end points, particularly in developmental toxicity studies. Some scientists believe that one type of abnormal outcome in one species may be predictive of a different outcome in another species. Others are more concerned about one outcome than another. For example, there is no general agreement about the relevance of certain minor skeletal abnormalities, which are sometimes detected in developmental animal tests.

Many substances show some evidence of reproductive or developmental toxicity to animals if exposures are large. But this leaves unanswered whether those results have any relevance to humans at lower doses, particularly with long-term exposures, which begin in the uterus and continue through the reproductive years.

Throughout the intricate processes of the menstrual cycle, egg and sperm production, fertilization, implantation, and growth and development of the fetus, there are specific and often short time intervals when there may be particular susceptibility to low-dose exposures, undetected if studies are not properly designed. For example, mice exposed to a single low dose of dioxin on day 15 of pregnancy give birth to male offpsring with delayed descent of the testes and decreased sperm production, prostate weight, and testosterone levels.[12] At other times in pregnancy, either larger amounts are needed, or the effects are not seen at all. Furthermore, if developmental testing does not include careful examination for neurological, behavioral, or immunological changes in test animals, a chemical may inappropriately be classified as safe for human exposure.

In recognition of these uncertainties, the Environmental Protection Agency and Food and Drug Administration continue to make important modifications to their animal testing protocols for pesticides and pharmaceuticals. Nevertheless, significant data gaps remain unaddressed. Neurotoxicity testing of pesticides is still rarely required despite increasing concerns. For most industrial and consumer product chemicals, including many that are pervasive in the environment, toxicity information is sketchy at best and often completely missing.[18] Many chemicals in commercial use have had no reproductive or developmental testing. Moreover, the emphasis continues to be on chemical-by-chemical analysis rather than the cumulative and multiple exposures that characterize the real world.

The Role of Science in Public Health Decisions

Many people assume that science provides clear answers to guide personal and public health policy decisions. In fact, although science is an important tool, it is often a blunt tool, capable of answering only certain narrow questions. Without an understanding of the strengths and the limitations of science, and an awareness of how both can be used and manipulated, we risk making personal and public policy decisions that will not best protect our children and ourselves.

This chapter first examines what science and scientific inquiry can and cannot do, what kinds of answers they can and cannot provide. Following is a look at animal testing and epidemiological research, two forms of scientific inquiry that are central to decision making concerning environmental reproductive toxicants. Finally, we consider quantitative risk assessment, a tool meant to improve our ability to make decisions by incorporating both scientific knowledge and uncertainty.

The Scientific Method and the Issue of Proof

Science is a way of looking at the world, a system for increasing our knowledge of the world, a tool for making decisions about how to act in the world. Fundamentally, science provides a structure for exploration: the scientific method, a well-defined system that scientists worldwide recognize as the basis for scientific understanding.

The scientific method, a system of hypothesis generation and testing, is the same whether used by laboratory scientists, epidemiologists, wildlife biologists, or others. A scientist begins with a curiosity about some particular topic and a question. Based on the scientist's previous observations and experience, she develops a hypothesis about how, why, or whether a particular phenomenon occurs. Next, she develops a test, or an experiment, to see how her hypothesis holds up in the face of the evidence. In fact, she is testing two hypotheses: the second, called the *null hypothesis,* is essentially the negative of the hypothesis she has developed. If, for example, her hypothesis is that (1) treating rats with the solvent toluene will result in birth defects in their offspring, the null hypothesis will state that (2) treating rats with toluene will *not* result in birth defects. Although it may seem counterintuitive, the scientist proceeds from the assumption that the null hypothesis is true.

Many nonscientists are not aware of the role of the null hypothesis in the scientific method, which can lead to misunderstandings between scientists and nonscientists. In the logic of the scientific method, a scientist looking for evidence to support a hypothesis must do so by disproving the corresponding null hypothesis. In other words, if she finds in her experiment that rats treated with toluene do in fact give birth to many more offspring with birth defects than do untreated rats, then she can no longer say that treating rats with toluene will *not* lead to birth defects. She then rejects the null hypothesis, and her experimental results support, but do not prove, the original hypothesis.

When deciding whether she can reject the null hypothesis, the scientist will use specific statistical techniques to assess the results of her experiment to determine whether the results are statistically significant. Statistical significance is a measure of the likelihood that a result did not occur due to chance alone. If the scientist decides that her result is not statistically significant, she fails to reject the null hypothesis. She states that nothing statistically significant happened in her experiment to support her original hypothesis, and she must go on to develop another hypothesis or refine the current one.

If, however, she decides that her result *is* statistically significant, she rejects the null hypothesis. Now she is in a curious position: she has shown that her original hypothesis has been supported, but *she has not proved* her hypothesis. She has simply shown that she has not found any result that disagrees with her original hypothesis. This is the scientists' dilemma: they can only support, but never prove, a hypothesis. With more research and further experimentation, the original hypothesis can be refined and further supported, elevated to the status of a theory, but in the end, the truth of a hypothesis is a matter of judgment rather than proof.

Nonscientists can be frustrated when scientists are unwilling to give absolute answers. The inability to provide such answers is one of the limitations of the scientific method. The scientist, knowing that her method of analyzing the world cannot absolutely prove a hypothesis, will tend to be conservative in making a statement about the truth of a hypothesis. She knows that further research may reveal that the hypothesis is inadequate in some way, and so will state that the science is incomplete. The scientist may feel fairly confident that her hypothesis is valid if there is a good body of research supporting it and numerous other researchers have confirmed her results, but she still cannot say that she is absolutely sure. Many of the difficulties in using science in the decision-making process thus have their roots in the very nature of the scientific method.

Statistical Significance and Statistical Power

As we have seen, scientific studies start with the assumption that the null hypothesis is true. For example, a scientist could assume, as a null hypothesis, that maternal exposure to toluene is not related to spontaneous abortion. After designing and conducting a study looking at the issue, the scientist would apply a test of statistical significance to determine whether the results justify rejecting the null hypothesis. A test of statistical significance assesses how much the results of a particular study deviate from those predicted by the null hypothesis. These statistical methods say nothing about whether a particular result is likely to be true or false. They simply indicate the probability that a certain result would be seen by chance alone, if there was really no association between the exposure and the effect in question.

One of the most commonly calculated statistics in epidemiology is the p-value, usually expressed as a percentage in decimal notation. The p-value indicates the probability of obtaining a result as extreme as the observed result if the null hypothesis is actually true. For example, a study may show many more spontaneous abortions in mothers exposed to toluene during pregnancy, and the p-value for this result may be less than 0.10. This means that if there is no systematic error in the study, and if there is really no association between miscarriage and toluene exposure, the probability of obtaining a result such as this would be less than 10 percent. To most nonscientists, this would suggest that the result supports an association between toluene exposure and miscarriage. Scientists, however, hold their research to a higher standard. By convention, scientists do not report that their results are "statistically significant" unless the p-value is less than 0.05, or 5 percent.

The decision to set the standard of statistical significance at less than 5 percent is arbitrary. One could argue that setting the level at less than 50 percent makes sense, since at that point it becomes more likely than not that a result is not due to chance alone. Alternatively, one could set the standard at less than 1 percent, arguing that only at such a high level can we be confident that the result is not due to chance. The scientific community has chosen 5 percent as its convention for a number of reasons—reasons that may not always apply in the context of public policy decision making.

Sometimes the results of a study may be presented with a "95 percent confidence interval." The confidence interval defines a range within which the true result is highly likely to fall. This statistical tool can tell a scientist that if there is no source of bias in the experiment, and she repeats it one hundred times, the results are likely to fall within this range ninety-five times. If a study finds that toluene triples the risk of spontaneous abortion, the investigators may report a 95 percent confidence interval, indicating that toluene may do anything from increasing the risk of spontaneous abortion by 20 percent to increasing it sixfold. In statistical notation, this would be reported as OR = 3.0 [1.2–6.0], where *OR* stands for *odds ratio,* a statistical measure of likelihood. An odds ratio of 1 indicates no increased or decreased risk, and represents the null hypothesis in this case. A 95 percent confidence interval that includes the number 1 in the range is therefore considered not statistically significant. If the number 1 is not within the range, such as in the example, the results are reported as statistically significant. The 95 percent confidence interval is more informative about the possible range of results than is the *p*-value, but it is still tied to the convention of requiring a *p*-value of less than 0.05 to be considered significant.

When the *p*-value is greater than 5 percent or when the 95 percent confidence interval includes the null value of 1, a study is usually reported as negative, or as not having found significant evidence of an association. A negative study that fails to reach the conventional criteria for statistical significance is often considered to be a justification for ending research in a particular area. In our example, a study that finds more cases of miscarriage in mothers exposed to toluene but fails to achieve statistical significance might be interpreted to the public as a negative study, showing no association between toluene and miscarriage. This is a misinterpretation of the science and of the statistics. In fact, there may or may not be an association between toluene and miscarriage. The study assumed that there was no association (null hypothesis). It then found a result that might be found by chance alone with a greater than 5 percent likelihood, and on that basis failed to reject the null hypothesis. Because these statistical concepts are confusing (and often confuse even scientists) it is easy to lose track of the fact that an association, even if it is not statistically significant may still have important public health significance.

When a scientist determines that a result is statistically significant, a number of possible interpretations exist. There is still the possibility, though slight, that the result is due to chance, or that a bias in the study design resulted in a spurious association. Alternatively, the result may indicate a real association between an exposure and a result. The association may be causal, meaning that the exposure actually causes the result observed. However, it is also possible that the relationship between the two factors is not so clear and, in fact, a third factor, unrecognized in the research, comes into play. For example, an association between coffee consumption and risk of heart attack might be assumed to indicate a causal relationship. However, if heavy coffee drinkers also tend to be cigarette smokers, then the connection with heart attacks may actually have to do with the cigarette smoking, not coffee consumption. Such a third, hidden variable is called a *confounder*.

If a scientist decides that a result is not statistically significant, again there are a number of possible interpretations. There may truly be no relationship between the exposure and the result. Alternatively, there may be an association, but because of chance or a bias in the study, no association is recognized. An important possibility to consider, however, is that the experiment did not have enough statistical power.

Statistical power refers to the ability of a study or experiment to detect reliably the effects that are being examined. For example, if an outcome is rare, then a scientist needs a large study population to determine whether the rate of occurrence changes. A study conducted on one hundred lab animals is unlikely to detect a doubling of birth defect rates if the defects normally occur only once in a thousand births. Similarly, in epidemiological research, studies must have a large enough study population to recognize small differences in rates of disease. A study on a small population may be able to detect major changes, such as a tripling of the disease rate, but a larger population would be required to detect a 20 percent increase in disease.

The Limits of Science

The scientific method is used for two distinct purposes. To many scientists, its primary use is in the quest for truth—the search for a more complete and accurate understanding of the phenomena of our world. In order to avoid false assertions that would lead away from the truth, scientists have rigorous standards for judging their results. In the scientists' view, it is better to avoid drawing a conclusion than to draw a false conclusion. Nonscientists, however, count on this same tool to generate information to help make decisions about how to act, whether on a personal or a societal level. Rather than seeking absolute truth, the decision maker seeks guidance—evidence that will help justify one course of action over another.

Since scientists choose to be extremely cautious in the reporting of their results, decision makers are often in a difficult position. The information they need will almost always be incomplete, but in the interest of protecting health, they must decide whether to take action. The scientist is cautious in the search for knowledge; the decision maker may wish to be cautious in protecting health. A better understanding of the limits of the scientific method will help us all use existing scientific data to make health protective decisions.

Aside from the limitations of the scientific method itself, a number of other factors can limit the effectiveness or relevance of scientific research. Scientific research is a process driven by people's curiosity and interest in certain areas, and that curiosity will shape the resulting research. From its first step, the scientific method is entirely dependent on a number of human factors. The questions that get asked are informed by, for instance, the scientists' previous work, their ability to make new connections between ideas, the established understanding of that area of inquiry, the interest of the department head, and the source of the funding for the lab. Science can attempt to address only questions that are actually being asked; questions that are not asked will not be answered.

Even when certain questions are asked, the interpretation of the answers can be colored by an individual's perspectives and preconceptions. Although the concept of scientific objectivity is a central aspect of the scientific endeavor, all human beings, including scientists, interpret their world based on what they already know and believe. What one scientist perceives as a benign change in organ function, another may perceive as an injury to the animal. A surprising result may be written off as a mistake, contamination, or a fluke instead of being fully evaluated. For this reason, scientists must be extremely explicit about their assumptions and observations so that others may make their own judgments of their work.

In the end, we must recognize that while science can inform us in many ways, science cannot itself make decisions for us. Science is a tool. When it is used appropriately, it can provide some of the information we need in order to make decisions likely to protect health.

Animal Toxicology

Animal studies have long provided a foundation for the understanding of chemicals' effects on reproduction and development. As early as the 1930s, some food additives and pesticides were studied using animal testing, but those tests were not sufficient to demonstrate the full spectrum of reproductive and developmental toxicity. Animal testing has improved considerably since then, but there are still major gaps in the testing of many chemicals in use today.

Scientists primarily use small mammals in their testing protocols. Rats, mice, and rabbits have relatively short life cycles and bear multiple young, simplifying the task of looking for reproductive and developmental effects. In some cases, animals have been specifically bred to have certain traits that may increase or decrease their susceptibility. Extrapolating effects from one species to another, or to humans, is a difficult and uncertain process. In order to reduce the likelihood of missing a potentially dangerous effect, tests should be conducted on at least two different species.

Types of Studies

In general, reproductive toxicity animal tests fall into two categories: segment and multigenerational studies. *Segment studies* look at specific portions of the reproductive process, assessing the impact of chemical exposure on some particular feature. For example, a test might look at the effects of a chemical on the female reproductive cycle or on the development of sperm in the male animal. In addition to examining female and male reproductive function, these tests can also be used to examine the development of the offspring of exposed parents.

In *multigenerational studies,* animals are exposed to a substance and then allowed to reproduce. Succeeding generations are observed, and various measures of reproductive success are tracked. These measures may include fertility, ability to carry offspring through full pregnancy, delivery and rearing of offspring, size and sex of litters, and examination of the offspring's organs. *Continuous breeding animal studies* are a form of multigenerational studies. In these, animals are dosed with the substance being studied and are then allowed to mate. Dosing continues during mating and production of successive litters. The last litter is then dosed with the substance from the time of weaning, and then mated to examine their ability to reproduce.

Occasionally substances are tested for toxicity to the developing neurological system. Offspring are examined after exposure during pregnancy or nursing for changes in motor activity, noise startle responses, learning and memory, or changes in the physical condition of the brain and nervous system. Although these effects are extremely important, these tests are not routinely performed.

Weaknesses of the Research

Data collected from animal testing are used to set standards for human exposure. In assessing reproductive and developmental toxicity, investigators generally assume that there is a certain exposure level, or dose, that will not cause a health effect seen at a higher level. This is a threshold below which exposures are considered "safe" for the animal. Regulators then decide what exposure level they believe is safe for humans. In practice, they generally apply uncertainty factors—usually a factor of ten for the uncertainty about differences between species, and another factor of ten to take into account particularly sensitive individuals. This gives a total adjustment of ten times ten, or one hundred, leading to the conclusion that humans will be safe from the effect of the substance if the exposures are one hundred times less than the no-effect level in animals.

On the surface, this appears to be a conservative approach, one that should protect human health. However, there are a number of problems with the assumptions used. First, not all scientists are certain that "safe" thresholds exist.[1-2] Levels of exposure believed to be safe for laboratory animals may actually produce effects that are so rare or subtle that they are not detected in limited animal testing. Large numbers of animals would have to be tested in order to detect a relatively rare effect, so testing with inadequate numbers of animals could lead to a false assumption about the safe threshold of exposure.

Another problem is that differences between species do not necessarily fall within a factor of ten. In some cases, slight differences in metabolism may lead to one species being relatively insensitive to a substance, while another is exquisitely sensitive. For example, in the case of thalidomide, a drug responsible for a terrible epidemic of birth defects in humans, tests in rats showed no sign of problems. Had the original investigators used rabbits, however, the potential risks of thalidomide would have quickly been recognized. Later research showed that rats, unlike rabbits and humans, metabolize the drug in such a way that it does not damage their offspring.

Other problems with relating animal testing results to humans have to do with differences between laboratory conditions and real life. For example, lab animals may be exposed to substances by different routes than are humans (e.g., injection). Lab animals are usually exposed to one chemical at a time, while humans may be exposed to a mixture of different chemicals. Human populations contain a great deal of genetic diversity, while animals used in toxicity testing are bred to be genetically uniform. Finally, humans vary in nutritional status, health, and living conditions. Lab animals are carefully maintained so that such variation is minimized.

Perhaps the most significant problem with current animal testing for reproductive and developmental toxicity is that many potential effects are simply not studied. Regulators seldom require testing for effects on development of the neurological system, effects that range from major brain damage to subtle changes in behavior and learning. Changes in the functioning of the immune system can significantly affect the health of an animal, yet are rarely assessed. While abnormalities in the structure of organs may be recognized, functional abnormalities go unnoticed. Regulators are just now becoming aware of the importance of chemical disruption of the endocrine system. There may well be effects that scientists have yet to imagine that will go unrecognized until some event brings them to scientific awareness.

In spite of these limitations, animal testing remains an important early warning system for detecting chemicals that may be particularly hazardous to humans and other animals in the environment. A positive animal test suggests reason for concern and a need for further investigation or action.

Using Animal Testing to Set Standards

Animal toxicity testing forms the basis for setting standards for human exposure to chemicals. The regulator is interested in determining at what level of chemical exposure there is no adverse health effect in the animal studied. From this, the regulator will extrapolate an exposure level that is expected to be safe for humans. The No Observable Adverse Effect Level (NOAEL) is the highest level of chemical exposure at which no adverse health effect is seen in the test animals. Sometimes researchers are unable to determine a NOAEL because their testing procedure does not include an exposure level so low that no adverse effect is found; in this case, they may report a Lowest Observable Adverse Effect Level (LOAEL).

While it might seem that determination of this "safe" level of exposure would be fairly straightforward, a question of interpretation complicates things. The determination of a NOAEL assumes that the researcher is able to recognize any potentially adverse effect. However, researchers frequently look only for certain kinds of effects and may entirely miss effects that are not anticipated. In some cases an effect on the animal may be noted, but it may not be clear whether the effect is actually "adverse." Some argue that the degree and nature of an effect should be considered before deciding if it is adverse, and point out the possibility that an exposure could result in a beneficial effect. Others argue that if a chemical alters the animal's normal form or function in any way, then it should be considered potentially toxic. Given our limited ability to extrapolate effects accurately from one species to another, it is more health protective to assume that any effect seen in an animal test is an indicator of potential toxicity of the chemical tested.

Dose-Response Issues

Scientists frequently refer to the concept of a *dose-response curve* when talking about the health effects of a substance. The curve describes the relationship between a certain dose, or amount, of a chemical exposure, and the health effect it has on an organism. The simplest type of dose-response curve is described in figure 2.1, in which the amount of the dose is directly proportional to the health effect that results. Another type of dose response curve shows little or no health effect until a certain level, or threshold, is reached (figure 2.2). A third type shows an effect increasing with dose until a certain point, then dropping off again, which may be characteristic of certain feedback dependent systems (figure 2.3). Other types of dose-response relationships also exist. Defining the dose-response curve of a particular substance and a particular effect is an important, and challenging, step in regulating chemical substances.

In general, most carcinogens are assumed to have a dose-response curve that does not have a threshold, as in figure 2.1, indicating that there is some risk of cancer at even the lowest doses. Most reproductive toxicants have been assumed to have a threshold dose-response curve, as in figure 2.2, meaning that below a certain exposure level, the substance poses no health risk. This assumption is based on limited evidence, and careful consideration of the evidence may support another model.

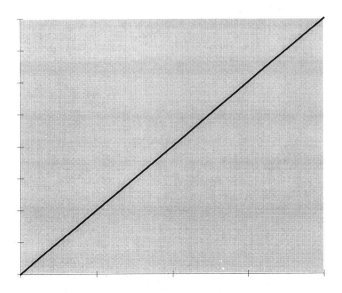

Figure 2.1   Linear dose-response curve

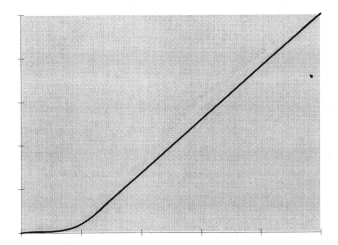

Figure 2.2   Threshold dose-response curve

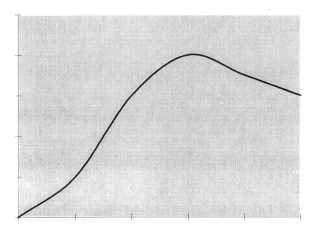

Figure 2.3 Nonmonotonic dose-response curve

Another factor to consider is that relatively large exposures may have very different effects than very small exposures. For example, a high level of lead exposure during pregnancy may lead to stillbirth or spontaneous abortion. Lower levels do not interrupt the pregnancy, but may interfere with the brain development of the child. Each of these health effects has its own dose-response curve. Animal testing that examines only the effects of high-level exposure, as is commonly the case, may entirely miss those more subtle effects that occur with much lower levels of exposure.

It is not enough to look at only varying doses of a substance in order to determine its potential health effects. Examining the timing and duration of exposure are particularly important in research on reproductive and developmental effects. Critical windows of vulnerability may make a fetus sensitive to small amounts of a substance at a critical time—amounts that have no detectable effects at other times.

In developmental toxicology, the time when embryonic cells become committed to a particular fate helps determine the likelihood, degree, and type of damage from an exposure. Early exposures, if they are not immediately lethal to the embryo, may be reparable, because the embryonic cells have the flexibility to replace those that have been damaged or destroyed. For example, the fungicide carbendazim causes birth defects in some rat embryos when given in mid- to late pregnancy.

However, when given earlier in pregnancy, it has no such effect, suggesting that the younger embryo may be able to repair or compensate for the damage more easily.[3]

On the other hand, once the embryonic cells have specialized, the precise timing of the exposure may determine the resulting type of abnormality. Children exposed to small doses of the drug thalidomide during the fifth to eighth week of gestation suffered horrible limb abnormalities; children exposed later had either no or entirely different health effects. During this window of vulnerability, the development of the limbs was taking place; the process proved to be extremely sensitive to the drug.

Epidemiology

In addition to animal studies, studies on exposed human populations are a major source of knowledge concerning reproductive and developmental toxicants in the environment. A variety of types of studies are used to assess the relationship between health effects and potential exposures. *Epidemiology* is the study of the patterns and causes of disease in human populations. It is useful for investigating associations between exposures and outcomes or in pinpointing groups at increased risk of an outcome. Because epidemiology studies populations, not individuals, it is of little use in predicting an outcome for a particular person. Thus, it allows us to say that exposure to organic solvents increases the risk of spontaneous abortion by twofold to fivefold, but it cannot say whether a particular spontaneous abortion in an exposed woman was due to solvent exposure or to other factors.

Most adverse reproductive and developmental outcomes have multiple causes, and some of these causes are still unknown. Epidemiology tries, but often fails, to tease apart multiple associations, to clarify how much each is likely to be part of the cause and how much each contributes to the overall burden of the disease in the population.

Types of Epidemiologic Studies

A variety of types of studies are used to examine relationships between disease and exposure (see figure 2.4). Each type of study has its strengths

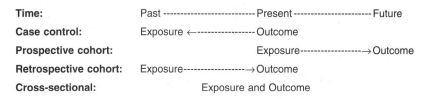

Time:	Past ------------------------ Present ---------------------- Future
Case control:	Exposure ←----------------- Outcome
Prospective cohort:	Exposure------------------→ Outcome
Retrospective cohort:	Exposure------------------→ Outcome
Cross-sectional:	Exposure and Outcome

Figure 2.4   Summary of types of epidemiologic studies

and limitations, which must be considered when assessing the significance of a particular piece of research.

Correlation studies look for evidence of strong relationships to generate theories about potential causes and effects. The change in a particular health effect is plotted against the change in a potential exposure and the strength of the association is assessed. For example, a study may graph a decline in sperm counts worldwide against the boom in chemical manufacture since World War II, demonstrating a striking correlation. Such studies are vulnerable to confounding because many variables tend to track together over time. Nevertheless, they are useful for generating hypotheses for further testing.

Cross-sectional surveys are frequently used because they are fairly quick and inexpensive, and they provide more information than correlation studies. In these studies, the health status and exposures of a group of individuals are assessed at one point in time. Studies on sperm counts in exposed workers, where sperm samples and exposure measurements are taken at the same time, are often of this type. Since they look at only a single point in time, cross-sectional surveys do not allow scientists to assess whether the exposure came before the outcome (i.e., if the men's sperm counts dropped after, and perhaps because of, the exposure).

Case reports and *case series* (a group of case reports) are not true epidemiologic studies, but are accounts of the particular health outcomes seen in one or more individuals. They are important because many serious medical problems first appear as case reports. For example, the effects of diethylstilbestrol (DES) exposure were first reported in a case series in the journal *Cancer,* which described a group of young women with a very rare vaginal cancer who were all seen at the Massachusetts General Hospital in Boston.[4]

Case-control studies are more useful because they attempt to identify possible causes of disease. Such studies identify people with a health outcome of particular interest, choose a comparison (control) group without the outcome of concern, and look back to see whether the "cases" were more likely to have been exposed to any particular risk factor than the "controls." An example is comparing a group of women who recently suffered a spontaneous abortion (cases) with an otherwise similar group of women who recently delivered healthy babies (controls). Both groups of women would then be asked about exposures during the pregnancy. This type of study is particularly useful in looking at rare outcomes, since it allows the use of a relatively small study population.

Cohort studies start with an exposed group and an unexposed comparison group and follow them over time, watching for the outcome of interest. A researcher can identify a group of children with high lead exposures in infancy and a similar group who had very little lead exposure. Both groups of children are then followed for years to observe whether there are differences in behavior and learning between the two groups. Some cohort studies are *retrospective cohorts;* they examine old records and identify a group of exposed people and a comparison group from many years ago. This is often done by looking through company records in an industry where workers were exposed to a chemical of interest. These people are then tracked down, where possible, and their current health status is determined.

Weaknesses of Epidemiologic Research

Epidemiologic studies trying to link exposure to a particular chemical with outcomes such as infertility, spontaneous abortion, behavioral problems, or childhood cancer suffer from a number of major difficulties. The most significant problems are those of bias and confounding.

Case-control studies and retrospective cohort studies, because they are interested in exposures that occurred in the past, can rarely identify the degree or precise pattern of chemical exposure that occurred. The result of this difficulty in quantifying exposure can be *exposure misclassification bias:* individuals are incorrectly assigned to the exposed or unexposed group. It is easy to see how this might happen, particularly if surrogates of exposure, such as job title or place of residence, are used to decide who was exposed and who was not. Not everyone who works in a plant nursery is heavily exposed to pesticides, while some office workers may be exposed. In most studies, misclassification of exposure is random: affected and unaffected individuals are equally likely to be misclassified. This tends to bias the study toward finding no association between the exposure and the outcome and can result in a falsely negative study or an underestimate of the risk.

Relying on the memory of those who are being studied may result in *recall bias*. For example, parents who had a bad pregnancy outcome will search their memories for any possible chemical exposure, while those who had healthy pregnancies may forget exposures that occurred months before. This is usually an issue only in case-control studies that rely on memory to determine exposure, and it can bias these studies toward finding associations between exposures and the outcome when in fact no such association exists.

Certain factors can muddy the ability to pinpoint particular risk factors and can create the appearance of an association that does not really exist. *Confounding factors* are those that are independently associated with both the exposure and the outcome. For example, if women who work in a particular industry are more likely to smoke than women who do not, and if women who smoke are more likely to have low-birth-weight babies, then it would be incorrect to assume that the industry work is responsible for the small babies, unless the difference in smoking is first taken into account.

Causation

An individual epidemiologic study generally cannot address the question of *causation,* that is, determining whether a particular exposure actually causes a particular outcome. Instead, each study will generally report an *association* between a specific exposure and a certain outcome. For example, a study may find a clear association between exposure to toluene and a certain type of birth defect. An association such as this could be due purely to chance, to confounders, or to bias in the study design. Alternatively, it may represent cause and effect.

An epidemiologist will generally move from saying that two factors are associated to saying that one causes the other when certain conditions are met. A set of criteria originally developed for the study of infectious disease has been refined for use in environmental epidemiology.[5-7]

1. There should be consistent, repeated studies in both animals and humans demonstrating the association.
2. A biologically plausible mechanism should exist to explain the association.
3. There must be a logical time sequence, in that the exposure consistently occurs before the outcome.
4. There should be evidence of a dose-response relationship with increasing risk or severity of effect with increasing dose.
5. The association must be strong enough to be persuasive.

A number of other factors may also be considered, such as the specificity of the association, the effects of removal, and the predictive performance of the hypothesis drawn from the association. If these conditions are met, a scientist may be willing to state that a particular exposure causes a certain outcome. However, deciding whether the conditions are adequately fulfilled is once again a matter of judgment, not proof.

It is not the place of scientists alone to make these decisions. We as a society must decide when the evidence is persuasive enough. We must decide when the weight of the evidence is sufficient to make it worth our while to take action that may limit a negative health outcome. In some cases, we may decide that the potential risks of exposure are great enough to take action even in the face of considerable scientific uncertainty. There will always be scientists who call for more studies in order to "prove" that some exposure results in an effect. Certainly research on an area of concern should continue, but if our interest is in protecting health, we cannot wait for science to provide proof. As one researcher pointed out, "In prevention, it is necessary to identify an exposure without necessarily identifying the ultimate cause of disease."[7] If we understand enough about the workings of science, we will be able to decide for ourselves whether the scientific data are adequate to suggest a health-protective course of action.

In the summer of 1976, an explosion at a chemical plant near Seveso, Italy, released a poisonous cloud containing toxic chemicals, including 2,3,7,8-tetrachlorodibenzodioxin. Over 120,000 people lived in the area. The 730 who lived in the most contaminated zone were evacuated two weeks after the accident, as were children and pregnant women from the remaining contaminated areas.

Immediately after the incident, teams of investigators began epidemiological studies to assess the impact of the exposure. The research results on reproductive and developmental effects were disappointing. Investigators reported an excess of spontaneous abortions, but because no baseline data existed, these reports were discounted. One research team that examined fetuses from spontaneous and elective abortions found no evidence of abnormalities attributable to exposure.[8] This research was limited because the abortion procedure often left insufficient tissue for evaluation. A birth defect registry established after the accident permitted researchers to compare birth defect rates in exposed and unexposed areas.[9] Although the rate was higher in the highly exposed population, the difference was not statistically significant. The small size of the population would have allowed detection of only an extremely large difference in rates with statistical confidence. A decreased number of births was recorded in the year after the accident, perhaps due to an increase in spontaneous abortions or to official suggestions to avoid conception.

To date, the only clear evidence of reproductive impacts following the accident comes from a study that examined changes in sex ratios of children born to exposed mothers.[10] In the seven years after the accident, almost two times more girls than boys were born to exposed mothers. In this case, the baseline measure was clear: the normal sex ratio is 106 males born for every 100 females.

The Seveso incident reveals how difficult it is to study reproductive and developmental effects and environmental exposures. Without baseline information, comparison of pre- and postexposure outcomes is impossible. Personal impacts of the accident may have affected choices regarding childbearing. Those who felt themselves at risk may have chosen to terminate pregnancies or not to conceive in the first place. Finally, although the number of people exposed was frighteningly high, the number is still too small for effective study of rare outcomes such as birth defects.

Surveillance and Birth Defects Registries

Surveillance is a process in which incidence (new cases) of certain illnesses or deaths from certain causes are tracked and tabulated in order to establish background rates of the disease in the population and to detect trends over time. Numerous infectious diseases are subject to surveillance by local and state health departments, the Centers for Disease Control and Prevention (CDC), and the World Health Organization (WHO). Such surveillance can greatly aid in the rapid detection of epidemics and quick response to disease threats. Surveillance has also been used to track cancers, homicides, occupational injury or disease, birth defects, and other noninfectious health problems.

Surveillance may be passive or active. *Passive surveillance* is characterized by requests or requirements that physicians or hospitals report cases, accompanied by efforts to educate the reporting community to improve compliance. *Active surveillance* involves a registry that goes out into hospitals or medical centers to identify and collect information about cases, generally done through review of medical records. Active surveillance is much more reliable than passive surveillance; however, it is more resource intensive, expensive, and difficult to implement.

If there is no surveillance system for a particular health problem, the likelihood of detecting a true increase in that problem over time is extremely slim. For example, in the United States, there is no surveillance system for spontaneous abortions. A true increase in this already common problem is unlikely to be noticed unless the increase is quite substantial. A weak surveillance system, such as a passive system with underreporting and poor reliability, may be little better. In the absence of a large and well-designed surveillance system, important changes in disease patterns may be missed.

In the United States, there are two federally administered birth defects registries: the Metropolitan Atlanta Congenital Defects Program (MACDP) and the Birth Defects Monitoring Program (BDMP).[11] There are also state and international birth defects monitoring programs. Perhaps the most comprehensive of these is the California Birth Defects Monitoring Program (CBDMP).[12] The MACDP and CBDMP are extremely well designed. Both are active registries involving case ascertainment through medical record review and review of numerous other information sources. Any defects diagnosed before the age of one year are included, as is information about fetal deaths after 20 weeks gestation. The BDMP, on the other hand, relies on voluntary reporting, is not necessarily representative of the population, and results in incomplete information on birth defects nationwide.

Maintaining an effective surveillance system for birth defects is challenging. On the one hand, data must be collected and analyzed quickly in order to detect any emergent health threats. But in order to collect and process the data quickly, some important information may be lost. Some systems review hospital records in detail, while others rely primarily on the birth certificate or discharge record, both notoriously unreliable sources of this type of information. Subtle or functional birth defects frequently remain undetected for some time after birth and may be missed. Minor defects may not be recorded, and birth defects found in stillborn babies are included in only some registries. Still, high-quality birth defects registries are critically important for detecting true increases in abnormalities and serve as a resource for epidemiologic studies investigating associations between exposures and birth defects.

In the early 1960s, the drug thalidomide was given to about 1 percent of pregnant women in Europe and Australia to treat morning sickness. This drug resulted in a 175-fold increased risk of a set of rare malformations involving extreme shortening of the limbs. Despite the immediately obvious nature of these defects and their previous rarity, six years elapsed and over 6,000 deformed babies were born before the drug-defect relationship was established.[13] At that time, there was no routine surveillance system in place for birth defects.

Unlike thalidomide, the environmental exposures discussed in this book do not involve increased risks of over 100-fold. In most cases, we are concerned about exposures that may double or triple the risk of a particular outcome. To understand the difficulty of detecting such subtle increases, let us assume that we are talking about a common birth defect, such as cleft lip/palate, occurring in about one in every thousand births; let us also assume that about half of the cases of this defect are caused by a mechanism that is susceptible to a new environmental toxicant. This toxicant increases the risk of this subtype of birth defect by fivefold, and 10 percent of the people in the population are exposed. A calculation of the resulting change in this defect indicates that, overall, it will increase by 1.4-fold. In the best U.S. birth defects registry, which represents about 40,000 births per year, the expected rate of this defect would increase from about 40 to about 56 cases per year, a change that would not be statistically significant. In fact, 200,000 births would need to be monitored in order to detect this increase within one year with the traditional cutoff p-value of 0.05.[13]

Because increases in adverse effects can "hide" in the background population disease burden, there is an urgent need for resource-intensive surveillance of disease trends. Without careful tracking of birth defects and other disease, we may miss increases that are significant from a public health point of view but otherwise would remain statistically hidden.

Birth Defect Clusters

In the early 1990s, Brownsville, Texas, health care providers became concerned when three babies were born without brains at one hospital in a thirty-six-hour period. A health department investigation revealed that the rates of this and related birth defects, known as neural tube defects (NTDs), were three times the average found in the United States, although they were lower than rates seen just across the border in Mexico.[14]

Brownsville sits on the Texas border, adjacent to Matamoros, Mexico. Early on, residents' concerns focused on the pollution released by factories, known as *maquiladoras,* located in Matamoros.[15] The *maquiladora* plants are mostly foreign owned and enjoy special tax status, along with lax environmental and occupational regulation. Numerous industrial sectors are represented, including electronics, textiles, plastics, and battery manufacturing.[16] In recent years, increasing international attention has been drawn to the dangerous occupational conditions and hazardous chemical emissions from some of these facilities.

A fairly thorough epidemiological investigation by the Texas Department of Health confirmed the statistical excess of NTDs in Brownsville but failed to find any significant associations. The researchers evaluated demographic, genetic, nutritional, occupational, and environmental factors. None appeared to explain the cluster. Neural tube defects are a tragic but interesting birth defect because, in contrast to many others, they have some well-identified causes.[17] The major known cause of NTDs is deficiency of the nutrient folate. Heredity, Hispanic heritage, and maternal age are other known risk factors, and valproic acid, used in the treatment of epilepsy, is known to cause this kind of birth defect. Maternal exposures to solvents and pesticides have been associated with NTDs in a few studies, as have paternal exposures to pesticides, solvents, and ionizing radiation.[18]

In Brownsville there are no clear answers. The Health Department's failure to find an association with any exposure may mean that the cluster of birth defects was due simply to chance. However, because the study was based on only twenty-eight cases and twenty-six controls, it may have lacked the statistical power to detect a meaningful association. The Brownsville residents' concerns about the *maquiladoras* may be well founded, or perhaps some other factor, such as genetics or nutrition, contributed to this cluster. Given the limitations of the science, we still do not know the answer.

Quantitative Risk Assessment

In the past twenty years, quantitative risk assessment (QRA) increasingly has become the tool of choice for regulators responsible for setting limits on potentially toxic exposures. This tool combines available scientific data with models and best guesses to generate estimates of risk from chemical exposures. In recent years, the limitations of this approach have become more clearly recognized, but QRA in one form or another continues to be a fundamental part of our regulatory system.

QRA is a process that attempts to quantify human health or ecosystem impacts of a particular chemical or pollutant. In most cases, the outcome of concern is cancer, although QRA can be applied to any health outcome. In order to quantify risk, first an assessment is made of the actual or potential exposure to the substance of interest. This is usually based on modeling rather than on existing exposure monitoring data, and the exposure assessment frequently considers only one particular source of exposure rather than all sources combined. Next, a dose-response model is developed so that the potency of the substance can be quantified. A different dose-response model must be developed for each potential health impact. Reproductive and developmental outcomes are assumed to fit a threshold dose-response model with a level below which there is no risk from the exposure. Then the estimated exposure is multiplied by the estimated potency of the substance to develop a measure of the risk of a particular human health impact.

The actual calculation is somewhat more complex, because a number of other factors may be included. For example, if the dose-response curve is based on animal data, an uncertainty factor may be added to take into account the unknown difference between animal and human response. If human exposure estimates are based on emission estimates, another factor may be included to account for the uncertainty of the data. Where data are limited, risk assessors must count on models to represent real-world conditions. By their nature, models will always be less complex than the real world, and therefore never entirely accurate.

Mathematical models used for QRA often create an illusion of scientific knowledge and certainty that is unjustified, while creating a confusing mass of mathematical data almost impossible for anyone but an expert to critique. Thus, the resulting risk calculation is based on limited scientific information and substantial uncertainty, and the unwary may be misled into believing that it represents scientific truth.

Limits of Quantitative Risk Assessment

Risk assessment as it is practiced today has a number of problems that limit its usefulness, particularly in the area of reproductive and developmental hazards. First, QRA rarely addresses potential reproductive and developmental threats, so any regulation based on this process may miss this area of concern entirely. But even if quantitative risk assessment were applied to these issues, there would still be problems. For example, the result of the QRA process is a numerical estimate, which implies a precision not justified by the data used in the process; too much uncertainty is incorporated in the process to allow for more than a general statement of risk. Yet regulators use the resultant risk estimates as if they were accurate assessments of the actual risk presented.

The uncertainty that comes into play in risk assessment has several distinct sources. First, there is the uncertainty arising from what we do not yet know scientifically—the gaps in our knowledge, which may eventually be filled with further investigation. We may know enough to estimate or model values for such factors, but these will be only best guesses, with a certain degree of inherent uncertainty. Then there are those things of which we are entirely ignorant—factors we do not even realize need to be considered. Sometimes we may be entirely unaware of factors relevant to assessing a particular risk, or changes may occur in the future that we are unable to anticipate. Finally, there are those things that are unknowable—factors that cannot be measured or defined by even the best scientific approaches.

Were it possible to produce a risk assessment with absolute certainty, there would still be important issues to consider. Once a risk level is determined, we are faced with the question of determining what the level of "acceptable" risk is. In the case of reproductive and developmental toxicants, it is difficult to come up with an "acceptable" increase in birth defects, infertility, or miscarriage. Some people would argue that it is unethical for regulators to impose any increased risk on any segment of the public without their informed consent, or when those exposed had nothing to do with creating the exposure in the first place. Another fact to be considered is that risks are not evenly spread across the population. Some groups bear a disproportionate share of the risks from toxic chemicals because of increased exposures, such as subsistence fishers or those living in particularly polluted communities. In addition, the impact of adverse reproductive and developmental outcomes is concentrated among infants, children, parents, and those who are trying to conceive.

Risk assessment is generally used to judge the potential effect of a single chemical or compound as emitted or released from one source. In the real world, people are exposed daily to a complex mixture of chemicals from a wide variety of sources. Risk assessment as practiced today does not take into account these multiple exposures, failing to consider even additive risk, much less the possibility that chemical mixes may synergistically increase risk. The determination that exposure to one particular chemical from one particular source poses an "acceptable" risk is meaningless in many ways.

Risk assessment, like any other tool for decision making, has certain limitations and biases that must be fully recognized and acknowledged. The assumptions must be revealed, the uncertainties in the science acknowledged, and the nature of the results described and explained. The utility of QRA as a tool for making public health policy is still under debate. Only when it is used properly will we be able to assess its ultimate usefulness.

Reexamining Scientific Tools

When science is conducted purely for science's sake, it is desirable and even necessary to hold to the strict standards established by the scientific community. But when science is asked to help make public policy decisions, it may be appropriate to apply a different set of standards that take into account our desire to act in a way that will be protective. Just as we have two systems of law, one to deal with criminal and the other with civil cases, it may be appropriate to have two systems of science. In criminal law, with its quest for absolute truth, responsibility for an action must be proved beyond a shadow of a doubt. In civil law, responsibility for an action must be shown to be more likely than not. In some cases, we might wish to act in a health-protective way if the science shows that it is more likely than not that a particular exposure would result in an adverse health outcome. When making real-life decisions, people base their actions on what is likely, not what is proved.

Unfortunately, this commonsense approach is frequently criticized by those who have reason to support the status quo. They argue that "sound science" requires that a cause and effect be proved beyond a shadow of a doubt, and that anything less is no more than "junk science." But as we have seen, science is an ongoing process, in which there is no gold standard of proof. Instead, there is a cautious weighing of evidence that gradually leads to growing consensus among scientists. If we wish to ask whether the scientific basis for a decision is sound, then what we really want to know is whether the *process* of the science underlying the decision is sound. We must ask whether the tool of science has been appropriately employed and the rules of the scientific method have been followed. If the rules have been followed, and if methods, assumptions, and results have all been clearly reported, then the science is sound. If the rules have not been followed and the reporting is inadequate, the science is not sound.

Those pursuing a particular political, social, or economic goal can quickly inject doubt into the minds of decision makers by using loaded but poorly defined terms such as junk science. If the public and policymakers are not familiar enough with the basic rules of science to assess the science themselves, they are more likely to be swayed by inaccurate and biased characterizations. It is perfectly valid to ask whether the scientific basis for making a decision is sound, if by "sound" we mean that the scientific process has been properly followed. Unfortunately, all too often the "sound science versus junk science" debate becomes nothing more than an exercise in name calling.

Instead of calling for absolute proof that a cause leads to an effect before proceeding to action, we would do better to consider the weight of the evidence and the consequences of being wrong or right. At any given time, we have a certain amount of scientific information available to work with when faced with making a decision. To consider the weight of the evidence, we need to assess the scientific work that exists on a question and judge whether it tends to support an association between an exposure and an outcome, taking into account both the soundness of the work and the results themselves. An examination of the weight of the scientific evidence can guide our decision making, even as we acknowledge that future evidence may change our understanding of the situation.

In the interest of protecting health, we may also choose to use scientific tools in different ways. Alternative ways of presenting the results of a study can provide much more information than current conventional data presentations. Graphical representations of a range of p-values or confidence intervals allow readers to interpret results in the light of their own scientific judgment. For example, an alternative presentation for confidence intervals involves presentation of not only the 95 percent confidence interval around the result but also the 50 percent, 75 percent, and 90 percent, in the form of "nested" confidence intervals. When results are presented in this way, we can look at the strength of an association and its statistical significance and decide for ourselves whether the evidence is strong enough for concern.

Another way to use scientific tools in a more health-protective way is to consider switching the null hypothesis. A scientist usually hypothesizes that a certain exposure may cause an effect; the corresponding null hypothesis states that the exposure does not cause an effect. Then the very high standard of scientific proof is applied to the original hypothesis. If we were to switch the hypotheses, then the high standard would instead apply to the hypothesis that the exposure has no effect. If applied to chemical testing, for example, we would require a high level of proof that a chemical exposure does *not* cause adverse health effects, rather than proof that it does. In this case, we are in effect arguing that a chemical is guilty until proved innocent, an inversion of the criminal law standard. But the standard of innocent until proved guilty was designed to protect the rights of people, not chemicals.

Quantitative risk assessment, a science-based tool, needs to be used in a thoughtful way. The focus in QRA should be on recognizing and acknowledging the uncertainty inherent in the process. Of course, acknowledging the uncertainties may mean that the answers provided by risk assessment are not so easy to use in decision making. We may find that, in the end, risk assessment is not able to provide as much useful information as we had hoped, and that the applications of this tool are in fact quite limited.

In order to participate in a meaningful way in the decisions facing us as both individuals and a society, we must become more familiar with the tools used to provide the bases for decisions. The process of science has a long and established history, and its usefulness in the quest for truth has been recognized. We need to consider further the best way to use this tool to protect public health. The quasi-science of risk assessment is a newer, less tested tool, whose efficacy is still under debate. We need to continue to assess this process, to determine whether it can indeed provide useful information. Not all of us can be scientists and researchers, but we all can learn to recognize the strengths and limitations of these tools, and to understand how to apply them to the questions we are asking. Only then can we hope to make decisions that truly serve our interests and protect human health.

Reproductive and Developmental Effects of Selected Substances and Human Exposures

II

Metals

3

Lead and mercury have been the most extensively studied reproductive and developmental toxicants. They are widely dispersed throughout the environment, and everyone is exposed to them. Three other common metals—cadmium, arsenic, and manganese—are also likely reproductive toxicants, and some animal studies suggest that chromium and nickel damage fetal development.[1] Other metals, such as tellurium, gallium, and indium, which only recently have come into widespread use as a result of high-tech applications, have some early indications of reproductive and developmental toxicity. These metals pose a potential hazard for future generations that cannot yet be quantified. Here we will concentrate on the effects of lead, mercury, cadmium, arsenic, and manganese.

Lead causes infertility in exposed males and spontaneous abortion in women exposed at high levels. Strong evidence suggests that lead exposure also leads to subtle neurological effects, developmental delays, and behavioral abnormalities in otherwise normal-appearing children. Mercury has been responsible for two major epidemics of spontaneous abortion and birth defects in human populations. Organic mercury compounds cause brain damage to the developing fetus and result in microencephaly (small brain), cerebral palsy, and mental retardation.

The reproductive effects of cadmium, arsenic, and manganese, by contrast, have not been well studied in humans. In animals, cadmium damages the testes and interferes with sperm production, and it may interfere with normal lung development and predispose to respiratory distress syndrome in newborns. Some evidence suggests that cadmium is toxic to the human placenta, and may thereby lead to spontaneous abortions and birth defects. Arsenic causes a characteristic set of malformations in lab animals exposed at high levels. In addition, some human studies suggest that arsenic exposure may lead to spontaneous abortion and stillbirth and may affect neurological development, particularly the development of hearing. Manganese is an important metal because it is a recent gasoline additive and may be even more widespread in the environment in the future. Animal studies and a few human studies indicate that manganese may interfere with hormone production and damage reproductive function in men. In addition, manganese is toxic to the fetal brain.

Lead

In its natural state, lead is found only in the earth's crust. Humans have mined and used lead ore for thousands of years, resulting in lead pollution of water, air, and soil. Lead is now found in the bodies of all living things on the planet and throughout the environment, including the polar icecaps. Most current environmental exposures to lead in the United States come from lead paint exposures, although people are also exposed through the water supply, usually from leaching of lead from pipes. Additional sources of exposure include lead-glazed pottery, certain medicinal and cosmetic preparations, which are used by a variety of ethnic groups, and food grown in contaminated soil. Some occupations and hobbies often involve exposure to lead—for example:

Painting
Removing old paint
Construction
Battery manufacturing or recycling
Automobile repair
Electronics
Ceramics and pottery
Printing
Welding and soldering
Firearm shooting and cleaning

Jewelry making and repair
Stained glass window making

The phase-out of lead in gasoline and the end of the use of lead solder in commercial food canning have greatly reduced lead exposures in this country. However, leaded gasoline is still used throughout the world and will continue to expose untold millions for years to come.

Distribution in the Body

When lead enters the body it distributes throughout the organs, including the brain, and crosses the placenta with ease.[2] Blood lead levels in the fetus are up to 90 percent of the maternal blood lead levels. Some lead is excreted, but the rest accumulates in bone, and can be released months or years later. Pregnancy is a time when a mother's body mobilizes nutrients for her fetus. Fetal calcium requirements are met, in part, by calcium from the mother's bones, a process referred to as *increased bone turnover*. If lead is also stored in bones, this toxic metal can be transferred to the fetus, or to the infant after birth in breast milk.[3]

Lead exposure can be measured through blood testing, urine testing, and x-ray fluorescence of bone. Blood testing is the most common, though it reflects exposure only over the past three months. Lifetime exposure to lead can be measured with either bone x-ray fluorescence or urine testing done after administration of a chelating medication, which increases excretion of lead. These tests are generally done at academic medical centers for research purposes.

Lead Dose and Health Effects

Over the past ten years there has been increasing evidence that lead may have serious health effects at exposure levels much lower than previously thought to be harmful. Most of the other substances discussed in this book are either disputed reproductive or developmental toxicants, or known reproductive or developmental toxicants with unclear dose-response ranges. Lead is a known toxicant with a well-studied dose-response relationship (see figure 3.1).

The average blood lead level in the U.S. population is now about 2.0 µg/dl (micrograms per deciliter) in women of childbearing age and about 4.2 µg/dl for men in the same age range.[4] Levels were much higher

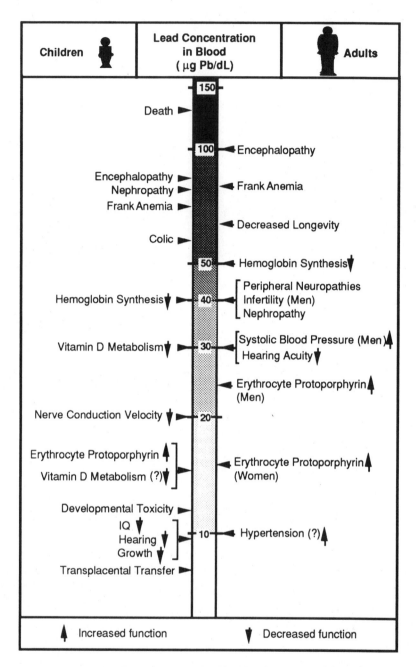

Figure 3.1   Health effects at various blood lead levels. Reprinted from the ATSDR Case Study on Lead.

in the 1970s: around 13.7 µg/dl in children aged one to five and around 11 µg/dl in women of childbearing age.[5] A 1990 governmental study stated that "every pregnancy potentially represents a risk if the mother has a blood lead level of 10 µg/dl or higher." The authors estimated that 4.4 million women of childbearing age have blood lead levels over 10 µg/dl and projected that during the next decade, over 4 million fetuses would be at risk because of maternal lead exposure in the United States.[6] The EPA has listed 10 µg/dl as the maximum acceptable blood lead level for fetuses and young children, and the Centers for Disease Control recommends action to monitor and lower lead levels in children with higher levels. Almost 22 percent of black children one to two years old currently have blood lead levels over 10 µg/dl.[6]

Reproductive and Developmental Effects at High Doses

Men

At blood lead levels over about 50 µg/dl, lead impairs fertility in males and females.[7-11] In men, lead may act directly on the testes to lower the sperm count; in fact, in the past, lead was used as a spermicide contraceptive. A recent study in male workers found effects on sperm function and quantity at blood lead levels near 40 µg/dl.[12] Blood lead levels of 40–50 µg/dl occur regularly in the workplace; employers are not required to remove workers from exposure until their blood level rises over 50 µg/dl. Evidence that lead may interfere with the endocrine system comes from studies showing an effect on testosterone levels and the hypothalamic–pituitary axis in men with severe lead poisoning.[13,14] Unfortunately, insufficient study size and few studies involving male exposure make it difficult to conclude at what dosage lead may affect male reproduction.[15]

Women

With exposure at or above levels sometimes encountered in the workplace, lead causes spontaneous abortions and stillbirths.[16] In the past, it was used to induce abortion. At lower blood levels, up to around 15 µg/dl, several studies have not found any increased risk of spontaneous abortion.[17,18] One study that tracked down women who had been lead poisoned as children forty years before and asked them about their reproductive history found a 60 percent increase in risk of spontaneous abortion.[19] Although this study was small and the results were not statistically significant, it suggests that early lead exposures may affect subsequent reproductive ability, not surprising given what we know about lead storage in bone.

Effects at Low Doses

The most worrisome effect of lead at exposure levels close to the U.S. average is developmental toxicity to the fetus, which may permanently affect neurologic and behavioral development. There is an apparent relationship between rising blood lead levels and preterm delivery, low birth weight, and fetal growth retardation.[16,20] This relationship is evident down to blood lead levels under 10 μg/dl. One study did demonstrate an association between minor birth defects and umbilical cord blood lead levels, but overall there is little evidence that lead causes birth defects.[21] The main effects of lead on the fetus are as a growth retardant and a neurologic toxicant.

Lead has long-term effects on behavior and intelligence in infants born to mothers with blood levels of 10–25 μg/dl. Developmental delays in lead-exposed children persist at least until five years of age. One study followed children into adulthood and found a sevenfold increased risk of nongraduation from high school, and a sixfold increased risk for reading disability in children exposed to lead as toddlers.[22-24] Although not all studies have found an effect on mental development at low doses, the studies that have found low-dose effects were well conducted and persuasive. Several recent reviews of the literature concluded that lead exposure, even at blood lead levels at or below 10 μg/dl, is linked with impaired neurobehavioral development, low birth weight, and intrauterine growth retardation.[25,26]

Two recent reports have also found that lead exposure is significantly correlated with aggressive, destructive, and delinquent behavior. One of these studies looked at bone lead as a measure of exposure in eleven-year-old boys; the second prospective study used blood lead in the mother during pregnancy and in the child until age three as an integrated exposure measure.[27,28] Effects of lead on the brain appear to occur after both prenatal and postnatal exposure. Monkeys exposed from birth to doses of lead that maintain their blood lead level at 15 μg/dl showed increased distractibility, inappropriate responses to stimuli, and difficulty in changing response strategy.[29] The evidence is persuasive that lead has subtle harmful effects on brain development even at quite low levels.

Summary

Lead is a well-recognized reproductive and developmental toxicant. In workers with high occupational exposures, it diminishes fertility and causes spontaneous abortion. Lower lead levels have not been well studied for their possible effects on the male reproductive system or on pregnancy in the partners of exposed males. Low lead levels result in fetal developmental delay, prematurity, and lasting deficits in concentration, learning, and behavior among children exposed in utero. There is no evidence of a threshold dose below which these effects do not occur. Despite the reduction of lead use in this country, the continued use of leaded gasoline around the world, the persistence of lead in soil, and the continuing problem of lead paint in houses make the effects of lead certain to persist in this country and to worsen worldwide.

Mercury

Mercury is found in the environment in three forms: elemental mercury vapor, inorganic mercury compounds, and organic (usually methyl) mercury. The three forms are produced and used for different purposes, are absorbed by the body differently, and have different effects on reproduction and development (see table 3.1).

Organic mercury is the most dangerous form of mercury because it is the most easily absorbed orally and crosses into the brain and into the fetus so easily. Levels in the fetal circulation are usually higher than levels in maternal blood, and methylmercury appears in significant levels in breast milk.[30] Bacteria in the environment transform other forms of mercury into organic mercury, which is taken up in algae and eaten by fish, and makes its way into the human diet (see figure 3.2). Contaminated fish, particularly carnivorous fish such as swordfish, tuna, shark, and pike, are the major source of organic mercury exposure for many people.[31]

Table 3.1　Profile of the three major forms of mercury

	Elemental mercury vapor	Inorganic mercury	Organic (methyl) mercury
Use	Used in dental fillings, thermometers, batteries A contaminant in coal and oil; used in gold mining and chlorine manufacture	Used in electrical equipment, some fungicides, antiseptics, and medications, and in skin-lightening creams	Fungicide in paints
Emission sources	Emitted by waste incinerators, oil and coal burning	Various industrial sources	Other forms of mercury are transformed into this in the environment.
Absorption	Absorbed through the lungs; poorly absorbed if swallowed	Not usually inhaled; absorbed slightly through skin or if swallowed	Rapid absorption if swallowed; some absorption via lungs, skin
Transport in the body	Crosses the placenta. Enters the brain.	Does not enter the brain or cross the placenta easily.	Crosses the placenta, enters the brain, is in breast milk.

Elemental mercury is a significant hazard only when inhaled, but the vapor pressure is low, so it can be inhaled at room temperature. For this reason, a broken thermometer should be disposed of by sweeping, not vacuuming, the mercury. The heat of the vacuum cleaner vaporizes the mercury into the air. People may be exposed to mercury in the air from waste incinerators that are burning batteries, switches, fluorescent bulbs, or medical waste, or oil or coal burning, since mercury is a contaminant of these fuels.[32] Once elemental mercury is in the body, it passes easily into the brain and across the placenta to the fetus.

Organic Mercury

Organic mercury exposure has resulted in two large epidemics of mercury poisoning in recent history. One episode, in the area around

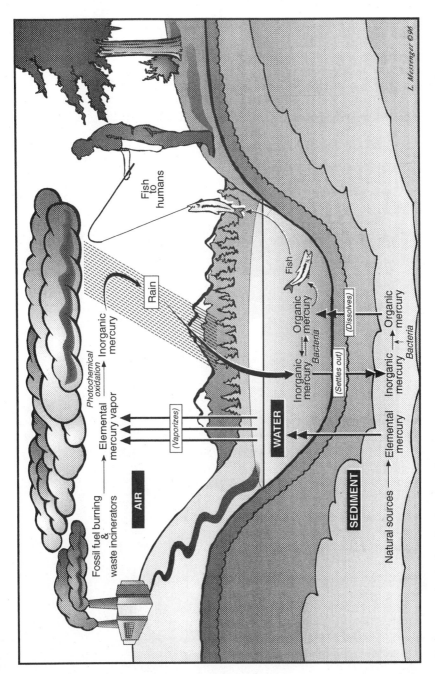

Figure 3.2　The cycle of mercury in the environment

Minamata Bay in Japan, occurred in the 1950s, and the second series of outbreaks occurred in Iraq in the late 1950s, early 1960s, and early 1970s, when grain imported for planting was treated with organic mercury to retard fungal growth. Instead of being planted, the grain was used for bread making, and thousands of people were poisoned. Although some adults developed symptoms, including constricted visual fields, numbness of the fingers and toes, and even poor coordination, the main victims of the exposure in both epidemics were children exposed before and after birth.

Organic mercury selectively damages the developing brain. In the outbreaks of poisoning in Japan and Iraq, infants had cerebral palsy, mental retardation, incoordination, weakness, seizures, visual loss, and delayed development.[33-36] Often a child exposed to organic mercury in utero appeared fairly normal at birth, with only slight abnormalities of reflexes and muscle tone, but later had seizures, long delays in learning to walk and talk, and severe clumsiness. At lower-dose levels, the only observed effects were abnormal muscle tone and reflexes and mild developmental retardation when retested at an older age.[37] The doses involved in these outbreaks were ten- to one-hundred-fold greater than doses most fish consumers are exposed to today.

Health effects of organic mercury are similar in animal studies and in human populations, and mercury is one of the best-understood developmental toxicants. Organic mercury interferes with cell division and migration of cells in the developing brain. Studies in mice have shown that cells in the developing brain stop in the middle of cell division when exposed to organic mercury.[38] In addition, methylmercury binds to DNA and interferes with the copying of chromosomes and production of proteins, processes that are essential to life.[30]

Two major ongoing studies of people who eat a lot of fish—one in the Seychelles Islands and one in the Faroe Islands—are attempting to evaluate the low-dose effects of methyl-mercury on brain development. Preliminary results are conflicting, with the Seychelles study showing little or no effect, and the Faroe study showing subtle but significant impairment of brain function.[39-41]

Based on an Iraqi study, the U.S. EPA projected that the highest chronic exposure to mercury tolerable without likely health effects is 1.0 µg/kg body weight/day, and on that basis set a reference dose (RfD) of 0.1 µg/kg/day. The *RfD* is the dose that is expected to be without any health effect even if exposure persists at that level over a lifetime.

Mercury in Minamata Bay: The Dangers of Bioaccumulation

During the 1950s, many residents living around the Minamata Bay in Japan developed a disturbing neurological disorder. These people suffered symptoms ranging from numbness of the extremities and tremors to paralysis, blindness, deafness, and even coma. Many children born during this period suffered from cerebral palsy, mental retardation, and microcephaly, a condition in which the brain is not fully developed.

It was several years before the cause of this mysterious outbreak was discovered. A nearby factory regularly discharged mercury into Minamata Bay. Inorganic mercury was not thought to be a serious health threat because it is poorly absorbed in the intestine of fish, animals, and humans. At that time, few people suspected that bacteria living in the sediments of wetlands and estuaries have the ability to transform inorganic mercury into the much more hazardous organic mercury. Organic mercury is easily absorbed by the intestine, and over time, mercury accumulated in the fish.

Because organic mercury has the ability to bioaccumulate in animals, the levels of organic mercury in fish can be up to 100,000-fold greater than the levels in the water. The people in the area depended on fish from the bay as a food source, and as they consumed the fish, the mercury accumulated in their bodies and moved easily across the placenta and into the fetal brain.

Chemicals that persist in the environment and bioaccumulate in the food chain can be especially hazardous to people. Humans are "top predators" because we eat animals and fish that in many cases are themselves carnivores. This means that we consume a dose of persistent toxic chemicals equivalent to that contained in all of the creatures eaten by the animal that we eat. It might have been easy to believe that the amount of mercury dumped into the bay was insignificant and would have quickly been diluted in the ocean. But because it entered the food chain, the mercury concentrated hundreds and thousands of fold and ended up ultimately in the bodies of the families living around Minamata Bay.

Elemental and Inorganic Mercury

The evidence of adverse effects regarding elemental and inorganic mercury is not clear. These forms of mercury do not appear to affect the developing brain like organic mercury does. Although animal studies indicate that elemental mercury can damage male fertility, men occupationally exposed to elemental mercury vapor did not have any apparent decrease in fertility compared to a group of unexposed men, nor did their children have a greater risk of malformations.[42-44] A different study of exposed male workers found a twofold increased risk of spontaneous abortion among their wives.[45]

Animal studies, however, have shown that elemental mercury can be toxic to the fetus.[45] And studies in women, mostly dental assistants, have found conflicting results as to whether elemental mercury increases the risk of spontaneous abortion.[46,47] One large cohort study demonstrated spontaneous abortion and other pregnancy complications in exposed women.[31] Several additional studies suggest that women occupationally exposed to elemental mercury may have an increased risk of menstrual disorders, particularly heavy bleeding and severe menstrual cramps.[42]

Inorganic mercury exposure in young children can lead to acrodynia, or "pink disease." Symptoms include a rash and peeling of the skin of the hands and feet, irritability, photophobia (being bothered by bright light), excessive hair growth, and profuse perspiration. This syndrome is seen when mercury is used as a disinfectant in diaper laundries or when mercuric salts are applied to the baby's skin as a disinfectant. This syndrome seems to be an allergic-type reaction to mercury.

Summary

Mercury is a known developmental toxicant that is particularly dangerous in the organic (methyl) form. It primarily attacks the developing brain, with effects ranging from mild developmental delays to severe cerebral palsy, blindness, and seizures. Organic mercury may pose a developmental danger to fetuses exposed at only slightly elevated levels of maternal fish consumption. Elemental mercury and inorganic mercury have less clear-cut reproductive and developmental effects on humans. Nonetheless, all mercury compounds should be considered probable reproductive and developmental toxicants.

Cadmium

Cadmium is toxic to the testes in animals at fairly low doses, and also concentrates in and damages the placenta. In comparison to lead and mercury, however, the effects of cadmium on human reproduction and development are poorly understood. Animal studies suggest both an increased risk of structural birth defects and an association between cadmium exposure and delayed lung development, with a possible increase in respiratory distress syndrome of the newborn.

People can be exposed to cadmium at work or through hobbies, including metal plating; semiconductor manufacture; wire, plastic, or battery manufacture; welding; soldering; ceramics; or painting. One other important source of cadmium is cigarette smoke; smokers typically have blood levels of cadmium approximately twice those of nonsmokers.[48] Cadmium can also be a contaminant of drinking water, air, and food, particularly shellfish. In the 1940s and 1950s there was an epidemic of poisoning in Japan due to contamination of water and rice crops with cadmium runoff from a zinc mine. Poisoned villagers experienced severe bone pain, a waddling walk, poor kidney function, and thinning of the bones.[49]

Everyone has cadmium in their bodies, where it concentrates in the kidneys, liver, pancreas, and adrenal glands and tends to accumulate slowly over time. Individuals with iron, calcium, or zinc deficiency or with protein malnutrition absorb cadmium more readily. A protein, metallothionein, binds to cadmium and is thought to help protect against the toxic effects of the metal. Normally very little cadmium is captured by metallothionein, but repeated low-level exposure to cadmium causes increased production of this protective protein. Thus, short-term, higher-level exposures may be more dangerous than low-level, chronic exposures.[49]

Testicular Toxicity

In male animals, cadmium severely damages the testes and kills the cells that produce sperm, even at low-dose levels that do not cause general toxicity to the animal.[50-52] In the few human studies done to date, the results are less clear-cut. Four men occupationally exposed to cadmium had one-hundred-fold higher levels of cadmium in their testes on autopsy compared to three unexposed men. Although the testes of the exposed men appeared essentially normal, almost no sperm were seen microscopically.[42] Another study showed no effects on the reproductive

---

hormones testosterone, luteinizing hormone, or follicle-stimulating hormone in a group of exposed workers, but no semen analysis was done.[53] Finally, recent research demonstrates an association between elevated cadmium levels in seminal fluid and varicocele-related infertility in men.[54]

Placental Toxicity

In both humans and animals, there is strong evidence for placental toxicity. Studies in female animals show that cadmium accumulates in the placenta.[55] Initially this accumulation was thought to be protective of the developing fetus, but there is now evidence that cadmium damages the placenta's ability to provide oxygen and nutrition to the fetus and can result in fetal damage or death.[56] Cadmium concentrates in the human placenta, and levels of exposure that cause placental toxicity are at least tenfold lower than those that result in other toxic effects in the adult, such as kidney damage. Cadmium leads to decreased production of a hormone, human chorionic gonadotropin (HCG), which is essential for maintaining the pregnancy; it also interferes with the transfer of zinc across the placenta and causes structural damage, initially to the blood supply and eventually to the rest of the placenta.[57] Cadmium does cross the placenta to some degree in humans. The level of cadmium in the skeletons of a group of stillborn infants was found to be ten times greater than levels in the bone of a comparison group of normal infants.[58]

Structural Birth Defects

Animal and human studies conflict regarding structural birth defects. Animals exposed to cadmium show birth defects, possibly due to damage to the placenta. Defects include decreased weight gain, abnormalities in the bony skeleton, damage to the central nervous system, and facial malformations, with the particular effect dependent on the timing of the cadmium dose during gestation.[59] In humans, two studies have reported a slight decrease in birthweight in infants of women exposed to cadmium during pregnancy, but one other study failed to confirm that effect. None of the three studies found an increase in congenital malformations.[60]

Other Adverse Effects on the Fetus

There is evidence of neurological effects, such as impaired reflexes and changes in activity level, in the offspring of exposed animals.[60] In one case, young rats exposed to cadmium during gestation had abnormally low levels of two essential metals, copper and zinc, in their brains, were less active than normal rats, and behaved poorly in neuropsychological testing.[61] In another study, prenatally exposed rats showed significant decreases in birth weight and growth rate, as well as hyperactivity and delays in development of instinctive cliff avoidance and swimming behaviors.[62] No human studies have been done in this area.

A series of other important animal studies exposed pregnant rats to cadmium and examined the lungs of the fetuses. All found that exposed rats have smaller lungs than expected. In addition, the important lung surfactants, which keep the air sacs in the lung from sticking together, were markedly decreased in the exposed rats. Not surprisingly, these exposed rats were found to have a high risk of respiratory distress syndrome and sudden infant death.[59] Again, no human studies have looked for an association between respiratory distress or sudden infant death syndrome (SIDS, or "crib death") in infancy and cadmium exposure.

Summary

Extensive evidence from experimental studies on rodents and on human placentas shows that cadmium can be toxic to the placenta at doses below those that cause other adverse effects of cadmium exposure. It is unclear whether this placental toxicity leads to adverse effects on the human fetus, although such effects were found in animals and would be expected in humans. The dramatic testicular toxicity found in animals has not been shown in humans exposed to low doses. There is worrisome evidence in animals that cadmium may affect neurological and behavioral development and lung development. These issues remain to be studied in humans and urgently require attention. While awaiting further research, this metal should be treated with extreme caution as a probable human reproductive and developmental toxicant.

Arsenic

Arsenic, like mercury, is found in organic and inorganic forms. In general, organic forms of arsenic appear to be of low toxicity, and different organic forms are found naturally in animals and plants.[63] Inorganic arsenic at very low doses is an essential trace element for some animals, but at higher doses is considerably more toxic, as its reputation as a tool of poisoners suggests.

The primary commercial use of inorganic arsenic is in wood preservatives, accounting for over two-thirds of commercial arsenic use. Arsenic-treated wood is resistant to decay and is widely used for outdoor building purposes. Agricultural chemicals account for most of the remaining commercial use.

Ingestion is a major route of arsenic exposure. In some areas, naturally occurring arsenic contaminates groundwater supplies.[64–66] Inorganic arsenic may also be ingested in nutritional supplements containing dolomite and bone meal, as well as in certain folk medicines.[67] Glassmaking and metal smelting are other sources of arsenic exposure. Finally, although the EPA has banned the production of inorganic arsenic pesticides for use on food crops, existing stores of these pesticides may still be used.

Distribution in the Body

Inorganic arsenic is well absorbed in the gastrointestinal tract, from the lungs, and to a lesser degree through the skin.[68,69] Animal studies have shown that arsenic distributes readily from the mother to the fetus and to all organs in the body.[70,71] In addition, these studies suggest that the placenta may selectively concentrate arsenic, although any effect this may have on the developing fetus has not been assessed. Limited evidence from human studies appears to show that arsenic distributes similarly in humans.[72] Finally, arsenic transfers into the milk of cows, goats, and humans.[73,74]

Arsenic in Wood Products

A family of eight living in a rural area of Wisconsin developed a series of health problems over a period of three years.[75] Their symptoms included recurrent rashes, respiratory problems, fatigue, muscle cramps, and diminished sensation in the hands and feet. Their symptoms were worst in the winter and spring.

The two infants in the family suffered worse effects than other family members. These children, who frequently crawled about on the floor wearing only a diaper, had red, peeling skin, bruising, bleeding, and seizure disorders. In addition, the children became completely bald. The infant, who was born prematurely, suffered recurrent severe pneumonia.

Health care workers suspected an environmental source for the family's illness. Biological monitoring revealed high arsenic levels in the hair and fingernails of all family members. Investigation of the home revealed high levels of arsenic in and around their wood-burning stove, the primary source of heat in winter. In fact, the father had been burning plywood scraps from a nearby construction site in the stove—scraps that had been treated with chromium copper arsenate, a common wood preservative.

Although we cannot be sure that all of the health effects that this family suffered are directly due to arsenic exposure, case reports such as this are important sources of information about the health effects of environmental exposures. When evaluated along with animal studies, we can begin to draw connections between controlled, experimental studies and actual life exposures. Cases such as these suggest areas for research and ways to avoid such exposures in the future.

Adverse Effects on the Fetus

High-dose exposure to arsenic has adverse effects on fetal development in animals. A distinctive pattern of malformations, including dose-related effects on brain and spinal cord development, malformed or missing eyes, failure of development of the kidneys and reproductive organs, and certain skeletal malformations, has been consistently reported.[76–78] In addition to malformations, arsenic causes significant reductions in litter size, increased intrauterine death and postnatal mortality, as well as growth retardation.[79,80] Finally, one study in mice suggested that maternal exposure to arsenic might lead to cancer in offspring.[81]

People living in an area with arsenic-contaminated drinking water had increasing risk of spontaneous abortion and stillbirth with higher levels of arsenic in the water.[82] Two case control studies found less clear evidence of arsenic's effects on the fetus—one suggesting a link between arsenic exposure and a particular heart defect, and the other showing a small but not statistically significant association between arsenic levels and spontaneous abortion.[83,84] Finally, a series of studies conducted on workers and residents exposed to smelter emissions in Sweden reported a variety of adverse reproductive outcomes, including spontaneous abortion, low birth weight, and malformations.[85–88] Because the smelter emissions contained a combination of arsenic, lead, cadmium, and mercury, it is impossible to assess what role arsenic played in the outcomes.

Neurologic Problems

A small but worrisome body of evidence suggests that arsenic may affect neurologic development. Mice exposed to arsenic before birth made more errors in learning a path through a maze.[89] In another study, rats from two to sixty days of age were given arsenic, while neurological development was still occurring. One hundred days after treatment, they had changes in both behavior and levels of neurotransmitters in the brain.[90,91]

Two human studies have reported that arsenic exposure may lead to hearing loss in children.[92] In one case, over 12,000 infants in Japan were accidentally poisoned with inorganic arsenic in dry milk. Fifteen years later, many of these children showed disturbances of central nervous system function, including severe hearing loss in 18 percent of the 415 children examined in follow-up studies. In another study, children living near a coal-fired power plant in Czechoslovakia that emitted large amounts of arsenic were found to have higher-than-expected rates of hearing loss.

Summary

In its inorganic forms, arsenic is highly toxic, widely used, and widely distributed in the environment. It is easily absorbed and distributed throughout the body, passing readily into the fetus, concentrating in the placenta, and passing into breast milk. A characteristic set of malformations occurs in animals exposed to high levels of arsenic. Human studies have been limited in both number and clarity, but suggest a connection between arsenic exposure and spontaneous abortion and stillbirth. A small number of animal and human studies suggest that arsenic may have effects on neurologic development, particularly affecting hearing. Although further study is needed, there is evidence indicating that arsenic may be a significant reproductive and developmental toxicant.

Manganese

Manganese, necessary to human growth and development at low levels, is naturally quite abundant in the environment. It is found in many foods, such as grains, cloves, and tea. But at high levels, manganese is toxic to the brain and the lungs, and inhalation of it appears to be much more hazardous than eating it in foods.

A major environmental source of manganese is emission from coal-fired power plants. Occupational exposure occurs in mining and metal products manufacturing (particularly iron and steel), dry-cell-battery manufacture, and the manufacture and use of certain paints, fertilizers, fungicides, and fireworks. Manganese, in the form of permanganate, is used in glass and ceramic manufacture. The neurologic and reproductive hazard of manganese is currently an extremely important issue because this substance is now being added to gasoline as an antiknock agent.

Manganese in Gasoline

A new octane enhancer and antiknock agent was developed in the 1970s for use in gasoline. This compound, methylcyclopentadienyl manganese tricarbonyl (MMT), was used briefly in the late 1970s during the oil crisis. Since then, it has been added to gasoline in Canada, where it was recently banned, but has not been used in the United States. The U.S. EPA refused to approve MMT for sale in this country until further investigation of the possible health effects. The Ethyl Corporation, however, challenged the EPA in court and in 1995 won the right to add MMT to fuel sold in this country. The corporation was founded in the 1920s to market tetraethyl lead, the infamous gasoline additive that led to dangerous lead exposure to the U.S. population and was phased out and finally banned as recently as December 31, 1995.

The court ruling allowing addition of MMT to gasoline stated that under the Clean Air Act, the EPA may concern itself only with MMT's effect on automobile pollution control systems and cannot prevent use of MMT on the basis of possible toxicity to humans. This additive may now be present in the gasoline sold throughout the country, except in regions where reformulated gasoline is required.

MMT is known to be extremely toxic at high doses. At low doses, the effects are essentially unstudied and therefore unknown. Many scientists are concerned that manganese may have subtle adverse neurologic effects, particularly in children, and that we may be witnessing a development similar to the original addition of lead to gasoline in the 1920s.

Toxicity to Adults

At high doses, such as found in some workplaces, manganese causes a degenerative neurologic condition similar to Parkinson's disease. This disease, known as manganism, begins as a loss of appetite, apathy, fatigue, leg weakness, and pain. It progresses steadily, and the final stages include an expressionless, masklike face, difficulty initiating movements, a shuffling walk, and tremors. Inhalation of manganese produces an inflammatory reaction, increasing susceptibility to pneumonia and bronchitis.[49]

Effects on Male Reproduction

Studies in mice and rats have found that male animals exposed to manganese during fetal development, at doses below those that caused other toxic effects, have retarded growth of the testes.[93] Further investigation revealed that testosterone concentrations are reduced in the exposed animals. Oral administration of manganese oxides to infant animals leads to accumulation of manganese in the hypothalamus and the pituitary. These two important regions of the central nervous system control a variety of hormonal systems, including the production of reproductive hormones, such as testosterone. The researchers who conducted these studies suggest that manganese may interfere with the male reproductive system by damaging hormone production.[94]

This research is supported by similar findings in a human study. Workers exposed at levels averaging one-fifth of the allowable workplace exposure limit had significantly fewer children during the period of exposure compared to similar unexposed workers.[43] A subsequent study in which workers were exposed at slightly lower levels, however, found no effect on birthrates.[95] Another study in male workers exposed to levels within the allowable workplace limits found effects on hormone levels.[96] Both prolactin and cortisone levels were significantly higher among exposed workers, suggesting the potential for interference with reproductive processes.

Absorption and Distribution in the Fetus or Newborn

In adult animals, only a tiny proportion of a manganese dose (about one-quarter of 1 percent) enters the brain. In contrast, in newborn animals, as much as 4 percent of a dose of manganese enters the brain.[97] To compound the problem, the newborn lacks the ability to eliminate manganese from its body.[98] In both humans and animals, gastrointestinal absorption of manganese is greatly increased in pregnant mothers and in newborns. For example, the newborn rat absorbs 70 percent of an oral manganese dose, while the adult absorbs only 2 percent.[99] Finally, manganese is known to pass from the mother to the infant across the placenta and in breast milk.[100]

Fetal Development

Mice exposed to manganese by one-time injection had growth-retarded fetuses with a high proportion of exencephaly, a birth defect in which the skull does not close. There was also an increase in fetal death. All of these effects occurred at the lowest doses tested, but doses were high compared to likely human exposures.[101] Another study in mice reported an increase in late resorptions, similar to spontaneous abortions, and delay in development of the bony skeleton at levels below those causing toxicity to the mothers.[102] Animals exposed prenatally to a very low-dose mixture of six metals, including manganese, cadmium, and lead, displayed severe growth retardations, suggesting a synergistic effect of various toxic metals in combination.[103]

The only human studies evaluating birth defects involve the population of a small island off the coast of Australia, where major natural manganese deposits contaminate water and food. Preliminary surveys of this exposed population revealed a higher-than-expected number of stillbirths and an apparent excess of the deformity clubfoot.[104] It is unclear if these observed outcomes are due to the manganese exposure or to other factors.

Neurologic Toxicity

Significant chemical changes in the brain and abnormalities in neurologic development have been reported in exposed infant animals. Reports have included reduced production of dopamine and excess of acetylcholine, important neurologic transmitters that must be maintained in delicate balance for normal function of the brain. These animals also have significantly lower activity levels and less exploratory behavior than unexposed animals, suggesting a neurotoxic effect.[93,105]

The population of the island off the coast of Australia with environmental exposure to manganese contains a large group of people with severe neurologic problems. One neurologic syndrome has its onset in infancy and progresses very slowly for a few years before remaining stable for a lifetime. It consists of weakness and muscle atrophy in the legs, leading to abnormal walking or, in severe cases, inability to walk or even sit up without assistance. All of these children also have clubfoot, scoliosis of the spine, and some other mild abnormalities of the joints and skin, but they are intellectually fairly normal.

A second syndrome occurs later in life, and consists of clumsiness, unsteadiness, staggering, tremor, weakness, and an expressionless face. This second syndrome is very similar to the Parkinsonian syndrome that affects workers exposed to manganese as adults. No other human studies exist on the possible neurologic effects of manganese in the developing fetus.

Summary

Although manganese is an essential mineral at low doses, overexposure may pose a hazard to human reproduction and development. It is likely that infants may experience overexposure to manganese at levels that are harmless to adults, and animal studies show evidence of growth retardation in fetuses, damage to the testes and sperm in young males, and some birth defects. There is evidence of neurologic damage to infants, which is not surprising, because manganese is known to be toxic to the brain, even in adults. Although human studies are insufficient, there is enough information to be concerned about the effects of manganese and to require further human studies before exposure increases from the addition of MMT as a gasoline additive.

Organic Solvents

4

Organic solvents are widely used in industry and at home. There have been many human studies on the reproductive and developmental effects of solvents. Although these studies are often unable to pinpoint specific solvents or specific doses of exposure, they have revealed a number of worrisome health effects.

Animal studies show variable effects on reproduction and development from one solvent to another, but many, if not most, of the solvents tested have been shown to be toxic to the fetus in animals. A few solvents cause birth defects in animals, and some have effects on male reproductive function. Unfortunately, animal studies almost always use a high dose of only one solvent, while humans are exposed to low or moderate levels of numerous solvents every day. Thus, most reports of effects in humans involve mixed solvents and may not allow us to identify one culprit, yet animal studies may not accurately reflect human risks.

In humans, there is consistent evidence that solvents may raise the risk of spontaneous abortion among exposed women by two- to four-fold. Two studies show an increased risk of spontaneous abortion among wives of men exposed to solvents. Solvents may increase the risk of certain structural birth defects in humans, particularly those of the central nervous system, urinary system, heart, lip, and palate. This area urgently needs further research. And one important study suggests that solvent exposure may predispose to preeclampsia, or toxemia of pregnancy. Finally, defects of the central nervous system and childhood cancers of the brain and urinary tract, as well as leukemia, may occur in offspring of exposed fathers at rates two to three times that of the general population, contradicting the previously accepted wisdom that only maternal exposures affect the fetus and child.

Solvents are characterized by their ability to dissolve other substances. They are generally liquids and can be water based or hydrocarbon (petroleum) based. The hydrocarbon-based solvents are known as *organic solvents*. Because of their tendency to evaporate at room temperatures, many organic solvents are also known as volatile organic compounds (VOCs). Organic solvents are used in an enormous variety of products. The most widely used organic solvents fall into several categories with varying possible reproductive toxicities:[1]

Aromatic hydrocarbons (benzene, toluene, xylene, styrene, phenol)
Aliphatic hydrocarbons (hexane, octane)
Chlorinated derivatives (trichloroethylene, perchloroethylene, 1,1,1-trichloroethane, methylene chloride, chloroform)
Alcohols (ethanol)
Aldehydes (formaldehyde)
Glycol ethers (ethylene glycol monomethyl ether, ethylene glycol monoethyl ether)
Complex solvent mixtures (gasoline)

People may be exposed to solvents at work in electronics, health care, dry cleaning, auto repair, laboratories, painting, and numerous other occupations. Household exposure to solvents may come from paints, strippers, glues, markers, cosmetics, correction fluids, and some cleaning agents. Pesticides frequently contain solvents as inert ingredients. Solvents contaminate drinking water in some areas, and airborne exposure may occur from dry cleaning shops or other facilities that emit large quantities of solvents. Toxic waste sites frequently contain solvents, and exposure may occur on or near the site through air, water, and soil contamination.

Organic solvents have physical properties that allow them to enter the human body easily: they evaporate in air at room temperature and are therefore easily inhaled; they penetrate the skin easily; and they cross the placenta, sometimes accumulating at higher doses in the fetus.[2] In addition, many solvents enter breast fat and are found in breast milk, sometimes at higher concentrations than in maternal blood.[3] Solvents contaminating drinking water enter the body through skin absorption and inhalation in the shower, as well as through drinking. In fact, the total exposure from taking a ten-minute shower in contaminated water is greater than the exposure from drinking two quarts of the same water.[4] Solvents are generally short-lived in the environment and in the human body, lingering for no more than several days. On the other hand, exposures may occur daily.

Reproductive and Developmental Effects in Humans

A large number of human epidemiological studies have examined the reproductive effects of solvents. In most, people were exposed to complex mixtures of these chemicals at work or in their environment, so the studies rarely allow us to pinpoint specific solvents as responsible for the observed reproductive effects. Animal testing has looked almost exclusively at one solvent at a time, and provides information about the variability of effects within this class of chemicals. The majority of the animal studies are discussed in the solvent profiles later in the chapter. The rich scientific literature on the reproductive effects in humans from exposure to solvent mixtures is the subject of the first part of this section.

Organic Solvents and Spontaneous Abortions

The increased risk of spontaneous abortion in women occupationally exposed to solvents was initially identified in Finland, where there is a nationwide database on births and spontaneous abortions. Finnish workers potentially exposed to organic solvents may undergo blood and urine testing for solvents at the Finnish Institute of Occupational Health.[5] In the Finnish studies, the biological measures were supplemented with questionnaire information about exposures.

Women who suffered a spontaneous abortion were consistently two to four times more likely to have been exposed to organic solvents during pregnancy.[5-8] Similar studies performed in the United States have come up with almost identical results. A group of California women who had spontaneous abortions were over three times more likely to report having been exposed to organic solvents at work compared with a group of otherwise similar women who had normal births.[9] Semiconductor workers and laboratory workers who are exposed to solvents also have an increased risk of spontaneous abortion.[7, 10–12] Specific solvents mentioned include perchloroethylene (PCE), trichloroethylene (TCE), glycol ethers, and aliphatic solvents, but almost all the women were exposed to complex mixtures.[13] The human studies on solvent exposure and spontaneous abortion are presented in table 4.1.

Although a variety of individual birth defects have been reported in association with organic solvents in one or more studies, for certain defects, the evidence linking them to solvent exposure is more significant and consistent. These defects include cleft palate and lip, and cardiovascular malformations. There is also some suggestive evidence regarding solvent exposures and defects of the urinary tract and central nervous system (see table 4.2). Because birth defects are fairly rare outcomes, there are few published cohort studies. Instead, most research on birth defects uses case control methods and relies on birth defects registries to identify children born with certain types of defects. Researchers then attempt to contact the mothers to ask them to recollect exposures during pregnancy. This study design, however, has limitations, which include weaknesses in some birth defects registries and differential recall of exposures engendered by the adverse outcome.

One interesting report noted a cluster of infants born with heart abnormalities in a neighborhood near Tucson, Arizona. The area had groundwater contaminated with the solvent trichloroethylene, and trace amounts of dichloroethylene and chromium. Investigators found a significant increase in heart defects in the contaminated zone.[20] Other studies have also shown associations between solvent exposure and cardiac malformations.[21,22] There are numerous questions about the degree of risk, the vulnerable time period, or the amount of exposure necessary to increase the risk, yet there is evidence implicating solvents as a potential cause of birth defects involving the heart, lip and palate, and perhaps also the nervous system and urinary tract.

Other Effects: Infertility, Low Birth Weight, and Preeclampsia

In addition to the increase in spontaneous abortions, one study showed a 25 to 50 percent decrease in fertility among women occupationally exposed to organic solvents. This apparent effect on fertility was particu-

Table 4.1   Studies on spontaneous abortion and solvent exposure in women

Location	Study type	Solvent	Spontaneous abortion incidence*
California[9]	Cross-sectional	Various unspecified	4.4 times more likely
Finland[5]	Case control	Various unspecified	2.2 times more likely
Finland[6]	Case control	Various unspecified	2.2 times more likely
		Methylene chloride	2.3 times more likely
Finland[7]	Case control	Toluene	4.7 times more likely
		Xylene	3.1 times more likely
		Formaldehyde	3.5 times more likely
Finland[14]	Case control	PCE	3.6 times more likely
Massachusetts[10]	Case control	Glycol ethers	2.2 times more likely
California[13]	Case control	Various unspecified	1.1 times more likely NS
		PCE	4.7 times more likely
		TCE	3.1 times more likely NS
California, Utah[11]	Retro-cohort	Glycol ethers	1.4 times more likely NS
Eastern United States[12]	Retro-cohort	Glycol ethers	2.8 times more likely
Singapore[15]	Retro-cohort	Toluene	2.8 to 5.7 times more likely
Santa Clara, CA[16]	Retro-cohort	1,1,1-TCA	2.3 times more likely
Santa Clara, CA[17]	Retro-cohort	1,1,1-TCA	1.4 times more likely NS
Italy[18]	Retro-cohort	PCE	4.0 times more likely NS
California[19]	Prosp-cohort	Trihalomethanes	1.8 times more likely

Notes: NS = not statistically significant; all other results statistically significant at the 0.05 level.
PCE = perchlorethylene. TCE = trichloroethylene. 1,1,1-TCA = 1,1,1-trichloroethane.
* In a case control study, this means that women who had a spontaneous abortion were more likely to have been exposed to organic solvents during pregnancy. In a cohort study this means that women who were exposed to organic solvents were more likely to have a spontaneous abortion.

Table 4.2 Maternal exposure to solvents and birth defects

Location	Study type	Solvent	Defect	Birth defect incidence★
Finland[31]	Case control	Various	Cardiac-VSD	1.5 times more likely
Finland[32]	Case control	Various	Cardiac-VSD	1.4 times more likely
Finland[27]	Case control	Various	CNS	No increase
Finland[25]	Case control	Various	CNS	Increased (no odds ratio provided)
Finland[24]	Case control	Various	CNS, cleft palate	5.5 times more likely
Finland[26]	Case control	Various	Cleft lip/palate	4.5 times more likely
Europe[29]	Case control	Glycol ethers	Cleft lip/palate	2 times more likely
France[28]	Case control	Various	Cleft lip/palate	8 times more likely
			Gastrointestinal	12 times more likely
			CNS	No increase
Canada[30]	Case control	Toluene/aromatics	Urinary tract	3.8 times more likely
New Jersey[23]	Case control	Trihalomethanes	CNS, cleft lip/palate	3 times more likely
		Trichloroethylene	CNS	2.5 times more likely
		Carbon tetra-chloride	Cleft lip/palate	2.2 times more likely
			CNS	3.8 times more likely
		Perchloroethylene	Cleft lip/palate	3.5 times more likely
Massachusetts[33]	Case control	Trichloroethylene/ perchloroethylene/ chloroform	CNS	4.5 times more likely
			Eye/ear	14.9 times more likely
			Cardiac	No increase
Maryland[34]	Case control	Various	Cardiac	1.6 times more likely
Arizona[20]	Cohort	Trichloroethylene	Cardiac	3.0 times more likely

Notes: NS = not statistically significant; all other results statistically significant at the 0.05 level. VSD = ventricular septal defect (a particular heart malformation). CNS = central nervous system (brain).
★ In a case control study, this means that mothers of babies with this birth defect were more likely to have been exposed to organic solvents during pregnancy. In a cohort study, this means that women who were exposed to organic solvents were more likely to have a baby with this defect.

larly strong among women working in dry cleaning, and those exposed to halogenated hydrocarbons such as perchloroethylene.[35] A more recent study found a 75 percent increased risk of infertility in women occupationally exposed to volatile organic solvents.[36] These results are strengthened by the fact that the study looked only at women with medically diagnosed infertility, but because researchers relied on subjects' memory to assess exposure, the results may have been biased toward finding an association.

An investigation of a New Jersey population exposed to solvent-contaminated drinking water revealed an association between exposure to certain solvents, particularly the trihalomethanes (volatile organic compounds including chloroform, which arise as a result of water chlorination) and carbon tetrachloride, with low birth weight and small size for gestational age.[23] This association was supported by a similar study in Iowa focusing on chloroform in drinking water.[37] People in New Jersey living adjacent to the worst U.S. Superfund toxic waste site, which emitted airborne volatile solvents for nearly a decade, were found to have a fivefold risk of having a low–birth–weight baby during the period of greatest contamination. These families also had twice the risk of having a premature infant.[38] Solvents implicated in this study included benzene, bis (2-chloroethyl) ether, methylene chloride, 1,2-dichloroethane, ethylbenzene, 4-methyl-2-pentanone, toluene, and xylene. This study is quite persuasive because the most likely sources of bias would tend to result in underestimating a true effect. A North Carolina study looking at trihalomethanes in drinking water and miscarriage, preterm birth, and low birth weight failed to find any consistent association.[39] This study did not look for full-term babies who were small for gestational age and did not consider exposures from showering. In summary, there is some evidence of an association between solvent exposure and smaller full-term babies, with conflicting evidence on the more nonspecific finding of low birth weight.

One well-designed prospective study showed a fourfold increased risk of preeclampsia, also known as toxemia of pregnancy, in women exposed to solvents.[40] Preeclampsia is a potentially life-threatening condition of late pregnancy consisting of hypertension, protein in the urine, generalized swelling (edema), and eventual seizures if untreated. Solvent exposure can lead to kidney injury, a presumed cause of preeclampsia.[41]

Reproductive Effects in Men

The effects of most solvents on men are not clearly understood, and research findings in this area have been contradictory except regarding the short-chain glycol ethers (see table 4.3). Animal and human studies on the glycol ethers indicate that they damage testicular function, lower sperm counts, and can cause infertility. In animals, short-chain glycol ethers lead to testicular atrophy (see the glycol ethers profile).[42]

There is some consistency in the finding that offspring of solvent-exposed men face increased risk of birth defects and low birth weight. Male spray painters and body shop workers have twice been shown to be at increased risk of fathering a low-birthweight baby.[43,44] Children of solvent-exposed men may also have an increased risk of birth defects including anencephaly (partial or complete absence of a brain).[45,46] One rare birth defect, the Prader-Willi syndrome (consisting of mental retardation, obesity, muscle weakness, and poor testicular or ovarian function), has been associated with paternal exposure to hydrocarbons and appears to be due to a chromosomal deletion transmitted by the father.[47] This is an important finding because it was previously believed that only maternal exposure could lead to adverse effects in the fetus.

An assessment of sperm quality among men exposed to perchloroethylene in dry cleaning found differences in sperm shape and swimming ability, but no overall difference in sperm count compared to an unexposed group.[48] Painters exposed to mixed solvents have increased rates of sister chromatid exchange, a chromosome abnormality. This finding is a measure of toxicity to genes on the chromosome and may imply a risk of birth defects in offspring.[49] Two studies have shown slight increases in spontaneous abortion in wives of solvent-exposed men.[6,50] Two others did not find any increased risk to wives, but one of these two did find an increased risk of infertility in these couples.[10,51]

Childhood Cancer

Fifteen of twenty studies that looked at parental solvent exposures and childhood brain tumors found an association between the two, though the finding was statistically significant only in ten.[55–57] The association was strongest when the father was exposed on the job to gasoline, trichloroethylene, methylethylketone, or freon.[58,59] Childhood

Table 4.3 Male exposure and adverse reproductive effects of solvents

Location	Study type	Solvent	Defect	Adverse effect incidence
United States[49]	Cross-sectional	Toluene, mixed solvent	Sister chromatid exchange	Positive (no odds ratio or risk ratio provided)
United States[48]	Cross-sectional	Perchloroethylene	Semen quality	Mixed
United States[52]	Cross-sectional	Glycol ethers	Semen quality	Positive
United States[53]	Cross-sectional	Glycol ethers	Semen quality	Positive
United States[54]	Cross-sectional	Glycol ethers	Semen quality	Mixed
Finland[50]	Case control	Organic solvents	Spontaneous abortion in wives	2.7 times more likely
			Birth defects	1.0 time more likely NS
Finland[6]	Case control	Various unspecified	Spontaneous abortion in wives	2.7 times more likely
United States[45]	Case control	Various unspecified	Anencephaly	2.5 times more likely
United States[47]	Case control	Various unspecified	Prader-Willi syndrome	1.9 times more likely
United States[10]	Retro-cohort	Glycol ethers	Spontaneous abortion in wives	2.4 times more likely NS
United States[51]	Retro-cohort	Perchloroethylene	Spontaneous abortion in wives	Negative
			Infertility	2.5 times more likely NS
United States[43]	Retro-cohort	Various unspecified	Low birth weight	1.6 times more likely
			Other effects	Negative
Sweden[44]	Retro-cohort	Toluene, mixed solvent	Low birth weight	Positive
			Other effects	Negative

★ NS = not statistically significant. All other results statistically significant at the 0.05 level.

cancers of the urinary tract show a similar association. All eight studies on this topic showed elevated risk of these cancers in offspring of solvent-exposed parents, although only four had statistically significant effects.[60,61]

The third childhood cancer showing an association with parental exposure is childhood leukemia. Parental exposure to gasoline has been particularly implicated in this malignancy.[62,63] This result is not surprising because of the known association of benzene, a constituent of gasoline, with adult leukemia. Several studies of childhood leukemia have pointed to other solvent-exposed occupations, including spray painting and beauty shop work.[64] Thus, there is some consistent preliminary evidence that solvent exposures to either parent may result in an increased risk of neurological and urinary cancers, and acute leukemias in their children.[55,65]

Childhood Leukemia in Woburn, Massachusetts

In the 1970s, residents of Woburn, Massachusetts became concerned about a large number of cases of childhood leukemia in their community. In May 1979, two of the wells supplying drinking water to part of the town were found to be contaminated with trichloroethylene, perchloroethylene, and chloroform, which leached into the well water from nearby toxic waste sites. Although the wells were promptly closed, they had supplied some portion of the town's water intermittently for fifteen years.[66]

Subsequent investigations confirmed a cluster of leukemia in children under the age of fifteen in the eastern section of town. In a town of that size, only six cases of childhood leukemia would be expected over twenty years based on national rates of disease; Woburn had twenty-eight cases in that time period. Extensive investigation has failed to reveal any other reason for the excess of leukemia.[67] The exposed community also had an increased risk of perinatal death and of certain birth defects, particularly of the eye and brain. Finally, exposure to the contaminated well water was associated with childhood diseases of the urinary tract and the lungs.[33]

Chlorinated solvents cause cancers in laboratory animals, and the time period of exposure to the well water fits with what is known about the time course of leukemia after chemical exposures. The increase in perinatal death fits with other epidemiological studies that found an increased risk of spontaneous abortion and fetal death in women exposed to solvents, although the risk of spontaneous abortion did not appear to be elevated in the areas of Woburn supplied with contaminated water. Although we may never know exactly what happened in Woburn to lead to this tragedy among the children, the evidence implicating solvent exposure is compelling.

> Solvents in the water supply are a hazard even to those who drink bottled water. Over half of the exposure to solvents in drinking water comes from inhalation and absorption through the skin in the shower or bath.[68]

Overall Assessment of Epidemiologic Studies

A large body of epidemiological literature addresses the question of whether organic solvents may have adverse reproductive effects in humans. Although some studies have failed to show any increased risk of adverse reproductive outcomes,[69,70] these negative studies are a very small minority.[71,72] The evidence that solvents raise the risk of spontaneous abortion among exposed women by two- to fourfold is consistent. Two studies even show an increased risk of spontaneous abortion among the wives of men exposed to solvents.

The evidence for structural birth defects in children of exposed women is also fairly consistent for defects of the central nervous system, heart, lip, and palate. This area urgently needs further research. Though there is only one study on preeclampsia, the study was well designed, and the conclusions are persuasive. Organic solvents should be considered among the potential causes of preeclampsia.

A disturbing body of evidence suggests that defects of the central nervous system and childhood cancers of the brain and urinary tract, as well as leukemia, may occur in offspring of exposed parents, particularly fathers, at rates two to three times that of the general population.

Most of the studies discussed concern solvents generally, and only a few have identified specific possible culprits. The following sections focus on certain solvents that may be particularly responsible for some of the reproductive effects noted.

Solvent Profiles

Benzene

Uses   Paint, rubber, degreaser, septic tank cleaner, ingredient in gasoline, range of chemical processes.

Routes of exposure   Occupational: some manufacturing jobs, gas stations, refineries; rubber manufacture. Environmental: contaminated drinking water, tobacco smoke, gasoline stations.

Reproductive effects In animals: damages fetal blood-producing cells; leads to bone deformities, reduced fetal weight. In humans: maternal and paternal exposures linked with neural tube defects, cancers; maternal exposures with cardiac defects and low birth weight; damages testicular function; may affect menstruation.

Benzene has long been recognized as a known cause of cancer in humans. Although its effects on reproduction and development have been less well studied, there is evidence in both animals and humans that benzene also interferes with these processes.

California conducted an extensive review of the scientific literature before concluding that benzene is a reproductive toxicant.[73] The review summarized studies in rabbits, rats, and mice that consistently found fetal growth retardation and delayed bone formation in animals exposed before birth. In some cases, these effects were seen at levels that did not produce maternal toxicity. Benzene does not appear to cause malformations in prenatally exposed animals. In mice, benzene exposure resulted in fetal chromosomal abnormalities, as well as changes in the blood-forming cells in the liver and spleen. Finally, benzene has adverse effects on testicular and sperm form and function in animals.

Data on human effects have been fairly limited but suggest a hazard. An early study from Eastern Europe reported menstrual disturbances in women who work with benzene, and another reported prolonged or heavy menstrual bleeding in women exposed to a mixture of benzene, toluene, and xylene.[74,75] More recently, researchers have found fetal effects after exposure through contaminated drinking water. In a study conducted in seventy-five New Jersey towns, mothers whose drinking water was contaminated with benzene were more likely to have a child with neural tube defects or major heart defects.[23] In Michigan, the presence of benzene and chlorinated solvents in drinking water was associated with an increased likelihood of low birth weight.[76] This association was as strong as the association between low birth weight and poor prenatal care, but did not reach statistical significance, possibly due to the small sample size. Finally, men exposed to benzene were more likely to father a child with anencephaly or spina bifida, malformations of the brain and spinal cord, respectively.[77]

Perhaps most worrisome is evidence that parental exposures may lead to childhood cancer. One study found that the mother's exposure to benzene in the year prior to the child's birth significantly increased the risk of childhood cancer.[78] Parental employment in industries where benzene is heavily used is associated with the development of a variety of childhood cancers, including leukemia, lymphoma, brain, urinary tract, and nervous system cancers.[57,79–82] Father's employment in gasoline-exposed jobs has also been linked with increased rates of childhood cancer.[59,60,83] It is impossible to say whether benzene exposure alone is responsible for these results, because people in these occupations may be exposed to a variety of chemicals. Still, given what we know about chromosomal damage from benzene and the fact that it is a known carcinogen in adults, this evidence is indicative of a real risk of childhood cancer from parental benzene exposure.

Benzene is an important hazard to reproduction and development. Its ability to damage chromosomes is unquestioned, and the probability that this damage can lead to adverse effects in the children of exposed individuals is supported by several studies. Less dramatic, but still troublesome, are the connections between environmental benzene exposure and low birth weight. Animal studies indicating testicular toxicity and limited human studies indicating menstrual dysfunction require further investigation.

Chloroform

Uses   Refrigerant, synthesis of fluorocarbons for plastics, propellants, synthesis of antibiotics, grain fumigant, general solvent, dry cleaning spot remover, dental work.[84,85]

Routes of exposure   Occupational: paper mills; wastewater, sewage and drinking water treatment plants; waste incinerators. Environmental: drinking chlorinated water, bathing, indoor pools and spas, breathing indoor air, eating some foods.

Reproductive effects   In humans: water chlorination by-products (including chloroform) are associated with smaller babies, spontaneous abortion, possibly some birth defects. In animals: decreased fetal weight, growth retardation, malformations, changes in testes and ovaries.

Chloroform, which can occur naturally, is widely distributed in the environment due to human activities. The most widespread human exposures are due to chlorination of drinking water. The addition of chlorine to water containing organic materials results in the production of chloroform.[4]

Animal studies suggest that exposure to chloroform can result in reproductive and developmental problems. Inhaled chloroform can result in decreased fertility, increased fetal malformations, fetal weight reduction, and developmental retardation in rodents, even at doses that are not toxic to the mothers.[86–88] Rats given chloroform orally also have decreased fetal weights.[89] Some reports found adverse effects in the fetus only at higher doses that were also maternally toxic. One study on oral intake of chloroform in rats and rabbits found no evidence of birth defects, and reduced fetal weights only at high exposure levels that were toxic to the mothers.[90] A number of animal studies have found other disturbing outcomes. In one, both male and female rats orally exposed to chloroform showed signs of atrophy of the ovaries and testes.[91] In another, mice who inhaled high levels of chloroform for five days had significantly more abnormal sperm than unexposed mice.[92,93] Researchers looking at the effects of chloroform exposure on two generations of mice found no effects on fertility, but did note changes in testicular tissue in some of the first-generation mice.[94] The significance of these studies is hard to judge because a number of other studies in mice, rats, and dogs found no effects on the reproductive organs.[95–97]

Only limited information is available on the effects of chloroform on human reproduction and development. Two cases of preeclampsia, a complication of pregnancy, were reported in laboratory workers exposed to chloroform at six to twenty times the recommended exposure limit, a level that caused liver problems in other exposed workers.[98] More worrisome perhaps are studies that show a possible link between chlorinated drinking water and developmental problems in infants. In an Iowa study, researchers found a connection between maternal exposure to chloroform in drinking water and both low birth weight and intrauterine growth retardation.[19] However, a number of other chlorinated and brominated organic chemicals were also found in the water. In most of the studies on drinking water, the role of chloroform is difficult to interpret. In some cases, researchers measured total levels of trihalomethanes, the class of chemicals to which chloroform belongs. One such study found correlations between trihalomethane exposure and reduced birth weight, small size for gestational age, central nervous system defects, cleft palate, and heart defects.[23] A second, more limited study found only a slight association with increased miscarriage at high levels of exposure. Unfortunately, the exposure measure was the number of glasses of water drunk per day, which is subject to recall problems and overlooks exposure from showering.[39] A more recent study using the same measure of exposure (reported number of glasses of water drunk each day, multiplied by level of trihalomethanes in the water) found nearly a doubling of miscarriage rates, which appeared to be due to bromodichloromethane, another trihalomethane found in drinking water.[19] Other studies have simply compared exposure to chlorinated versus nonchlorinated water. Researchers in Massachusetts linked exposure to chlorinated water to an increase in stillbirths, and researchers in Italy found a connection with small body and skull size, as well as an increased risk of neonatal jaundice.[99,100]

Chlorination of drinking water provides much of the United States with safer drinking water. An unfortunate result is that chloroform exposure in drinking water is widespread. The potential risk of reproductive and developmental effects must be viewed seriously, since both animal studies and the limited human evidence suggest reason for concern. A decision to reduce the use of chlorine disinfection would require the development of alternative techniques for providing safer water.

Water Disinfection By-products: A Complex Mixture and a Public Health Dilemma

Drinking water may be contaminated with pesticides and nitrates from agricultural runoff, metals from natural or man-made sources, and solvents from leaking storage tanks or toxic waste sites. Water can also be contaminated with microbes, and to prevent infectious disease, many water supplies are chlorinated. Chlorine kills most infectious organisms and is inexpensive. Unfortunately it reacts with organic compounds in the water to produce disinfection by-products (DBPs), a mixture of volatile chemicals, particularly trihalomethanes and haloacetic acids. People can be exposed to DBPs from drinking the water or through inhalation or skin absorption during showering or swimming.[4] Levels in indoor air rise any time hot water is run in the house.[101] Some bottled water has also been shown to contain DBPs, although some types of water filters can remove these compounds.

DBPs have been linked with bladder and colon cancer.[102] Now there is also evidence that these chemicals may be reproductive toxicants. Studies in Iowa, New Jersey, and California have implicated DBPs in low birth weight, small size for gestational age, spontaneous abortions, and even birth defects. These results are not conclusive, however, and at least one study failed to find evidence of harm. In animals, DBPs have caused reduced birth weight, heart malformations, and spermatotoxicity. It is not clear how to translate the high-dose effects seen in animals into the low-dose combinations found in water supplies.

Drinking water contaminants illustrate the complex mixtures that people are exposed to every day. These chemicals are present at low levels—far lower than any routinely evaluated in animal studies. Yet there is some evidence that they may have health effects, perhaps through interactive effects or because a certain constituent is toxic even at low doses during fetal life. Because of the health benefits of water chlorination, changes need to be implemented with great care. Improved filtration of water and ozone disinfection are a possible alternative. People should not have to choose between the risk of infectious disease and the risk of adverse pregnancy outcomes.

Epichlorohydrin

Uses   Manufacture of epoxy resins; semiconductor manufacture; chemical manufacturing; in some lubricants, adhesives, lacquers, paints, and pesticides.

Routes of exposure   Occupational: various manufacturing jobs. Environmental: some paints, lacquers, lubricants, adhesives. Contaminant of wines stored in epichlorohydrin resin vats, and of chemicals such as glycerin, food starch, and pharmaceuticals.

Reproductive effects   In animals: damages chromosomes; powerful male reproductive toxicant. Poorly studied in humans.

Although epichlorohydrin is not generally considered an organic solvent, its properties and uses allow consideration in this section. In rats and rabbits, epichlorohydrin decreases male fertility, causes abnormalities in sperm shape and swimming speed, and can lead to atrophy of the testes.[103,104] No effects are reported on fertility in female rats or on pregnancy outcomes in female mice and rats. Other research has shown that epichlorohydrin damages chromosomes in cell cultures, rats, and humans.[105] This raises the concern that the chemical may cause cancer, and makes it more likely to cause damage to the sperm, egg, or developing fetus.

Few reports exist on the effects of epichlorohydrin in humans, and these have involved only men. The two studies looking at human fertility found no effect of exposure on sperm count and reproductive hormone levels.[106,107] Both had low participation rates and failed to report on sperm shape, swimming speed, and other important aspects of the semen analysis. A number of human studies found chromosomal abnormalities in exposed workers. Epichlorohydrin is similar in structure to DBCP, a pesticide banned in the United States because it caused infertility in male workers.

Epichlorohydrin shows some of the characteristics of an important reproductive toxicant. As further research is done, epichlorohydrin should be treated as a chemical that is toxic to male reproduction, and because of the chromosomal toxicity, it should be considered likely to be toxic to pregnant women. Based on the animal evidence, epichlorohydrin is listed as a reproductive toxicant under California's Safe Drinking Water and Toxic Enforcement Act (Proposition 65), a right-to-know law that requires that people be informed about exposures to known carcinogens and reproductive toxicants and forbids discharge of such chemicals into sources of drinking water.

Formaldehyde

Uses Resins for particle board, plywood, insulation, table tops; rubber production; film manufacture; leather processing; dye production; cosmetics; hospitals; embalming.

Routes of exposure Manufacturing jobs, foam installation, funeral homes, hospitals, laboratories. Environmental: motor vehicle exhaust, new carpets, particle board, some furniture, wood stoves, cigarette smoke, emissions from resins, cosmetics.[108]

Reproductive effects Animals: damages reproductive organs at high doses. Humans: menstrual disturbances, increased risk of spontaneous abortion.

Many materials used in daily life emit formaldehyde for some time after manufacture, so many people are exposed to this chemical in their homes. Formaldehyde is a known irritant and a suspected carcinogen; evidence regarding its effects on reproduction and development is less clear, although human studies indicate reason for concern.

Formaldehyde damages the testes of rats after high-dose exposure, resulting in declines in sperm production, motility, and viability, as well as testicular degeneration.[109–111] Mice who inhaled near-lethal doses of formaldehyde had degenerative changes in the uterus and ovaries.[112] Rats

---

exposed to high levels of formaldehyde had disturbances of the estrus cycle, an effect that could have been due to stress from the irritant effects.[113]

Formaldehyde crosses the placenta in mice, and fetal animals eliminate it more slowly than adults.[113] However, even at doses that were highly toxic to the mothers, formaldehyde did not affect fetal size or increase the rate of malformations in mice, although it did slightly decrease litter size.[114] No malformations were reported in rats after inhalation exposure, but there was a slight increase in the length of gestation and an increase in average fetal weight.[115] Finally, beagles fed formaldehyde showed no physical effects on the mothers or pups, and no behavioral effects after birth.[116]

Although the animal evidence is ambiguous and generally negative, several human studies suggest reason for concern. Almost half of a group of women working in formaldehyde-exposed occupations reported having menstrual disorders, compared to less than 10 percent of women in unexposed occupations.[117] Another study also found increased rates of menstrual disturbances, though other factors may have been involved.[118] Cosmetologists, who are often exposed to formaldehyde, have an increased rate of spontaneous abortions.[119] Finally, lab workers exposed to formalin, a water-formaldehyde mixture, had more than a threefold increased risk of spontaneous abortion.[7] The evidence is clear of menstrual abnormalities with formaldehyde exposure, consistent with animal evidence of disturbed estrus and ovarian injury, although the latter studies were at high-dose levels.

Human studies have not looked at testicular function, although high-dose animal studies indicate a possible effect. Further study is also needed to confirm the observation of increased risk of spontaneous abortion, which was seen only in the human studies.

Glycol Ethers

Uses Jet fuel deicing, brake fluid, ink, dye, varnish, paint, printing, photography, circuit board production, cleaning solutions, some pesticides,[120] perfumes and cosmetics.[121]

Routes of exposure Occupational: where used as deicers, in cleaning solutions, or as additives in inks, dyes, or photographic chemicals. Environmental: home use of cosmetics, perfumes, paints, inks, varnishes, or stains.

Reproductive effects In animals: testicular toxicity, infertility in males, birth defects and toxicity to the fetus. In humans: damage to male reproduction, possible risk of spontaneous abortion, possible birth defects.

The glycol ethers are a class of related compounds. The short-chain glycol ethers—including ethylene glycol monomethyl ether (EGME), ethylene glycol monoethyl ether (EGEE), ethylene glycol monomethyl ether acetate (EGMEA), and ethylene glycol monoethyl ether acetate (EGEEA)—are reproductive toxicants. Other glycol ethers may also be hazardous to reproduction based on limited animal studies.[42] Animal studies demonstrate reproductive toxicity at low doses, close to those encountered in occupational settings.[121]

In male animals, glycol ethers cause microscopic testicular damage, testicular atrophy, spermatotoxicity, and infertility.[122-125] In female animals, these compounds cause infertility, prolonged pregnancy, and increased reabsorptions.[42] These solvents lead to decreased fetal weight, abnormalities in the bony skeleton, and birth defects in the offspring, including defects of the heart, kidneys, and urinary system.[126-128] In addition, there is some evidence that exposure to some glycol ethers during development affects later neurologic function in offspring.[129] Similar effects have been found in five animal species, increasing the likelihood that humans are also affected.

In humans, two studies show lowered sperm counts in exposed workers.[52,130] Another smaller study found no effect on sperm count, but did find decreased testicular size in occupationally exposed men.[54] There is one case report of a woman who used a cleaning product containing EGMEA throughout two pregnancies and had two sons with hypospadias, an abnormality of the penis.[131] Women in the semiconductor industry have a significantly increased risk of spontaneous abortion and reduced fertility; these effects have been attributed to exposure to glycol ethers.[10-12,132-134]

A large multicenter study in Europe using six regional birth defects registries identified women who had a child, a stillbirth, or an aborted fetus with a birth defect and matched these women with controls who had healthy babies. All women were contacted and questioned about their occupation, and experts ranked the probability of occupational exposure to glycol ethers. Women who had a child with a birth defect were 44 percent more likely to be rated occupationally exposed to glycol ethers. The risks increased to 94 percent for central nervous system defects and over twofold for cleft lip and multiple anomalies. Most of the sources of bias in this study would tend toward underestimating actual risk. In this case, exposures were not confined to the four short-chain glycol ethers but encompassed the entire class of these compounds.[29]

The short-chain glycol ethers may lead to reduced fertility, spontaneous abortion, a variety of birth defects, and behavioral changes in the offspring. The National Institute of Occupational Safety and Health[135] and the state of California[136] have designated the four short-chain glycol ethers as known reproductive and developmental toxicants.

Methylene Chloride

Uses Paint and varnish remover, degreaser, aerosol propellant, decaffeination of coffee, food processing, fumigant for grains and fruits, urethane foam production, pharmaceutical manufacture, acetate film production.

Routes of exposure Occupational: various manufacturing jobs, some food processing jobs, furniture refinishing. Environmental: home use of paint and varnish removers, some aerosol products.

Reproductive effects Due to metabolism to carbon monoxide. In animals: decreased fetal weight, brain damage. In humans: malformations of the limbs and face, psychomotor disturbances, subnormal mental development, central nervous system damage.

In the human body, methylene chloride is quickly metabolized into carbon monoxide. The amount of carbon monoxide found in the body is directly related to the amount absorbed. Exposure thus may result in health problems due to the toxic effects of carbon monoxide.[137] Health effects are due to an inability to provide sufficient oxygen to body tissues, a condition known as hypoxia.[138]

Fetal animals are less able to increase blood flow to compensate for low blood oxygen levels and are more likely to suffer damage from hypoxia than is the mother.[139,140] Relatively low maternal exposures to carbon monoxide result in decreased fetal weight gain and neurobehavioral problems in rodents.[141-143] Higher exposures result in lower fetal survival.[144] Mice chronically exposed to moderate levels of carbon monoxide had increased incidence and severity of cleft lip and palate in their offspring.[145] Monkeys exposed to carbon monoxide at levels well tolerated by the mothers had moderate to severe fetal hypoxia. The least hypoxic fetuses survived without significant injury; the severely hypoxic fetuses suffered brain damage and early death.[146,147] One important study looked at the combined effect of protein deficiency and carbon monoxide exposure in mice. Although protein deficiency did not influence the effect of carbon monoxide on the mother, it did worsen the hypoxic effect on the fetus, suggesting greater susceptibility.[148]

The few animal studies that have looked at the effects of methylene chloride itself did not find any evidence of birth defects or fetal toxicity, although one found reduced fetal body weight in rats exposed to methylene chloride at levels that affected the mother's liver.[149-151] Little is known about the effects of methylene chloride itself in humans. Among thirty-four men exposed to methylene chloride, eight were infertile.[152] Four of these men submitted semen samples, and all had abnormal sperm movement, shape, and density. Female pharmaceutical workers exposed to methylene chloride had a slight increase in spontaneous abortions, although other job factors may have contributed.[153]

More is known about the impact of hypoxia on the human fetus. A review of case reports of pregnant women exposed to carbon monoxide found that fetuses either died or developed significant problems when their mothers experienced unconsciousness or coma as a result of the exposure.[154,155] Outcomes included malformations of the limbs and face, psychomotor disturbances, subnormal mental development, and central nervous system damage.

Methylene chloride exposure should be considered a potential threat to the health of the fetus. Although the chemical itself is not known to have any direct effects on the fetus, its metabolism to carbon monoxide can result in low oxygen levels, potentially leading to deformities, functional problems, and death. Since the fetus is even more susceptible than the mother to hypoxia, any exposure to methylene chloride that causes symptoms in the mother may threaten the fetus.

N–Methyl–2–Pyrrolidone (NMP)

Uses Microelectronics, petroleum production, paints, paint strippers, and cleaners; production of resins such as Kevlar, wire coating, graffiti removal; topical veterinary products, cosmetics; under consideration as an absorption enhancer for topical pharmaceuticals for human use.[156]

Routes of exposure Occupational: in labs, semiconductor work, factories. Environmental: paints, strippers, graffiti removers, cosmetics.

Reproductive effects In animals: fetal resorptions, stillbirth, low birth weight.

N–Methyl–2–pyrrolidone (NMP) is a popular new solvent marketed as a safer alternative to chlorinated solvents. Little is known about the reproductive and developmental effects of NMP in humans, but animal studies have shown toxic and even deadly effects on fetuses at doses at or below those causing maternal toxicity.

Mice fed or injected with NMP at a range of doses suffered increased rates of fetal resorption.[157,158] Surviving offspring had lower birth weights, decreased size, an increase in cleft palate, and delayed bone formation, yet the mothers did not exhibit any toxic effects. Other researchers exposed rats to NMP orally, dermally, and through inhalation. Each route of application led to significantly increased fetal resorption, increased stillbirths, and in some cases delayed bone formation in surviving offspring.[159–162] These studies generally showed no, or mild, evidence of maternal toxicity at these doses, as shown by reduced weight gain during gestation in one study and dry skin at the application site in the dermal study.[159] A multigenerational rat reproduction study found fetal death and reduced body weight at a dose that did not affect the mother.[163] Fetal death and some malformations were also found in rabbits, and some maternal toxicity occurred.[164,165]

Researchers have looked at postnatal physical and behavioral development in rats exposed to NMP in utero. The mothers inhaled NMP at a dose that did not cause significant fetal loss. The exposed pups had lower body weight throughout the preweaning period and delayed physical development. Neurobehavioral studies revealed abnormalities in dealing with difficult tasks.[166]

Information on human reproductive and developmental impacts of NMP is extremely limited. One case report suggests a connection between NMP exposure and stillbirth. A young laboratory technician who was regularly exposed to NMP at work through her twentieth week of pregnancy subsequently developed intrauterine growth retardation and ultimately delivered a stillborn fetus with no evidence of malformation.[167]

NMP has consistent fetotoxic effects on animals at or slightly below levels that cause mild toxicity in adult animals. The results are stillbirth, low birth weight, some skeletal malformations, and perhaps neurologic impairment. The mechanism for these effects is unclear, but the finding across species, with different routes of exposure, and in a dose-dependent fashion is fairly convincing. On the basis of the animal evidence, NMP should be considered fetotoxic in humans.

Perchloroethylene

Uses Dry cleaning, vapor degreasing, machining, auto paint, assembly plants, electroplating.

Routes of exposure Occupational: in dry cleaning, facilities using degreasers. Environmental: location near dry cleaners, manufacturing and repair shops;[168,169] from recently dry cleaned clothes;[170] drinking water contaminant in some areas.[171]

Reproductive effects In humans: probably increases the risk of spontaneous abortion by two- to fivefold in those exposed at workplace levels; may increase the risk of infertility in both men and women; concentrates threefold in breast milk and can lead to jaundice in infants. In animals: decreased fetal weight, spontaneous abortion in some studies.

Perchloroethylene (PCE) is widely used and relatively well studied in humans. According to one study, men who work in dry cleaning shops had more sperm abnormalities than men working in laundries.[48] The findings are hard to interpret because both the exposed and unexposed group had high percentages of men with low sperm counts, and it is not clear if the abnormalities have any significance for reproductive function.

A partner study looked at fertility in male dry cleaners and their wives, compared with laundry workers. The dry cleaners' wives were twice as likely to report unsuccessfully attempting to get pregnant for more than twelve months or seeking medical care for infertility.[51] However, both groups had similar numbers of pregnancies, and both had fertility rates above the national average.

Women exposed to PCE in dry cleaning shops were found to take twice as long as an unexposed group to become pregnant.[35] In another study, women seeking care at an infertility clinic were almost three times more likely to report exposure to dry cleaning chemicals than were women without fertility complaints.[172] This study carries little weight because it may suffer from biased selection and recall.

Two studies found that exposure to PCE increases the risk of spontaneous abortion by two- to fivefold.[13,14] Two others also found an increased risk of spontaneous abortion, but the finding was not statistically significant; moreover, one failed to find an association between PCE exposure and spontaneous abortion.[18,173,174] There was no increased risk of spontaneous abortion among the wives of PCE-exposed men.[50] A study of solvent exposure in drinking water found a weak association between PCE exposure and oral cleft defects,[23] but there is little other evidence that PCE exposure increases the risk of birth defects.[30]

PCE may pose a risk to newborns. A significant case report concerned a nursing mother who visited her husband at a dry cleaning plant during his lunch breaks. Their six-week-old infant, who never entered the plant, developed liver damage and jaundice, which resolved after cessation of breast-feeding. After a thirty-minute plant visit, the mother had detectable PCE in her blood, and levels in her breast milk were over three times greater than in her blood.[175] Exposure modeling indicates that women occupationally exposed to PCE at levels below the workplace standard and women living in apartments over dry cleaners may have enough PCE in their breast milk to risk health damage to their infants.[176]

Animal studies have shown that PCE can cross the placenta and enter the developing fetus. A few studies in chickens and rodents showed decreased survival, decreased fetal body weight, and increased resorption of fetuses.[177] Most animal studies showed no effect on development and no increase in malformations.[177]

Overall, animal testing on the reproductive effects of PCE has not demonstrated significant reproductive toxicity, although human studies have shown toxic effects including spontaneous abortion and possible effects on human fertility. The presence of PCE in breast milk is a worrisome finding; this solvent is classified as a possible human carcinogen,

---

and infant exposures during breast-feeding could lead to harmful effects later in life.[176]

Phenol

Uses Synthesis of resins, nylons, plasticizers, aspirin, herbicides; used as a disinfectant, analytical agent; by-product of leather tanning, timber products manufacture, pulp and paper production, textile manufacture, iron/steel production.

Routes of exposure Occupational: in factories, laboratories. Environmental: contaminated drinking water; emissions from wood and gasoline combustion; in some consumer products, including disinfectants, mouthwash, medicated skin products.

Reproductive effects In animals: reduced fetal weight, sperm chromosome damage, possible changes in the estrus cycle. In humans: infant jaundice.

Despite phenol's widespread usage in consumer products, information concerning its effects on reproduction and development is limited. The few animal studies show mixed results. In one study, pregnant rats inhaled phenol at levels that humans might encounter occupationally and suffered increased fetal and neonatal loss.[178] Other researchers injected rats with phenol on specific days of gestation. They found no evidence of birth defects or increases in fetal resorptions, but did note fetal weight reduction in rats treated with the highest dose.[179] In another set of experiments conducted in both mice and rats, phenol exposure led to low birth weights, at doses at which the mothers showed no evidence of harm.[180,181]

Two other studies are harder to interpret. In one continuous breeding study, several generations of mice were constantly exposed to phenol. The researchers found a dose-related increase in damage to sperm cell chromosomes in all generations of offspring.[182] It was not clear whether paternal, maternal, or fetal exposures led to the effect; however, the fact that there was a marked increase in chromosome damage at even the lowest levels of exposure is cause for concern. Finally, a study looking at the effects of phenol inhalation on female rats found changes in the estrus cycle, the rodent equivalent of the menstrual cycle.[183] In this case, the dose was so high that these changes may have been simply due to the general toxicity of phenol.

In humans, phenol is produced naturally in the intestine, and the human body seems able to process low levels without difficulty.[184] There are, however, reports of newborns developing jaundice as a result of exposure to phenol in disinfectant detergents used in hospitals.[185,186] This implies that newborns are particularly sensitive to phenol exposure, a question that needs further examination.

Phenol is widely distributed in the environment and used in a broad range of products and processes. Animal evidence suggests that it may damage chromosomes and lead to fetal toxicity, and several human case reports indicate that low-level exposures to phenol may result in infant jaundice. Given these facts, phenol urgently needs further study to determine its impacts on humans.

Styrene

Uses Reinforced plastics manufacture, polystyrene manufacture, polyester resins, rubber manufacture.

Routes of exposure Occupational: various manufacturing jobs, boat building, firefighting. Environmental: burning of plastic and polystyrene; common water contaminant; leaches from polystyrene cups in small amounts; present naturally in cinnamon.[187]

Reproductive effects Very conflicting evidence. In animals: reduced weight gain, hypoactivity, birth defects in chicks. In humans: possible testicular toxicant; may interfere with endocrine function and disrupt menstruation.

Although styrene is not generally used as an organic solvent, structurally it is related to the other aromatic solvents discussed in this chapter (benzene, xylene, toluene, and phenol). Styrene has been studied extensively in animals and humans. Two animal studies suggest an effect on hormone function. Rats exposed to styrene vapor at levels much lower than those allowed in the workplace showed a lengthening of their estrus cycle.[188] In another study, investigators removed one ovary from each experimental rat, then exposed half of the group to styrene orally. After removal of one ovary, normally the other ovary grows in size as it takes over responsibility for hormone production. In the exposed rats, the remaining ovary did not grow to compensate for the loss, suggesting an effect on hormonal function in rats.[189]

One animal study using levels of styrene vapor tenfold lower than the allowable workplace average showed significant increases in embryonic death, but another study did not confirm this finding.[190,191] Injection of styrene into chicken eggs consistently causes developmental abnormalities in the chicks.[192] These studies are the only ones that suggest that styrene might cause birth defects.

Evidence that styrene may affect development and behavior comes from a study of rats exposed to low levels of styrene vapor for seven weeks after birth. These rats had significantly reduced weight gain and delayed ear and tooth development. The exposed rats displayed a dose-related reduction in exploratory and avoidance behavior.[193]

Human studies, for the most part, have not found a consistent effect on reproduction and development.[194] The largest human epidemiological studies were performed in Finland, and these generally found no significant effect of styrene exposure on pregnancy outcome.[195–198] A series of Russian studies found an association between styrene exposure and self-reported menstrual abnormalities.[199] A small study in Italy found greater menstrual irregularity and reduced fertility in women occupationally exposed to styrene, but any conclusions are limited by the small sample size.[200] Similar studies in Finland and the United States did not find any effects of styrene on menstrual function.[201] Interestingly, a small study of women occupationally exposed to styrene found significantly elevated levels of the hormone prolactin and elevated levels of human growth hormone.[202] Elevated prolactin levels can lead to menstrual dysfunction and could explain the findings of abnormal menstrual cycling in women workers and abnormal estrus cycling in rats.

The evidence on whether styrene affects male fertility is conflicting. One group of styrene-exposed workers had a significantly lower proportion of normal sperm than a comparison group who sought care at an infertility clinic.[203] A styrene-based chemical, styrene maleic anhydride, is under consideration for use as a male contraceptive.[204] Animal studies on the effects of styrene on the testes are conflicting, with some studies showing reduced sperm counts and changes in the microscopic appearance of the testes and other studies finding no effect.[205] Researchers have found an increase in damage to DNA in both human and rat testicular cells exposed to styrene in vitro.[206]

The conclusions of studies on the reproductive toxicity of styrene have been extraordinarily conflicting. The evidence for an effect on male testicular function is strongest, but needs further study. Despite a large number of studies, there is no clear answer to whether styrene affects female menstrual function, though this would not be surprising in the light of evidence that the solvent may affect the endocrine system in rats and humans.

Toluene

Uses Glues, coatings, inks, paint, cleaning agent, gasoline additive; used in manufacturing, cleaning, chemical production, coke ovens, dye making.

Routes of exposure Occupational: widespread in painting, assembly work, cleaning, general industry, chemical plants. Environmental: consumer products such as stain removers, nail polish, paint thinners, dyes, inks, adhesives, and some cosmetics; lower-level exposure from automobile exhaust, cigarette smoke, gasoline, sometimes in drinking water.

Reproductive effects In animals: fetotoxic, learning and behavioral deficits. In humans: 2.5-fold increased risk of spontaneous abortion, birth defects of the head, face, urinary tract, and limbs. May disrupt hormones, particularly in men.

Animal studies show that toluene has a fetotoxic effect in rats and mice, including a reduction in fetal weight, delayed development of the skeleton, spontaneous abortion, and fetal resorption.[207,208] In addition, some studies have found evidence of learning impairment and behavioral changes in rodents exposed during the period of brain development.[209–212] Effects on the fetus occur at doses below those causing toxicity to the mother. Extrapolation from the animal studies shows that human occupational exposure levels are near levels shown to have adverse effects on fetal development in rats and mice.[210]

Several studies of spontaneous abortion in solvent-exposed women have particularly implicated toluene, with risks up to ninefold higher than among unexposed women.[5] Women exposed to toluene alone experienced five times more spontaneous abortions than unexposed women.[15] Wives of men exposed to high and/or frequent quantities of toluene had a twofold increased risk of miscarriage.[50]

A large questionnaire-based case-control study found that exposure to aromatic solvents (toluene, xylene, benzene) was significantly associated with birth defects. Odds of toluene exposure, in particular, were almost fourfold higher among cases than controls. The defects included urinary and cardiac abnormalities and congenital cataract in the group reporting toluene exposure. Numerous case reports describe serious congenital defects among children of women who sniffed glue or paint containing toluene during pregnancy. These infants suffered from intrauterine growth retardation; neurologic abnormalities; abnormalities of the head, face, and urinary tract; and malformations of the arms and legs. The resemblance to babies with fetal alcohol syndrome led some investigators to propose the existence of a fetal solvent syndrome.[213,214] Solvent sniffing leads to higher exposures than occupational or home use of toluene.

Men exposed to toluene had dose-related decreases in luteinizing hormone, follicle-stimulating hormone, and testosterone, hormones that regulate the reproductive system.[215] A young man who died from sniffing a toluene-based paint thinner had testicular atrophy and suppression of sperm production.[216] At least one animal study found a reduction in sperm counts and reduced epididymal weight in rats exposed to high levels of toluene.[217] These reports indicate a probable effect on male hormonal and reproductive function.

Toluene increases the risk of spontaneous abortion in exposed women. High doses in humans, though not in animals, have been associated with a syndrome of severe congenital defects. One study found hormone-suppressive effects in exposed men. More research needs to be done on the possibility of hormone suppression in humans from toluene exposure. Based on the evidence, toluene is currently regulated by California as a developmental toxicant.[218]

Trichloroethylene

Uses Vapor degreasing, textile processing, refrigerant; production of polyvinyl chloride, pharmaceuticals, insecticides; in stains, finishes, lubricants, adhesives, rug cleaners.

Routes of exposure Occupational: vapor degreasing, various production processes. Environmental: contaminated drinking water, inhalation indoors from building materials, consumer products.

Reproductive effects   In animals: cardiac abnormalities, impaired brain development. In humans: possible association with miscarriage, cardiac abnormalities.

Trichloroethylene (TCE) is a common indoor air pollutant, widely used in building materials and consumer products.[219,220] The most common organic contaminant in groundwater, it appears in one-tenth to one-third of all samples tested.[221,222]

In animals, TCE appears to target the reproductive organs, concentrating in the ovaries and spermatocytes.[223,224] Mice exposed by inhalation had an increase in abnormally shaped sperm, suggesting genetic damage.[224] However, rats exposed orally had no changes in sperm count, shape, or movement.[225] Two studies in rats showed an association between TCE inhalation and reduced fetal weight; one used extremely low levels of TCE.[226,227] However, other studies in rats, rabbits, and mice found no significant increases in birth defects after maternal exposure to TCE.[227–231] In the frog embryo, the developmental toxicity of TCE appears to be mediated by a short-lived metabolic by-product, trichloroethylene oxide, formed by metabolism of this solvent by the mixed-function oxidase enzyme system.[232] It is possible that genetic variability in the function of this enzyme system may explain the differences in susceptibility to the developmental effects of TCE.

Cardiac abnormalities are a recurring theme in the developmental toxicology of TCE. Direct intrauterine instillation of TCE or its breakdown product, dichloroethylene, in rats leads to a variety of cardiac defects in the absence of other types of birth defects.[233] Rats exposed to TCE in drinking water during pregnancy at doses that did not cause maternal toxicity had offspring with more cardiac deformities than expected at the higher dose. Interestingly, when maternal rats were exposed starting before conception, the offspring had heart deformities even at the lower dose.[234] Earlier rodent studies that did not report cardiac defects may not have specifically dissected and examined the heart.[230] Investigators also found increases in heart deformities in chicks from eggs injected with TCE.[235] Chicks exposed to TCE developed electrocardiographic abnormalities almost immediately following TCE injection and then had arrest of cardiac development.[236] In human adults, TCE is known to cause electrical cardiac abnormalities.

Finally, some evidence suggests that maternal exposure to TCE in drinking water may affect brain development and behavior in offspring. In rodents, maternal exposure leads to structural and functional changes in the brain, as well as behavioral change, although not all studies reached statistical significance.[227,237–240]

In humans, an early study found an increase in miscarriages among nurses exposed to TCE in the operating room, but concurrent exposure to other chemicals makes it impossible to specify TCE's role.[241] A comparison of women who had spontaneous abortions with those who did not found that affected women were more likely to report exposure to TCE during pregnancy.[13] This study design was prone to recall bias. A study focusing on parents exposed to TCE and other chemicals at work found no increases in malformations in their children.[242] Two studies of male workers exposed to TCE found levels of testosterone and sex hormone binding globulin that decreased with increasing years of exposure.[243,244] Male workers exposed to TCE also had sperm abnormalities.[245]

Researchers have tried to assess the effects from TCE in drinking water, but the results are far from clear. One Massachusetts population exposed to TCE and other solvents in drinking water had an apparent increase in eye, ear, central nervous system, chromosomal, and oral cleft abnormalities.[33] However, this research has been criticized for lumping the anomalies together in ways that may not be scientifically valid. Researchers studying the occurrence of certain congenital heart defects in Arizona found an association with parental exposure to TCE-contaminated drinking water.[20] Maternal exposure before pregnancy and during the first trimester was associated with a threefold increase in the risk of congenital heart defects. Although this study too had limitations, the result is particularly interesting in connection with the animal studies showing that TCE exposure can lead to heart abnormalities. The Massachusetts population with TCE-contaminated water also had an unusually high incidence of childhood leukemia, leading some investigators to implicate TCE.[246]

TCE exposure is widespread in this country, but human and animal studies of possible health effects have shown conflicting results. Given the associations of other solvents with spontaneous abortions, this finding with regard to TCE is plausible and should be taken seriously. The consistency of the animal and human studies showing an increase in heart defects from TCE exposure prior to and during pregnancy is of great concern and implies a hazard that requires further action.

Xylene

Uses Paints, lacquers, varnishes, insecticides; in rubber, plastic, and leather manufacturing; an ingredient in gasoline.

Routes of exposure Occupational: various manufacturing jobs, painting and varnishing. Environmental: home use of paints, lacquers, varnishes, gasoline exposure, water contamination.

Reproductive effects In animals: toxic to the fetus, may cause certain birth defects, may interfere with endocrine function. In humans: association with spontaneous abortions.

Some animal studies involving xylene are particularly troubling because of the toxicity and birth defects at low doses. One study found lethal effects at late stages of fetal development, abnormal bleeding, abnormalities in skeletal development, and growth retardation in rats exposed by inhalation to levels between 50 and 500 mg/m^3.[247] The workplace standard is 435 mg/m^3 for eight hours.[248] Higher-dose animal studies have found increased fetal resorptions, fetal death, delayed fetal development, and low birth weight.[248–251] An interesting study exposed rats to xylene alone or to xylene and aspirin. The xylene alone was found to be somewhat toxic to the embryo, inhibiting normal growth. When xylene and aspirin were combined, the effects were much more serious, with dramatic fetal toxicity and malformations, particularly of the skeletal system and the kidneys.[252] Aspirin is known to cause birth defects, and it appears that xylene may act synergistically to worsen this outcome.

Prenatal xylene exposure may lead to changes in development and behavior.[253] Rat pups exposed prenatally at fairly low levels showed a decrease in brain weight, delay in reflex development, and impairment in tests of neuromotor ability, learning, and memory. The effects were most marked in the female pups.

Two important studies looked at the effects of xylene on sex hormones. Rats exposed to high levels of xylene have significantly lower blood levels of progesterone and 17ß estradiol (a form of estrogen), two of the hormones responsible for regulating the female reproductive cycle.[254] In addition, xylene prevents ovulation in rats.[255] Alterations in maternal hormone levels may be responsible for the toxicity to animal embryos.

There have been few human studies of the reproductive toxicity of xylene. One early investigator reported five cases of a rare birth defect, caudal regression, in mothers exposed to solvents; this defect involves incomplete development of the pelvic region and legs. Of nine reported cases of this rare defect, five mothers were exposed to solvents. In a companion study, the investigator exposed chicken embryos to xylene and found many deformities. About half of the deformities included "rumplessness."[256] Other human studies found a fivefold increase in spontaneous abortions in women exposed to xylene.[7] Mothers of children born with central nervous system defects were also more likely to have been exposed to aromatic solvents, particularly xylene, during pregnancy.[25]

There is clearly evidence of toxicity and neurodevelopmental effects in the rat fetus at inhaled levels similar to those encountered in the workplace, as well as suppression of maternal sex hormones in rats. This conclusion is of considerable concern because human exposures to xylene are common. The evidence that xylene causes birth defects is based on animal studies with large doses of xylene and on a few human reports. The fact that caudal regression was reported in both humans and chickens is important and implies that xylene might be involved in the causation of this unusual birth defect.

Pesticides

Throughout the world, enormous quantities of pesticides are used on crops, forests, nurseries, golf courses, lawns, gardens, and pets, and in public spaces and homes. In the United States about six hundred active ingredients are used in over twenty thousand pesticide products as insecticides, herbicides, rodenticides, and fungicides. Most formulations contain inert ingredients, with their own toxicity and health risks. In 1995, the United States used approximately 1.2 billion pounds of pesticide active ingredients, or about 5 pounds for each person in the country, accounting for 20 percent of world use.[1] Repeated year after year, the environmental and health effects of this volume and mixture of chemicals are extraordinarily important.

Chemical pesticides are designed to kill insects, fungi, plants, or other unwanted organisms, usually by interfering with some essential biochemical process in the target. However, their acute and chronic toxic properties also pose risks to the health of exposed humans, pets, wildlife, and entire ecosystems. In most cases, pesticides are classified according to the mechanism of their toxic action. But even within each class, there is a wide range of chemical structures and potential health effects. Pesticides may cause cancer; adverse reproductive, developmental, neurological, or immune system effects; or other organ damage at varying exposure levels. Each of these outcomes must be considered for each chemical.

Institutional protection from toxic effects depends largely on pesticide registration and regulation (see chapter 8). But there are significant gaps in the registration and regulatory processes, which government agencies have only partially addressed. Toxicity testing for many pesticides that have been in use for years is inadequate. One source estimates that complete toxicologic data are available for only about one hundred of the approximately six hundred active pesticide ingredients.[2] Reproductive and developmental toxicity data are often particularly deficient.

Active Ingredients and "Inerts"

A final pesticide product includes a mixture of active and inert ingredients. An *active ingredient* is a chemical "that can kill, repel, attract, mitigate or control a pest, or that acts as a plant growth regulator."[3] An *inert ingredient* is one that is not defined by the manufacturer or EPA as active. About twelve hundred inert ingredients are present in twenty thousand pesticide formulations. Inerts may serve as vehicles, assisting in the transport of the active ingredient to the target pest. They also give certain properties to the final recipe useful in mixing and application or in affecting the length of time the product remains active in the environment. Although inert ingredients may not have pesticidal properties, many have their own toxicity. For example, some of the solvents used as inert vehicles, like epichlorohydrin and trichloroethylene, have harmful reproductive effects. (See chapter 4.) Skin absorption is a significant route of human exposure to pesticides, and some inert ingredients penetrate protective clothing and skin, carrying active ingredients and increasing the risk of toxic effects.

Inert ingredients often comprise over 90 percent of the final product formulation. Until recently, the EPA did not require that the names of these ingredients be listed on any pesticide label, accepting manufacturers' claims that the identity of the inert was a trade secret. However, in 1994 the Northwest Coalition for Alternatives to Pesticides charged that the EPA wrongfully accepted claims of confidentiality without first determining that the inerts were actually trade secrets. A federal district court ruled that pesticide manufacturers must disclose information about inerts in the six different products that were at issue. Although the ruling did not apply to all pesticide products, the EPA is now faced with having to apply the court decision more broadly.

Table 5.1 Characteristics of pesticides

Pesticide	Persistence (days)	Solubility (ppm)
Herbicides		
Atrazine	60–100	33
Bromoxynil	11	0.08
Cyanazine	12–108	170
2,4-D	10	890
Dicamba	8–25	8,310
Diuron	30–400	42
Molinate	12	800
Insecticides		
Acephate	3–6	650,000
Chlorpyrifos	11–141	2
Cypermethrin	7–82	0.004
Diazinon	3–13	60
Parathion	7–30	12
Carbaryl	7–28	50
Endosulfan	4–200	0.32
Dicofol	16–60	0.8
Lindane	400	7
Methoxychlor	7–180	0.1
Permethrin	6–106	0.006
Fungicides		
Mancozeb	7–139	6
Vinclozolin	14	3
Pentachlorophenol	50	14

Note: *Persistence* refers to the soil half-life, in days—the amount of time required for the pesticide to break down to one-half of its initial concentration. Large ranges indicate a variability in half-life depending on soil type, pH, aerobic, or anaerobic conditions.

A high *solubility* means that the pesticide is more likely to be carried by water throughout the environment and into groundwater aquifers (ppm = mg/liter).

Pesticide Fate and Transport

The distribution and life history of pesticides in the environment are largely determined by the chemical and physical properties of each agent. The following properties are of particular interest:

Environmental persistence   Indicates how long it takes for a pesticide to break down in soil, sunlight, surface- or groundwater, or indoors (table 5.1)

Water solubility Determines the degree to which a pesticide will run off in rainwater or be transported into groundwater, measured in mg/liter or parts per million (ppm) (table 5.1)

Volatility Determines the extent to which a pesticide will evaporate into the air and be transported through the atmosphere

Soil binding Influences environmental persistence and runoff into water bodies. A chemical that binds to soil particles is less mobile in the environment and less likely to run off into water bodies or move into groundwater.

Tendency to bioaccumulate Indicates how much the concentration of a pesticide is likely to build up in a living organism over time

Some pesticides persist for long periods, tightly bound to soil particles, while others readily evaporate and are dispersed over great distances through the atmosphere. They may be biodegraded by soil organisms and sunlight or persist unchanged as they cycle through ecosystems. Several pesticides whose chemical structure includes chlorine (organochlorines) have been banned for years in the United States but are so persistent that they are still detected in homes, and residents continue to be exposed.[4] Sprayed pesticides drift to nearby land and water. Applications from airplanes are sometimes windblown many miles. Some pesticides bind to water droplets and are commonly found in fog and rainwater.[5] Pesticide atmospheric dispersion is global and penetrates the food chain at all levels.

Rachel Carson: The Woman Who Blew the Whistle on Pesticides

Marine biologist Rachel Carson's interest was captivated by an angry letter to the editor of the *Boston Herald* in 1958. Author Olga Owens Huckins claimed that the synthetic pesticide DDT had caused the death of songbirds on her land and complained that the pesticide did not even accomplish its purpose of mosquito control. The letter resonated with Carson's own concerns, and she embarked on a systematic investigation of the biological repercussions of DDT and other synthetic chemicals. Publication in 1962 of her book *Silent Spring* was the result of those efforts.

In *Silent Spring* Rachel Carson approached the human health risks of pesticides from a public health and ecological perspective. Her long-standing interest in predation and food chains became central to her work. Many years of painstaking ecological research coalesced into an urgent public health message: synthetic chemicals may pose a serious threat to human health and the environment.

Industry scientists and representatives attacked the arguments in *Silent Spring* vehemently, trying to discredit Carson's scientific methods. Because she was at the vanguard of a movement requiring new methods and statistical evaluations, many fellow scientists also withheld their support. Carson faced the even broader challenge of changing a deeply rooted societal outlook concerning pest eradication that had emerged from World War II. The postwar era saw an explosion into the marketplace of pesticides developed during the war, where total eradication was the ideal. In the subsequent agricultural applications of DDT, the same uncompromising effects were desired. Carson criticized this shortsighted vision as not only destructive to many organisms and to human health, but also as ineffective because insects quickly evolve resistance to chemicals.

Silent Spring heralded much of what we now know about the health effects of toxic persistent organochlorine chemicals. Carson's work brought environmental protection into the arena of public debate, and over time her views were vindicated. Legislation eventually banned DDT and several other organochlorine pesticides. Unfortunately many environmentally persistent toxic chemicals are still produced, and we are living with residues in our bodies from those banned decades ago. Nevertheless, Rachel Carson's influence has been profound, and her vision continues to guide those willing and able to take the broader view.[6–9]

Pesticides that persist in the environment and accumulate in living organisms tend to concentrate at the top of the food chain. For example, some organochlorine pesticides are dispersed worldwide, contaminating oceans, sediments, bottom-dwelling clams, mussels, sea urchins, and fish.[10] Persistent pesticidal and nonpesticidal organochlorines, along with other long-lasting and fat-soluble chemicals, concentrate in the fat tissue of marine mammals. Inuit mothers, whose diet at the top of the Arctic food chain is rich in marine mammal fat, have the largest known body burden of organochlorines, some of which are pesticides. They pass these chemicals on to their developing fetuses and nursing infants.[11]

Groundwater used for drinking in large areas of the United States is contaminated with pesticides. For example, some of the hundreds of millions of pounds of herbicides applied to corn and soybean fields in the Midwest each spring migrate into the drinking water supply of nearly 14 million people in this area of the country.[12] Among them, triazines, the most commonly used herbicides in the United States, tend to persist for long periods of time in underground aquifers. Elsewhere, ground-water on Long Island in New York is contaminated with aldicarb, a carbamate insecticide, which kills insects by interfering with transmission of nerve impulses. Spray drift or pesticide runoff from treated land enters surface water and large aquatic ecosystems. Concentrations in surface water rise dramatically with heavy pesticide use in the spring.

Exposure to Pesticides

Pesticides contaminate air, soil, food, water, and the indoor environment. A focus on only one source will seriously underestimate total human and environmental exposure levels. Unfortunately, with only a few exceptions, accurate pesticide use and exposure information, necessary for studying health effects, is not routinely collected.

The EPA estimates that U.S. consumers in 1993 spent approximately $1.2 billion for 71 million pounds of insecticides, herbicides, and fungicides for home, lawn, and garden use.[13] Some of these agents are also routinely used in and around schools and other public buildings. A nine-month study of 238 families in Missouri in 1989 disclosed that 98 percent used pesticides at least once annually and two-thirds more than five times per year. More than 80 percent used pesticides during pregnancy and 70 percent during the first six months of a child's life. Pesticide use in the home was most common (80 percent), followed by herbicide use in the yard (57 percent), and flea and tick control on pets (50 percent).[14]

About 771 million pounds of pesticides were used on U.S. farms in 1995, and at least half of the millions of farmworkers in the United States come into direct contact with farm chemicals.[1,15] Pesticide labeling instructions specify the amount to be used on specific crops. Farm owners or employees are responsible for complying with instructions, but in most states there is little monitoring, though there are usually applicator licensing and training programs. For an estimated 5 million migrant and seasonal farmworkers, most of whom belong to an ethnic minority, the extent of pesticide exposure and resultant health effects is largely unknown.[16,17]

Pesticide residues on the food supply are monitored by the Food and Drug Administration (FDA). For each pesticide ingredient registered for a particular food crop, the EPA has established a tolerance value that represents the maximum residue allowable on that food. Historically, pesticide food tolerances were based on the results of field trials conducted by manufacturers in which pesticide residues on food were determined after the pesticide had been used in the amount necessary to have its intended effect. Tolerances based primarily on health considerations are more recent, but since a number of pesticides in use for years have not had adequate toxicity testing, existing tolerances are often based on incomplete data.[18] In 1996, the U.S. Congress passed the Food Quality Protection Act (FQPA), which requires the EPA to begin setting health-based standards for all pesticide residues in food. This will require a review of approximately nine thousand existing tolerances.

Indoor carpets, dust, and furniture are also sources of ongoing human exposure, particularly for crawling and playing infants and children. The largest number of chemicals and the highest concentrations are often found in household dust, compared to air, soil, and food.[4] Children living in homes near sites of agricultural pesticide use are likely to be exposed to pesticides that are not registered for residential use. Levels of pesticides in their homes are higher than in homes more remote from agricultural operations.[19]

Skin absorption, inhalation, and ingestion are important potential routes of exposure. Fat-soluble chemicals, purposely or accidentally applied to the skin, such as lindane, used to treat body lice, are readily absorbed into the body. The types of spray equipment, spray velocity, and pesticide volatility determine the extent of inhalation exposure.[20] Product-labeling requirements that emphasize the need for protective clothing or equipment may be the only regulatory safeguard against excessive exposure to farm chemicals. Yet in warm climates, this equipment is often intolerable because of the heat and is rarely used.[21] Even when pesticides are used as recommended, exposures may be excessive. In a study of air and surface residues after chlorpyrifos (an organophosphate pesticide) had been used for indoor flea control according to directions, total absorbed doses for infants were estimated at up to five times the no-observable-effect level.[22] Another study found chlorpyrifos levels peaking on furniture, toys, and other indoor surfaces thirty-six hours after the pesticide had been applied to the floor of a room with subsequent ventilation according to directions.[23] The researchers concluded that skin contact, ingestion, and inhalation were likely to cause unsafe exposures to children playing in the room. Some chemicals banned years ago, like chlordane and aldrin, are still present in recent testing of indoor air and carpet dust.[24]

Pesticide ingestion from food depends on dietary patterns and details of food preparation. Infants and children consume more fruits and vegetables, such as apples, bananas, tomatoes, and squash, per unit of body weight than adults do. They also have less variety in their diets than adults, and consequently are sometimes excessively exposed to pesticide residues on those foods. Some pesticides banned or restricted in the United States continue to be manufactured domestically for export. They may then return as residues on the billions of pounds of fruits and vegetables imported into the United States annually and are not routinely tested for by the FDA. In 1990, for example, U.S. pesticide manufacturers exported over 465 million pounds of pesticides, and of those, 52 million pounds were banned, restricted, or unregistered for use in the United States.[12]

In order to estimate the extent of pesticide exposure in the general population and as part of the 1994 National Health and Nutrition Examination Survey III (NHANES III), urine samples were collected from about a thousand adults selected from a broad spectrum of the U.S. population.[25] Specimens were analyzed for twelve different chemical compounds that result from the metabolic breakdown of about thirty different pesticides with a detection limit of 1 μgm/liter urine.[26] More than half of the individuals tested had at least six of the pesticide residues in their urine. Chlorpyrifos residues were detected in 82 percent of the study group, pentachlorophenol in 64 percent, lindane in 20 percent, and 2,4-D in 12 percent. This widespread exposure in the general public justifies concern about health effects and supports arguments for more comprehensive toxicity testing.

Reproductive and Developmental Toxicity of Pesticides

In addition to their effect on target pests, pesticides may also harm non-target organisms like beneficial insects, earthworms, soil fungi and bacteria, fish, wildlife, domestic animals, and humans. Features of ecosystems such as predator-prey relationships, wildlife distribution, biodiversity, and the organic quality of soil are also altered by pesticide use.

Information about the health effects of pesticides comes from animal testing, epidemiological data, and case reports. For registration purposes, the potential for harm is estimated from the results of animal testing for cancer and effects on reproduction, development, the nervous system, and other organs.

Epidemiological Studies

Epidemiological studies are not used in the pesticide registration process but are useful for examining health effects of actual exposures. A number of these studies have evaluated the risks of spontaneous abortions, delayed pregnancies, birth defects, retarded intrauterine growth, some childhood cancers, spermatotoxicity, and chromosome damage in people exposed to pesticides. Agricultural workers exposed to multiple pesticides often serve as the study population, but several analyses also attempt to assess the risk to the general population.

Epidemiological studies are often limited by inaccurate or inadequate exposure assessment or inadequate data on health outcomes, potentially masking any true relationship between exposure and health effect. A large agricultural health study underway in North Carolina and Iowa may partially address these concerns.[27] Investigators estimate that seventy-five thousand people will be questioned about or monitored for pesticide exposures and a variety of health outcomes including cancer and reproductive effects. The results of this study will not be available for years.

Tables 5.2 and 5.3 summarize many of the available epidemiological studies. Each table indicates the criteria used to classify the subjects as exposed to pesticides, the reproductive outcomes studied, and the observed effects—whether the exposed subjects were more or less likely to experience the health effect than a control group. Collectively, the studies demonstrate a range of adverse reproductive outcomes, primarily among agricultural workers.

Spontaneous Abortions and Fertility Problems

A number of studies report an increased incidence of spontaneous abortions and stillbirths among women agricultural workers (table 5.2). Some

---

Table 5.2   Spontaneous abortion and fetal death in women with agricultural occupation and potential pesticide exposure

Exposure	Reproductive outcome	Observed effect
Agricultural occupation[29]	Spontaneous abortion	1.3 times more likely
Agricultural occupation[30]	Spontaneous abortion	2.8 times more likely
Agricultural or horticultural occupation of more than 30 hours per week at beginning of pregnancy[31]	Spontaneous abortion	No effect
Gardener[32]	Spontaneous abortion	2 times more likely (NS)
Grape garden spraying (both parents)[33]	Spontaneous abortion	5.5 times more likely
Floriculture[34]	Spontaneous abortion, stillbirth	2.2 times more likely
Agricultural work for at least 2 weeks at beginning of pregnancy and pesticide exposure estimated by interview later[35]	Stillbirth without major malformation	3.1 times more likely
Agricultural or horticultural occupation at any time of pregnancy[35]	Stillbirth without other birth defect	5.7 times more likely
Male pesticide mixers and sprayers[28]	Spontaneous abortion, stillbirth	1.7 times more likely

Note: Refers to women exposed to pesticides except where otherwise noted. All increases in adverse outcomes are statistically significant ($p < 0.05$), unless otherwise indicated (NS). "More likely" means "more likely than in a control group in the study."

of these studies have inherent limitations. For example, when agricultural occupation is used as a surrogate for pesticide exposure there is always the possibility of exposure misclassification and underestimation of the true risks. Some are subject to recall bias, which may tend to exaggerate the risks. Studies that use hospital discharge summaries to document health outcome and occupation avoid recall bias but capture only women who were treated in a hospital. Self-reports of spontaneous abortions are likely to underestimate their true incidence when they occur

Table 5.3 Studies of birth defects and low birth weight in offspring of women and/or men exposed to pesticides

Exposure	Outcome	Observed effect
Male pesticide applier[37]	Birth defects in offspring (from state birth registry)	All defects: 1.4 times more likely Circulatory or respiratory defects: 1.7 times more likely Urogenital defects: 1.7 times more likely
Agricultural occupation as farmer's wife or gardener[39]	Nervous system, musculoskeletal defects, oral clefts	Musculoskeletal defects 5 times more likely for gardeners
Agricultural occupation at least 15 hours per week at beginning of pregnancy[40]	Chromosomal, developmental, musculoskeletal defects	Developmental defects 4.5 times more likely
Agricultural occupation—either or both parents[41]	Malformations, premature birth, low birth weight	Limb defects more likely (NS)
Agricultural occupation—either or both parents[42]	Limb defects	1.6 times more likely (NS)
Agriculture, fishing, forestry occupation[43]	Congenital malformation	No effect
Exposure to pesticides in first trimester as estimated by occupational hygienist on basis of interview[36]	Oral clefts, nervous system, skeletal defects	Oral clefts: 1.9 times more likely Any defect: 1.4 times more likely (NS) Nervous system defect: No effect
Agricultural exposure to pesticides estimated from occupation and industry, reported on birth certificates of child[47]	Limb defects	1.4 times more likely (NS)
Exposure to pesticides based on interview of mother (China)[48]	Birth defects (hospital diagnosis) Intrauterine growth retardation	No effect 2.9 times more likely
Floriculture[34]	Birth defects (parent report) Prematurity	1.3 times more likely 1.7 times more likely

Table 5.3 (*continued*)

Exposure	Outcome	Observed effect
Floriculture[44]	Birth defects (confirmed from medical data)	Birthmarks only—6.6 times more likely★
Paternal occupation pesticide exposure, estimated[45]	Birth defect—anencephaly	No effect
Agricultural work, more than 30 hours per week until thirteenth week of pregnancy, pesticide exposure estimated by interview later[46]	Congenital defects (from medical records)	No effect
Municipal water contaminated with herbicides, Iowa[38]	Intrauterine growth retardation	1.8 times more likely
Agricultural occupation at beginning of pregnancy[32]	Low birth weight	No effect
Agricultural occupation at any time in pregnancy[49]	Low birth weight	No effect

Note: Refers to mothers unless otherwise noted.

All increases in adverse outcomes are statistically significant ($p < 0.05$) unless otherwise indicated (NS). "More likely" means "more likely than in a control group in the study."

★ In this study information about congenital defects was collected through maternal interview and proved to be unreliable when checked against hospital records. When repeated with confirmed defects from medical record, the association with floriculture work was positive only for birthmarks.

early in pregnancy and often go unrecognized. Nevertheless, considered collectively, there appears to be an increased risk of spontaneous abortion in women occupationally exposed to pesticides, and it may be up to five times the risk in control groups.

An investigation of a group of Indian men employed as pesticide mixers and sprayers in cotton fields showed that their wives also experienced more miscarriages and stillbirths than a comparison group.[28] The men applied a variety of pesticides, often without the use of protective equipment.

In the Netherlands, investigators studied time-to-pregnancy and occupational exposure to pesticides in male fruit growers.[20] Increased time-to-pregnancy depends on a number of biological factors: frequency of intercourse, egg and sperm production, fertilization, embryo transport and implantation, and early fetal survival. Pregnancy was delayed among farm owner couples who were trying to conceive when the farm owner was the only pesticide applier. This was most noticeable in the period from March to November, when pesticides are applied. During that time, in the high-exposure group, time-to-pregnancy more than doubled, and 28 percent of the pregnancies were preceded by a visit to a physician because of fertility problems compared with 8 percent in the low-exposure group. These results indicate a potential adverse effect of pesticide exposure on fertility and may be related to very early spontaneous abortions.

Developmental Abnormalities: Birth Defects and Low Birth Weight

A series of studies have been conducted investigating the association between parental pesticide exposure and birth defects or growth retardation in their offspring (table 5.3). Parental agricultural occupation is used as a surrogate for pesticide exposure in most cases. However, in one well-conducted Finnish study of women in agricultural occupations, trained industrial hygienists estimated the amount and duration of pesticide exposure in an attempt to assess exposure more precisely. They found that exposure to pesticides during the first trimester of pregnancy nearly doubled the risk of cleft lips and palates in offspring (OR 1.9, 95 percent CI 1.1–3.5).[36] (Refer to chapter 2 for definitions of odds ratio and confidence interval.)

Two studies suggest that risks may extend to the general population as well as agricultural workers. Using statewide data from birth certificates in Minnesota, investigators determined that the birth defect rate was significantly increased for pesticide appliers and included circulatory, respiratory, skin, musculoskeletal, and urogenital abnormalities.[37] Further analysis showed that the birth defect rate was highest in the western part of the state, where chlorophenoxy herbicides (e.g. 2,4-D) and fungicides are most heavily used. Moreover, families from the general population living in western regions were 85 percent more likely to have a child with a birth defect than those from other parts of the state. And both the general population and pesticide appliers were more likely to have a child with birth defects when the child was conceived in the spring, the time of heaviest pesticide use. This seasonal effect was not seen in other areas of the state. The use of birth certificates to identify birth defects is a weakness of this study, inasmuch as abnormalities identified after birth were not included in the analysis. It is also unfortunate that the investigators did not consider neural tube defects (spina bifida) separate from other central nervous system defects since other evidence suggests that this subclass may have a unique relationship to pesticide exposure.

In Iowa, residents of a community where municipal drinking water is contaminated with commonly used herbicides were at increased risk of having children with retarded intrauterine growth.[38] This study is limited by its ecologic design in that there was no attempt to determine whether individual women whose offspring suffered from retarded growth were drinking more of the contaminated water than women whose children developed normally.

The weight of evidence from these studies supports the conclusion that there is some increased risk of birth defects in the children of parents exposed to pesticides before or during pregnancy. More accurate exposure assessment will be necessary in future studies to quantify the excess risk and assess its relevance to the general population.

Childhood Cancer

Childhood cancer is the second leading cause of death of children between ages one and 14 years of age in the United States. Moreover, the incidence of childhood cancer has been steadily increasing for twenty years, most markedly for leukemia and brain tumors.[50] Fortunately, more effective treatments have reduced the mortality from these diseases. Epidemiological studies of environmental factors that may contribute to childhood cancer are limited by small numbers of cases, making it difficult to achieve statistical significance. Nonetheless, a number of studies demonstrate an increased risk of these malignancies with parental occupational pesticide exposure or home use of pesticides.

A review of the published literature that examines the link between pesticide exposure and childhood brain cancer finds eight of nine studies showing an increased risk, with three reaching statistical significance.[51] Of particular concern are the residential use of pesticide bombs and no-pest strips during pregnancy, where the risk of brain cancer may be increased five- to sixfold.[52] This study also reported a strong association with childhood use of lice shampoo (lindane) and childhood contact with pesticides used on pets.

Five of nine studies found an increased risk of childhood leukemia with parental occupational exposure to pesticides.[51] Increasing frequency of home or garden use of pesticides was also reported associated with an increasing risk of childhood leukemia.[53]

Of particular note are studies performed within the Children's Cancer Group Epidemiology Program. Participants diagnose and treat more than 90 percent of childhood cancer in the United States, providing an opportunity to design studies with substantial statistical power. Completed studies of this program consistently find a statistically significant association between reported pesticide exposure and childhood acute myeloid leukemia.[54]

The mechanisms by which parental pesticide exposure may increase the risk of certain childhood cancers are not well understood. Possible explanations include mutations in the chromosomes of eggs, sperm, or the developing fetus and alterations in the immune system, hormone function, or DNA repair mechanisms of offspring.

Spermatotoxicity

Dibromochloropropane (DBCP), a nematocide, and ethylene dibromide (EDB), a fumigant, are toxic to sperm and have been banned from agricultural use in the United States, although EDB is used for other industrial purposes.[55,56] DBCP and EDB still contaminate groundwater in some areas where they were previously used.

2,4-D is a heavily used chlorophenoxy herbicide that may also be toxic to sperm. Sperm counts declined and abnormal sperm increased with exposure to 2,4-D in a study of farm sprayers.[57] Many weed killers for large-scale, commercial use, as well as over-the-counter preparations for home and garden use, contain 2,4-D. The urine of an estimated 12 percent of the United States population contains 2,4-D residues. The health significance of this finding is uncertain.[26]

Dibromochloropropane was first produced in the 1950s by Shell and Dow Chemical Companies for use as a soil fumigant. The chemical was used to protect a variety of fruit crops from damaging nematodes. At the time, its persistence in the soil, with a half-life of decades,[58] was considered a positive feature.

Widespread use of the chemical in vineyards, citrus orchards, and banana plantations in both the United States and abroad, began in the late 1950s. At about this same time, tests conducted by the manufacturers began to show that DBCP exposure led to testicular damage in at least three different animal species. The chemical companies chose not to act on this new information and did not submit it to the government agencies responsible for registering the chemical.[59]

In the 1970s, a number of workers at a chemical plant working with DBCP discovered that they were sterile or had suffered reduced fertility. One worker who had been trying without success to conceive a child decided to have his fertility tested. He mentioned his positive sterility test to one of his coworkers, who wondered about the possibility of a relationship with the chemicals they used in the plant. As word of their coworker's condition spread, the men discovered that a number of them had been unsuccessful at conceiving children. The men were tested and discovered that fourteen of them were sterile, and thirty-four had reduced fertility.[60] As a result of this discovery, by 1977 the California Department of Food and Agriculture had banned DBCP, and in 1979 the EPA followed suit, banning DBCP for all uses except on pineapples. A complete ban followed in 1985.

The banning of DBCP has not been the end of the story, however. During the 1980s, DBCP was discovered to be contaminating groundwater in large parts of California.[61] The chemical has made its way into the water supply of many of the agricultural communities in the state, leading to concerns about potential health effects. Not long after, the plight of workers on banana plantations outside the United States came to light. When DBCP was banned in the United States, its manufacturers continued to find willing markets in many Central American countries. As a result, thousands of plantation workers there have suffered sterility and reduced fertility; in many cases, the effects have been permanent.[62] Both California municipalities and the banana workers are engaged in ongoing legal action against the companies that produced DBCP.

Chromosome Abnormalities

Several investigations have examined the effect of pesticides on chromosomes of exposed agricultural workers, usually assessing lymphocytes from the blood. Chromosomal damage in those cells suggests that similar damage may be occurring in other cells, including sperm, raising concern about mutations and inheritable disorders. In a group of floriculture workers in Argentina who were using organochlorine, organophosphate, and carbamate pesticides, the frequency of some types of chromosomal abnormalities was four times higher than in a control population.[63] Similarly, lymphocytes from a group of pesticide-exposed workers in Hungary showed a 31 percent increase in damaged cells when compared to controls.[64] Pyrethroid exposure in a group of pesticide dealers and workers in Syria was associated with up to three times the frequency of chromosome breaks compared to controls.[65]

Summary

Spontaneous abortions, delayed pregnancies, birth defects, retarded intrauterine growth, some childhood cancers, spermatotoxicity, and chromosome damage are associated with pesticide exposure in a number of epidemiological studies. From this evidence, it is difficult to determine the magnitude of increased risk for each outcome with precision because of varying study design, accuracy of exposure assessment, and outcome measurement. Other factors may also influence the risks in agricultural work. But, recalling that more than 80 percent of families surveyed in Missouri in 1989 used pesticides during pregnancy and that large numbers of the general population have pesticide residues in their urine, obvious concerns are raised by this body of evidence. Although animal testing drives the regulatory process, these epidemiological studies should not be ignored. The data are essential to agricultural workers, employers, consumers, regulators, and pesticide manufacturers for more informed decision making.

Animal Studies

The pesticide registration process is intended to ensure that every active ingredient proposed for manufacture and use will be subject to a standard battery of animal tests. There are, therefore, considerable toxicological data available for newly proposed pesticides. Historically, however, regulators did not require rigorous toxicological evaluation of pesticides before allowing their commercial use. Consequently, for some widely used chemicals, the data are sparse and inadequate to meet current standards, not to mention proposed refinements.

Many pesticides have not been adequately tested for a range of developmental effects. New understanding of subtle and delayed expressions of toxicity, such as developmental neurotoxicity and endocrine disruption, indicates that reevaluation of many currently registered pesticides is necessary. Reregistration of chemicals "grandfathered" when current regulations became effective is underway but will not be complete for at least another ten years.

The EPA uses animal test data, usually from at least two mammalian species, to determine what they believe to be safe exposure levels for humans and the need for use restrictions and warning labels. An oral reference dose (RfD), intended to be without adverse health effects in exposed individuals, is calculated from the data. When animal tests are conducted, different health effects occur at different levels or timing of exposure. For example, for one pesticide, birth defects in test animals might occur only with a higher exposure at a different time of pregnancy than spontaneous abortions or kidney toxicity. For another chemical, it might be the opposite. Regulators typically attempt to discover the highest oral dose that fails to elicit any adverse health effect in the test animals—the no observable adverse effect level (NOAEL). They then usually divide that dose by an uncertainty factor of 100, to account for species differences and particularly susceptible individuals, calling that the RfD—the oral reference dose for humans that they believe is "safe"—that is, protective of health. Therefore, the lower the RfD, the more toxic the chemical is in animal studies—for some adverse health effect. Occasionally the uncertainty factor used is only 10 when there is considerable information about species differences in metabolism of the chemical and therefore less uncertainty. Inhalation or skin absorption is not considered in establishing an RfD. Regulators sometimes attempt to acknowledge important gaps in the data used to calculate the RfD by indicating a level of confidence in the final figure. For some pesticides in current use the level of confidence is low.

Pesticide Profiles

The following profiles summarize the reproductive and developmental toxicity data from animal studies of some members of various pesticide classes. Many of the approximately six hundred active ingredients currently in use are not mentioned, but this does not imply that they have no important toxicity. Our intent is to review the reproductive and developmental toxicity of some commonly used, high-volume chemicals.

Of particular note, the EPA identified numerous chemicals as reproductive toxicants when establishing reporting requirements for the Toxics Release Inventory (TRI). For a chemical to be added to the TRI list, it must be known or reasonably anticipated to cause, in humans, cancer, teratogenic effects (birth defects), serious or irreversible reproductive dysfunction, neurological disorders, genetic mutations, or other chronic health effects. The following list contains the chemicals that are produced and used in the United States in excess of 1 million pounds annually *and* appear on the EPA TRI list because of reproductive or developmental toxicity:[66]

Herbicides

bromoxynil	linuron
cyanazine	metribuzin
dicamba	molinate
diclofop	prometryn
diuron	simazine
EPTC	

Insecticides

diazinon	propargite

Fungicides

benomyl	maneb
mancozeb	ziram

Fumigants

methyl bromide	metam sodium

Organophosphates and Carbamates

Acephate: Used on vegetables, peanuts, tobacco, forests, ornamentals

Chlorpyrifos: Used on fruit, vegetables, nuts, cotton grain, ornamentals, turf, in the household

Diazinon: Used on fruits, vegetables, tobacco, forage, field crops, nematodes, turf, seed treatment, fly control, in the household

Dimethoate: Used on fruits, vegetables, grain, tobacco, cotton, ornamentals

Malathion: Used on fruits, vegetables, ornamentals

Naled: Used on ornamentals, poultry houses, kennels, in food processing plants, mosquito control

Tetrachlorvinphos: Used on fleas, ticks, mites, houseflies, animal feed, as larvicide

Carbaryl: Used on fruits, vegetables, forage, and field crops, nuts, ornamentals, lawns, forests, in the household

Reproductive effects   Vary among individual agents and include fetal deaths, abnormal sperm, abnormal ovarian follicles and eggs, hormonal changes, DNA damage, birth defects, neurobehavioral disorders

Organophosphates, originally designed as nerve warfare agents, are widely used in many pesticide products. Most are much less toxic than the original chemical weapons although acute toxicity is still their most commonly recognized adverse effect.

Organophosphates and carbamates disable the enzyme cholinesterase, which breaks down a naturally occurring neurotransmitter, acetylcholine. The result is runaway transmission of nerve impulses along certain nervous system pathways. Symptoms of acute intoxication include excessive salivation, tremors, muscle twitching, nausea, vomiting, and diarrhea. Chronic exposure to lower doses of some organophosphates may also lead to delayed neurological symptoms.

Since many different organophosphates and carbamates are used for various purposes, total human exposure to these pesticides is likely to be higher than predicted from consideration of individual agents and single routes of exposure. Indeed, some farmworker exposures, many in violation of state and federal regulations, are sufficient to depress cholinesterase enzyme levels.[67] Low enzyme levels may be associated with symptoms that often go unreported or are unrecognized by health professionals as associated with pesticide exposure. Indoor use of organo-phosphates according to label directions may also lead to excessive exposures.[22,23]

Animal studies sometimes show dose-related adverse reproductive and developmental effects from dosing at levels that do not cause obvious evidence of acute toxicity (table 5.4). These chemicals cross the placenta and depress cholinesterase levels in the fetal blood and brain, though not always to the same degree as in maternal tissues. Although dose thresh-olds for easily recognized reproductive effects in animals are generally above likely human exposure levels, the large majority of animal tests have not examined subtle long-term effects on the developing fetal brain after exposure during pregnancy.

In a study of pregnant rats exposed to chlorpyrifos at 6.25, 12.5, or 25 mg/kg/day by injection on days 12 to 19 of a twenty-one-day pregnancy, the investigators concluded that marked neurochemical and behavioral alterations occur in the developing organism following repeat exposures in the absence of overt maternal toxicity.[68] Cholinesterase lev-els were reduced in maternal and fetal brains in all exposure groups. Behavior testing, limited to the high-exposure group, included observing the newborn rat's ability to right itself when placed on its back and purposeful avoidance of the edge of a table ("cliff avoidance"). Young chlorpyrifos-exposed rats had markedly reduced performance in these two tests, yet the animals had no visible evidence of birth defects. They would have been judged normal by toxicity tests that are routinely used to assess the safety of pesticides for regulatory purposes.

Another study of rats injected with 0.03 to 0.3 mg chlorpyrifos/kg/day during days 7 to 21 of pregnancy found a dose-related increase in fetal deaths and birth defects in the highest-dose group.[69] The abnor-malities included small limbs and lack of spinal development. Moreover, healthy-appearing animals were abnormal on neurological and behav-ioral testing. In the Dow Chemical research laboratory, investigators found that similar doses of chlorpyrifos administered directly into the stomach of rats on days 6 to 15 of pregnancy did not cause fetal death or obvious birth defects. However, these animals did not undergo neuro-logical or behavioral testing.[70]

Rats given daily doses of parathion (1.0 mg/kg/day) on days 6 to 20 of pregnancy at doses that showed no evidence of maternal toxicity

Table 5.4 Reproductive and developmental toxicity of selected organophosphates and carbamates in animal tests

Chemical	Reproductive and developmental toxicity	RfD (mg/kg per day) and EPA confidence in RfD
Organophosphates		
Acephate	Mouse: reduces luteinizing hormone[77] Rat: fetal losses, decreased litter weight	0.0003, high confidence
Chlorpyrifos	Mouse, rat: increased birth defects at 25 mg/kg/day[78,69] Rat: behavioral neurotoxicity[68] Rat: fetal deaths, birth defects, neurobehavioral toxicity at 0.3 mg/kg/day by injection[69]	0.003, medium confidence
Diazinon	Rat, mouse: decreased sperm motility, increased abnormal/dead sperm, decreased testosterone level, increased fetal deaths, increase in some birth defects[79,80] Mice: neurotoxicity in offspring[81]	No RfD set by EPA; under review
Dimethoate	Rat: decrease in testes weight and sperm motility, abnormal sperm, decreased testosterone (6–12 mg/kg/day for 65 days)[82]	0.0002, medium confidence
Malathion★	Cows: decreased progesterone (1 mg/kg) Rats: smaller litters, reduced pup weight[84]	0.02, medium confidence
Naled	Rat: decreased survival, litter size, pup body weight (18 mg/kg/day)[85]	0.002, medium confidence
Parathion	Rat, mouse, hamster: DNA damage Rat: fetotoxicity Chick: birth defects[86]	Under review by EPA
Tetrachlorvinphos	Mouse: ovarian follicles show poor growth, premature ovulation, poor egg development[81]	0.03, medium to high confidence
Carbamate		
Carbaryl (Sevin)	Dogs: birth defects at 5–6 mg/kg/day (not in monkeys at 20 mg/kg/day) Rat, gerbil: decreased reproductive capacity, trend to sterility with increasing dose[87]	0.1, medium to low confidence

★ This pesticide is under special review by the EPA.

gave birth to offspring with altered postnatal development of neurons and subtle alterations in behavior.[71]

Pregnant mice given daily doses of diazinon (0.18, mg/kg/day, 9.0 mg/kg/day) gave birth to normal-appearing offspring.[72] However, even mice in the low-exposure group showed impaired endurance and coordination on neuromuscular testing as they developed into adults. This pesticide is under review by the EPA.

Recent research provides insight into mechanisms by which fetal exposures to organophosphates and carbamates may have long-term effects on brain function in offspring. Acetylcholine is but one of a number of different neurotransmitters that transmit nerve impulses across the connections (synapses) in established networks of nerve cells (neurons). During fetal and early infant brain development, these same neurotransmitters serve the very important additional function of signaling information for further development of the brain.[73] Abnormal fluctuations in neurotransmitter levels during fetal and early infant life interfere with differentiation of maturing brain cells and the development of normal nerve connections in the brain. The number and distribution of receptors, to which the transmitters attach, may also be altered. These are distinctly unlike effects in adults, whose brain connections are already established, where neurotransmitters temporarily alter nerve impulse traffic rather than permanently affecting the connections themselves.

One study found that a single low dose of an organophosphate given to mice on day 3 or 10 after birth caused increased activity in the animals when measured at four months of age and permanent alterations in neurotransmitter receptor levels in the adult brains.[74] In another study, when chlorpyrifos was administered to neonatal rats at doses that showed no other evidence of toxicity, both protein and DNA synthesis were inhibited in the brain.[75] It is important to note that the first ten days of postnatal life in the rodent represent stages of brain development corresponding to the last trimester of gestation in humans.[76]

Effects on neurological development and behavior at low doses in animals are of more concern at current human exposure levels. Animal studies demonstrate the need to redesign required toxicological testing of these pesticides to include better examination of neurodevelopmental effects as called for in the FQPA.

Organochlorines

Dicofol: Used for mite control on fruit, vegetable, ornamental, field crops

Dienochlor: Used for shrubs, trees, and in greenhouses

Endosulfan: Used for fruits, vegetables, coffee, tea, forage, and field crops, grains, nuts, ornamentals, tobacco

Lindane: Used for seed and soil treatment, nurseries, tree farms, tobacco, human louse control

Methoxychlor: Used for fruit and shade trees, vegetables, dairy and beef cattle, home gardens, around farm buildings

Reproductive effects   Endocrine disruption, including effects on estrogen, androgens, prolactin and thyroid hormone, fetal loss, and reduced sperm counts in animal tests

Organochlorine insecticides are used in agriculture, forestry, and building and human protection from insects. DDT was among the first of this class of chemicals to be developed in the 1930s. Organochlorines were of particular concern to Rachel Carson, who, in *Silent Spring,* protested the growing use of pesticides with harmful effects that cascaded through the food chain, decimating populations of birds and threatening other species. Years later, heightened scientific, governmental, and public awareness of the environmental persistence of these chemicals with harmful effects on nontarget organisms finally prevailed over entrenched industry resistance and led to withdrawal of or bans on DDT, heptachlor, kepone, aldrin, dieldrin, and chlordane in the United States. Many organochlorines, including DDT, continue to be widely used in other parts of the world, particularly in developing countries, for controlling insects responsible for crop loss and human disease (e.g., malaria). Short-term benefits and established manufacturing and trade practices perpetuate their use. In the United States, endosulfan, methoxychlor, and dicofol are still used on the food supply.

Normal nerve cell function depends on the transport of electrically charged ions of sodium, potassium, and calcium across the cell membrane. Organochlorines exert their toxic effects by altering the normal transport of sodium and calcium across nerve cell membranes. The net result is an increase in the sensitivity of the neurons to small stimuli that would not otherwise elicit a response in an unexposed nerve. Studies in wildlife and laboratory animals at exposure levels not acutely toxic show hormonal and other biochemical (enzyme-inducing) properties of organochlorines. Developing animals are more sensitive than adults, and there is considerable concern about the organochlorines' long-term effects on human and wildlife fertility, reproduction, and development.

Organochlorines in use in the United States are not as persistent in the environment as older members of the class. Half-lives are generally measured in weeks, but lindane may be detected in pine needles and forest soil years after spraying, with a typical half-life of four hundred days.[88] All have some tendency to bioaccumulate so that small exposures result in much larger tissue levels over time. Bioaccumulation sometimes occurs in the middle of the food chain where, for example, methoxychlor concentrations in mussels and snails are about 10,000-fold higher than levels in the surrounding water or soil, but not in fish, which tend to metabolize the chemical rapidly.[89] Lindane, however, does tend to bioaccumulate in mammals at the top of the food chain. (Table 5.5 summarizes the reproductive and developmental effects of some organochlorines.)

Lindane acts as an antiestrogen, weakly interfering with the effect of naturally occurring estrogen on target tissues. Chronic treatment of newborn rats delays vaginal opening, disrupts normal ovarian cycles, and reduces pituitary and uterine weight.[90,91] Lindane given orally to pregnant mice at various stages of pregnancy in amounts sufficient to cause fetal death may be prevented from having this effect by simultaneously giving estrogen and progesterone.[92] In adult male rats, lindane retards testicular growth when given at 4 and 8 mg/kg/day over forty-five days.[93] Fetal exposure to lindane also alters development of the immune system. Pregnant mice exposed to lindane at 10 mg/kg/day throughout gestation produced offspring with overactive immune responsiveness.[94] In an interesting demonstration of synergy, investigators exposed a group of pregnant rats to lindane (20 mg/kg/day on days 6 to 14 of pregnancy), another group to cadmium in their food (approximately 4.2 mg/rat/day), another group to both lindane and cadmium (same doses), and a control group to neither substance.[95] Cadmium or lindane alone did not produce significant malformations in twenty-day-old fetuses, but the combination increased fetal deaths and produced a marked increase in skeletal abnormalities.

Table 5.5 Reproductive and developmental effects of selected organochlorine pesticides in animal tests

Chemical	Reproductive and developmental toxicity	RfD (mg/kg/day)
Dicofol	Rat: prenatal administration alters behavior in offspring—10 mg/rat, days 4–15 of pregnancy[105] Bird: eggshell thinning, reduced hatchability, abnormal gonads (male), submissive behavior (male), infertility of offspring[102] Reptile (see chapter 6)	No data (none established)
Endosulfan	Humans: Estrogen-like effects on estrogen-sensitive breast tumor cells[100] Rat: Shrinkage of testicles—inhibits hormone synthesis (follicle-stimulating hormone, luteinizing hormone) at 7.5 mg/kg/day[106] Mouse: reduces sperm count[107]	0.006, medium confidence
Lindane	Rat: testicular degeneration and androgen deficiency at 4 mg/kg/day[93] Mouse: early pregnancy, absence of implantation of fertilized eggs in uterus at 40 mg/kg/day; in mid-pregnancy, loss of fetuses; in late pregnancy, newborn deaths[92] Mouse: single dose on day 9 of pregnancy (25 mg/kg/day) toxic to fetuses[109] Rabbit: reduced ovulation[110] Rat: delayed vaginal opening, disrupted ovarian cycles, reduced uterine weight[90]	0.0003, medium confidence (persists for years after spraying)
Methoxychlor	Birds, mammals: Has estrogenic activity[99] Female rat: accelerated vaginal opening, abnormal estrus cycle, inhibited luteal function, blockage of implantation, reduced fertility and litter size[111] Male rat: elevated prolactin levels, suppression of Leydig cell function (some at 25 mg/kg per day)[111] Rat: aggressive behavior in male offspring[98] Rat: birth defects with higher doses[111]	0.005, low confidence

Both endosulfan and methoxychlor are estrogenic, as shown in a large number of animal and other laboratory studies. Unlike endosulfan, methoxychlor must first be metabolized to a by-product, which is the estrogenic substance. Investigators injected fertile gull eggs with either DDT or methoxychlor at levels found in eggs from southern California in the early 1970s and demonstrated feminization of developing male embryos.[96,97] As discussed in chapter 1, male-type brain development depends on the chemical transformation of testosterone to estrogen in the brain by the enzyme aromatase. When mice are exposed orally to methoxychlor during pregnancy, their male offspring show more aggressive territorial behavior. This effect is also seen with estrogen and diethylstilbestrol at much lower doses than those required for methoxychlor.[98] Mice treated with methoxychlor or estrogen on days 6 to 15 of their twenty-one-day pregnancy have female offspring whose vaginal opening (evidence of sexual maturation) occurs earlier than normal. When these same mice are mated again, female offspring from their second pregancies show a similar result, indicating a residual effect from previous treatment.[99]

Endosulfan interacts directly with the estrogen receptor, as demonstrated in cultures of estrogen-sensitive breast cancer cells.[100] Endosulfan, kepone, DDT, dieldrin, and toxaphene, all organochlorine pesticides, directly stimulate growth of these estrogen-responsive cancer cells. In this assay, however, each compound is thousands of times less potent than estrogen in producing the effect. It is therefore a matter of significant debate as to whether this observation has any relevance to human health.

Dicofol has been used in the United States since 1955 and thereby escaped any thorough assessment of its toxicity, having been grandfathered for continued use as new testing requirements evolved. It is in the process of reregistration by the EPA.

Dicofol is manufactured from DDT. In the 1980s dicofol manufactured in Europe contained as much as 20 percent DDT contamination, with somewhat lesser amounts in U.S. preparations.[101] This contamination not only complicates toxicity testing but provides ongoing release of DDT into the environment, where dicofol is used. Currently the EPA requires manufacturers to use techniques that minimize DDT contamination.

Studies of the effects of dicofol on reproduction and behavior of captive kestrels show that maternal exposure by oral intake leads to egg-shell thinning, feminization of male embryos, abnormal submissive behavior in male offspring, and impaired reproductive capacity of the offspring after they mature.[102,103] In widely publicized studies at Lake Apopka in Florida, where alligators were exposed to dicofol contaminated with DDT, along with other pollutants associated with agricultural activity, male juveniles had significantly depressed testosterone levels, abnormal testes, and small penises when compared to control animals from another lake.[104] Exposed females had significantly elevated estrogen levels and abnormalities of their ovaries.

Pyrethrins and Pyrethroids

Cypermethrin: Used for cotton, fruit, vegetables, cockroaches, household insects, termites

Fenvalerate: Used for a wide range of crops, Christmas trees, pine seed orchards, tree nurseries

Permethrin: Used in the household, garden, and greenhouse; a broad spectrum used for a wide range of commercial crops

Resmethrin: Used in the household and greenhouses, for indoor landscaping, mushroom houses, stored products, mosquito control

Reproductive effects   In animal tests, some pyrethroids decrease offspring weight, increase fetal losses, and interfere with brain development; potential for endocrine disruption

Pyrethrins are naturally occurring pesticide compounds derived from chrysanthemums. Pyrethroids, chemically similar to pyrethrins, are synthesized for commercial use. These chemicals are widely used throughout the world and are found in many home-use pesticide products.

Pyrethrins and pyrethroids have a paralytic action on insects. They cause repetitive nerve discharge and interfere with enzyme levels in the brain. Like organophosphates and carbamates, the developmental neurotoxicity of pyrethrins and pyrethroids is not routinely assessed for regulatory purposes, and currently established tolerances do not consider the particular vulnerability of the developing brain to permanent effects from low-dose exposures. (Table 5.6 summarizes the reproductive and developmental effects of the pyrethrins and pyrethroids.)

Table 5.6 Reproductive and developmental health effects of selected pyrethrins and pyrethroids

Chemical	Reproductive and developmental toxicity	RfD (mg/kg per day) and EPA confidence in RfD
Cypermethrin	Rat: decreased offspring weight at 5 mg/kg/day; decreases brain neurotransmitter receptors in offspring when given on days 5–21 of pregnancy (15 mg/kg per day); delays maturation of cerebral cortex at 3–6 weeks of age;[112] developmental delays (given on days 5–21 of pregnancy at 15 mg/kg per day)[116]	0.01, high confidence
Fenvalerate	Rat: decreases brain enzymes in offspring when given on days 5–21 of pregnancy (10 mg/kg per day)[112]	0.025, high confidence
Permethrin	Rat: Three-generation reproductive study; liver and eye abnormalities in offspring at lowest dose tested (25 mg/kg per day)[117]	0.05, high confidence
Resmethrin	Rat: multigeneration study; stillborn pups, low birth weight at lowest dose tested (25 mg/kg per day); same result with one-generation study[114]	0.03, high confidence

The offspring of rats treated with fenvalerate or cypermethrin, (10–15 mg/kg/day) during days 5 to 21 of pregnancy, have abnormal brain levels of chemical neurotransmitters.[112] Neonatal mice given 0.21–0.42 mg/kg of bioallethrin for seven days soon after birth had permanent changes in brain neuroreceptor levels and increases in their level of activity.[113] But when bioallethrin was administered at one hundred times the doses that caused these effects, the animals showed decreased activity and no change in receptor levels. This observation raises important questions about the appropriateness of using high-dose testing when studying the toxicity of pesticides for registration purposes.

The RfD for resmethrin, one of the synthetic pyrethroids, is established on the basis of its developmental toxicity in a three-generation rat study.[114] The effects at the lowest dose tested (25 mg/kg daily) included an increased incidence of stillborn offspring and decreased body weight at weaning.

Some pyrethroids bind to androgen receptors and also displace testosterone from its carrier protein in the circulation.[115] The importance of this to humans or animals is not clear since the binding is weak, but inasmuch as most testosterone is protein bound, even a small decrease in binding could result in a significant increase in free testosterone levels in the blood. Free testosterone can enter cells and activate the testosterone receptor in testosterone-responsive cells, whereas protein–bound testosterone is confined to the circulatory system and does not have the same biological activity.

Fungicides

Dithiocarbamates: Used for fruits, vines, hops, vegetables, potatoes, ornamentals, tobacco

Benomyl and thiabendazole: Used for fruits, nuts, vegetables, grains, nuts, turf, bulbs, flowers, ornamentals

Vinclozolin and iprodione: Used for grapes, strawberries, soft fruit, vegetables, ornamentals, hops, oilseed rape

Reproductive effects Birth defects, testicular toxicity, and endocrine disruption in animal tests

Fungicides are used to prevent fungal growth on agricultural and various consumer products. Foliar fungicides, applied to the leaves of plants, and soil fungicides, applied to the soil as liquids, powders, or granules, may be taken up into the plant. Dressing fungicides are applied after harvest to protect crops like cereals and grains. There is a long history of controversy surrounding the use of fungicides since most cause gene mutations in bacterial test systems, raising concerns about carcinogenicity.[118] Some, like hexachlorobenzene, are no longer used in the United States because of their toxicity and long life in the environment. Others are being reinvestigated because of new findings of toxicity in animal studies. Chemicals used as fungicides fall into several classes. (Table 5.7 summarizes the reproductive and developmental effects of fungicides.)

Table 5.7 Reproductive and developmental health effects of fungicides

Chemical	Reproductive and developmental toxicity	RfD (mg/kg per day) and EPA confidence in RfD
Benomyl	Rats and rabbits: birth defects[134] (no effect at 30 mg/kg per day with rat); interferes with mitosis Rats: damage to testicles, Sertoli cell toxicity/sperm toxicity[135]	0.05, high confidence; California Proposition 65 birth defects list
Thiabendazole	Rats: fetal death, birth defects at 60 mg/kg per day[127]	0.1, high confidence
Dithiocarbamate fungicides		
Maneb, nabam, zineb, mancozeb	Rat: birth defects, single dose on day 11 or 13 of pregnancy (>0.5 g/kg needed)[136]	
Thiram	Rat: blocks LH surge, interferes with ovulation[137] Testicular toxicant at 5 mg/kg per day[138]	0.005, low to medium confidence
Vinclozolin	Rat: feminization of male offspring exposed in utero[133]	
Iprodione	Rabbit: abortions at 60 mg/kg per day	0.042, high confidence

Dithiocarbamates

The dithiocarbamates include maneb, mancozeb, nabam, thiram, ziram, and zineb, which are used on a variety of fruit and vegetable crops. These fungicides are broken down into ethylene thiourea (ETU) in the environment and in mammals. ETU causes mutations, birth defects, and cancer and may be formed by cooking food contaminated with the fungicides.[118,119]

Since 1977 the various uses of and tolerances for dithiocarbamates have been the subject of ongoing negotiation between the EPA and manufacturers, based largely on concerns about carcinogenicity and thyroid effects. Tolerances and crop uses of dithiocarbamates have frequently changed and may be influenced further by provisions of the 1996 FQPA, which requires the EPA to issue health-based tolerances after considering total exposure to agents with similar mechanisms of action.

In rats, maneb and mancozeb cause birth defects or fetal death after fairly high levels of maternal exposure (500–1,300 mg/kg/day during eleven days of pregnancy or inhalation of 500 mg/m^3/day for 6 hours daily during five days of pregnancy).[120–122] Brain abnormalities are among the birth defects in rats and hamsters. There is conflicting evidence of birth defects in chicks following exposure to maneb in the egg, with some studies showing lower limb deformities.[123,124] Zineb, maneb, and mancozeb are also toxic to sperm and damage the testes of rats at fairly high exposure levels.[125,126] However, toxic effects on the thyroid and the potential for carcinogenesis, rather than reproductive effects, drive current tolerances of dithiocarbamates on food.

Benzimidazole Fungicides

The benzimidazole fungicides, benomyl and thiabendazole, are used before and after harvest on different foods, bulbs, flowers, ornamentals, and shade trees. Thiabendazole is used not only as a fungicide but also to treat certain parasitic diseases in humans. In mice, rats, and rabbits, it causes fetal toxicity, death, and birth defects with maternal exposures above 60 mg/kg/daily during nine or more days of pregnancy.[127,128]

Benomyl is metabolized into carbendazim, which is thought to be the chemical responsible for most of the toxicity of the parent compound.[129] Benomyl causes birth defects and testicular toxicity in rats and rabbits.[130–132] It is on the California Proposition 65 list of reproductive hazards.

Dicarboximide Fungicides

Vinclozolin and iprodione are fungicides used to control a variety of crop diseases. Vinclozolin, an androgen antagonist, causes demasculinization of male offspring when given to pregnant rats. Abnormalities include reduced anogenital distance (more female-like), nipple development, and abnormal penises with hypospadias.[133] In rabbits, iprodione causes increased abortions at 60 mg/kg/daily throughout pregnancy.[84]

Herbicides

Triazines (atrazine, cyanazine, simazine, prometryn): Used for grasses and weeds in field crops, orchards, vineyards, turf

Chlorinated phenoxy herbicides (2,4-D, diclofop, dicamba): Used for wild oats and annual grassy weeds

Substituted urea herbicides (linuron, diuron): Used for annual and perennial broadleaf and grassy weeds, field and vegetable crops, sugar cane.

Bromoxynil: Used for postemergent control of annual broadleaved weeds in corn, cereal, sorghum, onions, flax, mint, and turf

EPTC: Used for control of annual and perennial weeds in beans, legumes, potatoes, corn, and sweet potatoes

Metribuzin: Used for control of grasses and broadleaved weeds in field and vegetable crops, turf

Molinate: Used for control of weeds in rice paddies

Reproductive effects In animal studies: Spermatotoxicity, fetal losses, decreased fetal weight, birth defects. In humans: Evidence of birth defects and spermatoxicity

Herbicides are used to control unwanted vegetation and often replace mechanical cultivation. They are applied on large tracts of forest, farmland, and tree farms; along roadsides; beneath power lines; and on lawns and gardens. Their chemical structures and toxicities vary considerably. Herbicides are often referred to as pre- or postemergent herbicides, depending on whether they are applied to soil to prevent weed growth or directly to weeds after sprouting. Monoculture favors the emergence of particular weeds, which are often treated with herbicides.

These chemicals may contaminate the soil for long periods, migrate to groundwater, or run off in surface water to lakes, streams, and rivers. Aquifers beneath much of the nation's farmland contain a mixture of agricultural chemicals, including herbicides.

Of the sixty pesticidal active ingredients that are produced and used in excess of 1 million pounds annually in the United States, thirty-four are herbicides.[66] Total U.S. production of herbicides in 1993 has been estimated at more than 750 million pounds.[139] Their use on corn, soybeans, small grains, and cotton accounts for 71 percent of the U.S. herbicide market. In the early 1990s, nearly 150 million pounds of five herbicides were applied just to corn and soybean fields annually in the

Midwest. These include three triazine herbicides plus alachlor and met-alachlor, each of which is commonly found in drinking water sup-plies.[12,18] In 1993, the EPA estimated the annual use of home, lawn, and garden herbicides to be 27 million pounds.[13] (Table 5.8 summarizes the reproductive and developmental effects of herbicides.)

Triazines

Atrazine, simazine, cyanazine, and prometryn are triazine herbicides. Cyanazine causes fetal toxicity in rabbits at 2 mg/kg/day and birth de-fects in rats at 25 mg/kg/day. It is on the California Proposition 65 list of reproductive hazards, and manufacturers say that they intend to eliminate its production by 2002.

Atrazine causes birth defects when pregnant rats are dosed at 70 mg/kg/day on days 6 to 15 of pregnancy.[140] Recent studies in rats show that high doses of atrazine inhibit some estrogen-induced responses.[141,142] Others show that atrazine accelerates the onset of breast tumors in one strain of rats but not in another, suggesting that it may also have an estrogenic effect.[143] There is considerable debate about the relevance of these observations to humans, particularly in the light of the steady in-crease in breast cancer incidence in the United States over the past thirty years.

A series of experiments in cell cultures and in juvenile animals sug-gests that it is unlikely that atrazine exerts its observed effects by attaching to the estrogen receptor and mimicking or blocking naturally occurring estrogen.[141,144] However, in a study using yeast containing the human estrogen receptor, triazines appear to block estrogenic activity by in-terfering with receptor binding, but only in the presence of low concen-trations of estrogen.[145] Other mechanisms, unrelated to receptor binding, may explain atrazine's interference with normal estrogen function, such as resetting regulatory feedback loops between the hypothalamus and ovaries or changing metabolic breakdown pathways of estrogen. For ex-ample, studies in human breast cancer cells show that atrazine alters met-abolic degradation of estrogen, resulting in a larger amount of a more persistent estrogenic by-product.[146] One view holds that this mechanism may increase breast cancer risk in women exposed to atrazine. These matters, far from settled, are of considerable concern since triazine herbi-cides are heavily used, and large numbers of people are exposed.

Table 5.8 Reproductive and developmental health effects of herbicides

Chemical	Reproductive and developmental toxicity	RfD (mg/kg per day)
Atrazine*	Rat: decrease in male offspring weight in second generation at 25 mg/kg per day; skeletal birth defects at 70 mg/kg per day; increased prostate weight at 120 mg/kg per day;[152] androgen antagonist interferes with receptor formation and testosterone conversion[153] Rabbit: increased fetal death at 75 mg/kg per day[154]	0.035, high confidence
Cyanazine	Rabbit: fetal toxicity at 2 mg/kg per day Rat: birth defects at 25 mg/kg per day[155]	RfD withdrawn; manufacturer agrees to phase out; on California's Proposition 65 list
Simazine	Sheep: testicular toxicity, spermatotoxicity (1.4 mg/kg per day)[147] Rat: birth defects, decreased fetal weight (200 mg/kg per day)	0.005, high confidence
Dicamba	Rabbit: offspring with less weight gain, reduced fetal body weight, increased fetal loss (pregnant rabbits given 10 mg/kg per day) Rat: heart abnormalities in offspring, skeletal malformations[147,111]	0.03, high confidence (1980 estimate); 2.3 million U.S. residents with dicamba in urine[156]
2,4-D	Rat: fetal deaths (maternal dosing at 50 mg/kg per day before mating and throughout gestation)[148]	RfD under review; estimated 12 percent of U.S. population with 2,4-D in urine
Diclofop	Rat: fetal losses, birth defects (1.6 mg/kg per day)[147]	RfD under review
Bromoxynil	Rat: reduced fetal weight, increased fetal deaths (35 mg/kg per day)[84] Rat: birth defects (extra ribs) (5 mg/kg per day)[147] Rabbit: brain, eye, skull abnormalities (60 mg/kg per day)[147]	0.02, medium confidence
Paraquat	Rat: fetotoxic (weight loss and bone abnormalities) (5 mg/kg per day)[157]	0.0045, high confidence; adheres to soil particles, persists for years

Table 5.8   (*continued*)

Chemical	Reproductive and developmental toxicity	RfD (mg/kg per day)
Linuron	Rat: fetal losses; reduced pup survival, pup weight, liver and kidney weight (6.25 mg/kg per day) Rabbit: decreased fetal weight, litter size; increased skull malformations (5 mg/kg per day)[147]	0.002, high confidence
EPTC	Rat: fetotoxicity (300 mg/kg per day, days 6–15 of pregnancy);[147] decreased pup weight (40 mg/kg per day)	Medium confidence

* A pesticide/fertilizer mixture of alachlor, atrazine, cyanazine, metolachlor, metribuzin, and ammonium nitrate at 1, 10, and 100 times the concentrations found in groundwater in Iowa was evaluated for reproductive toxicity in mice. There was no significant reproductive toxicity at any of the concentrations tested.[158] However, in a study of chromosome damage, N-nitrosoatrazine, readily formed from atrazine and nitrate in an acid environment such as that found in the stomach, was thousands of times more damaging to chromosomes than atrazine and nitrates separately or combined.[159]

The toxicity database for prometryn is old, and few reproductive and developmental data are available. One study reports fetal toxicity in rabbits at 72 mg/kg/per day. The EPA has low confidence in the established tolerance and lists prometryn as a developmental toxicant subject to TRI reporting.[147]

Chlorinated Phenoxy Herbicides
Chlorinated phenoxy herbicides have been in extensive and uninterrupted use since 1947.[118] 2,4-dichlorophenoxyacetic acid, otherwise known as 2,4-D, is widely used by commercial appliers and home owners to kill weeds. A mixture of 2,4-D and another member of this class, 2,4,5-T, is known as Agent Orange and was sprayed as a defoliant over vast areas of Vietnam during the 1960s and early 1970s. This mixture was inevitably contaminated with dioxin as a result of the production process. Consequently, epidemiological studies intended to reveal the health effects of exposure to this class of chemicals have had to contend with the potential contribution of dioxin to the observed results.

In animal studies, 2,4-D causes toxicity to the blood, liver, and kidneys at 5 mg/kg/day.[148] Larger doses are necessary to elicit reproductive toxicity. For example, there was an increase in the mortality of offspring of rats given 2,4-D in a dose of 50 mg/kg/day for three months before mating and throughout gestation.[148] Musculoskeletal, nervous system, urinary system, and head and face abnormalities appear at still higher doses in animal tests.

In a study of male farm sprayers exposed to 2,4-D as determined by measuring residues in their urine, significantly lower sperm counts and increases in abnormal sperm were seen in the exposed group when compared to controls.[149] This study has been criticized by a pesticide industry–sponsored review of the toxicity of 2,4-D for failing to describe how samples were handled and how controls were selected.[150] Yet an epidemiological study in Minnesota showed significantly higher rates of birth defects among the offspring of pesticide appliers and the general population in areas of the state with the highest use of chlorophenoxy herbicides and fungicides.[37] The increase was most pronounced for infants conceived in the spring, the time of highest herbicide use.

Diclofop is another heavily used chlorophenoxy herbicide. Rats given this agent at 5 mg/kg/day during pregnancy experienced increased fetal losses and gave birth to offspring with reduced body weights and abnormalities of the urinary tract.[147] Increased offspring mortality occurred at 5 mg/kg/day in a three-generation rat study.

2,4-D and other chlorophenoxy herbicides are under review by the EPA primarily because of concern about carcinogenicity. In particular, a body of evidence indicates a relationship between exposure to these chemicals and development of malignant lymphoma.[118] However, there are also studies that show no relationship. It is unlikely that this controversy will be resolved anytime soon.

Substituted Urea Herbicides
Linuron and diuron belong to a chemical class called substituted urea herbicides, which work by inhibiting photosynthesis. In rats given linuron during pregnancy at doses of 6.25 mg/kg/day, pup weights and survival were reduced. Rabbits show the same adverse effects at 5 mg/kg/day. Rabbit offspring also show evidence of skull abnormalities.[151] Diuron causes decreased body weight in the offspring of a three-generation reproduction study in rats and rib abnormalities in a teratology study. Both diuron and linuron are listed as developmental toxicants subject to TRI reporting.[147]

Other Herbicides

Bromoxynil, a nitrile herbicide that inhibits photosynthesis, it is on the California Proposition 65 list of reproductive toxicants. In rats, this herbicide causes fetal toxicity and rib abnormalities in offspring at 35 mg/kg/day. In rabbits, it causes brain, eye, and skull defects at 30 mg/kg/day.[84]

Metribuzin is a selective herbicide that causes maternal and fetal toxicity in rabbits when given at 45 mg/kg/day during days 6 to 18 of pregnancy. In rats, abnormalities of the spinal column and decreased pup body weight occur at 85 mg/kg/day. The EPA lists metribuzin as a developmental toxicant subject to TRI reporting.[147]

Molinate is a selective herbicide that causes fetal losses, decreased fetal and pup weight, and skeletal abnormalities when given to pregnant rats at 35 mg/kg/day. When given to male rats at 4 mg/kg/day, molinate causes abnormal sperm, decreases fertility, and causes fetal death. The EPA lists molinate as a reproductive and developmental toxicant subject to TRI reporting.[147]

EPTC (S-ethyl dipropylthiocarbamate) is a cholinesterase inhibitor used as a selective herbicide. When given to pregnant rats at 40 mg/kg per day, it causes reduced pup weight.[84] At even lower doses, pregnant females develop degenerative heart disease.

Acaricide

Acaricides are used to control mites, which are parasites on plants and animals. Among these are dicofol (see under Organochlorines) and propargite. Propargite is used on and around citrus crops. According to the California Department of Pesticide Regulation, propargite ranks highest among pesticides as a candidate for evaluation as a toxic air contaminant in that state.[160] In a developmental toxicity study in which rabbits were given propargite (6 mg/kg/per day) during days 6 to 18 of pregnancy, there were an increase in fetal losses, decreased fetal weight, and delayed bone development in offspring.[157] Bone developmental abnormalities also occur in rats at similar doses. The U.S. EPA lists propargite as a reproductive toxicant subject to TRI reporting.

Fumigants

Ethylene dibromide: No current pesticidal uses in the United States; was used as a soil and spot fumigant of grain milling machinery, to control infestations in fruits, vegetables, and grain; now used as a lead scavenger in gasoline and as a solvent

Ethylene oxide: Used in the manufacture of antifreeze, polyester fiber, and film, many organic chemicals; fumigant and fungicidal sterilizing agent for medical supplies, drugs, books, leather, clothing, and furniture.[161]

Methyl bromide: Pesticidal gas that is injected into soil before planting strawberries, grapes, almonds, tomatoes, tobacco, and other crops; as a grain fumigant; to treat imported produce and timber at ports of entry; in industrial chemical manufacturing; as a solvent for extraction of oils from nuts, seeds, and wool

Metam sodium: Used to sterilize soil before planting by killing seeds, weeds, bacteria, nematodes, fungi, and insects

Reproductive effects Spermatotoxicity, chromosome damage, mutations

Fumigants are used to kill insects, nematodes, weed seeds, and fungi in soil, as well as in stored grains, fruits, vegetables, and clothing. Most fumigants are volatile and are applied in closed spaces. These chemicals are generally highly toxic and may cause considerable damage to living tissues with inhalation or direct contact. The reproductive and developmental health effects of several of the more heavily used fumigants are summarized in table 5.9.

Ethylene Dibromide

Ethylene dibromide (EDB) was widely used for many purposes until it was discovered to cause chromosome damage, cancer, and toxicity to sperm. An EPA review of its use as a pesticide began in 1977. Most agricultural uses were cancelled in 1983, when it was discovered in stored grain and wells. Traces of EDB have been found in some Connecticut soils up to twenty years after their last known fumigation.[162] Improper disposal of EDB and fuels led to contamination of groundwater as well. Concentrations of EDB over 100 μg/liter contaminate a groundwater aquifer near the Massachusetts Military Reservation on Cape Cod. Maximum contaminant levels (Safe Drinking Water Act standards) are set at 0.05 μg/liter.

Table 5.9　Reproductive and developmental health effects of fumigants

Chemical	Reproductive and developmental toxicity	RfD (mg/kg per day)
Ethylene dibromide	Bull: lower sperm counts and abnormal sperm[164]	No data on RfD
Ethylene oxide	Rat: low birth weight[165] Monkey: lower sperm counts, chromosome damage[166] Mouse: birth defects[168]	0.03 mg/m^3 (inhalation exposure)
Methyl bromide	Mouse: testicular toxicant[177] Rat: lower testosterone levels[172]	
Metam sodium	Rat: Fetal deaths, birth defects (10 mg/kg per day) Rabbit: birth defects[174]	

Both human and animal studies demonstrate EDB's toxicity to sperm. Bulls exposed to dietary EDB develop lower sperm counts and sperm with diminished motility.[163, 164] Sperm maturation is affected but recovers over a period of days to months when the EDB is removed from the diet. Agricultural workers exposed to EDB have also had decreased sperm counts, decreased viable and motile sperm, and increased numbers of abnormally shaped sperm when compared to an unexposed group.[56] Most uses of EDB have been cancelled, but groundwater contamination persists in some areas.

Ethylene Oxide

Ethylene oxide (EtO) is a highly toxic, explosive chemical and is usually kept in tightly closed, automated systems, with little opportunity for worker exposure. However, improperly operated or malfunctioning sterilizing systems in hospitals may result in brief but significant exposures. The hazards of EtO are widely known. Most hospitals, for example, are equipped with elaborate sterilizing systems, continuous EtO monitors, and gas recovery systems. There are less toxic alternatives, which are gradually replacing EtO in some hospitals.

EtO is a potent chromosome toxicant, causing mutations and other forms of damage even at low and intermittent exposure levels. Animal studies demonstrate that EtO is carcinogenic and causes harmful reproductive effects. Rats inhaling small amounts of the gas during and after mating produce smaller litters with lower birth weights.[165] EtO exposure has lowered the sperm counts of monkeys exposed to small amounts seven hours a day for five days each week, for two years[166] and has produced sterility in male mice.[167] Chromosome damage insufficient to cause fetal death may result in genetic damage transmissible to the next generation. Given to mice intravenously at doses thousands of times above occupational standards, EtO causes birth defects in off-spring.[168]

A study of hospital sterilizing staff in Finland demonstrated a significant increase in miscarriages among those exposed to EtO when compared to those unexposed.[39]

Methyl Bromide

Pesticide-intensive agricultural practices in Florida and California make them the largest methyl bromide–consuming states. In 1993, California used nearly 15 million pounds of methyl bromide, mostly for soil fumigation.[169] Like other fumigants, it is extremely toxic and must be used with great care. Furthermore, methyl bromide is a major depleter of the stratospheric ozone layer, with phase-out called for in the Montreal Protocol. However, as a result of pressure from agricultural interest groups, the Clinton administration has led a successful effort to shift the cutoff date for production and use of methyl bromide from 2001 to 2010, with provisions for "essential uses" after that date.

The toxicity of methyl bromide is well known. Large short-term exposures may rapidly cause death. Smaller nonlethal exposures over a period of weeks damage the brain, kidneys, nasal cavity, heart, adrenal glands, liver, testes, esophagus, and stomach. The reproductive and developmental toxicity of methyl bromide has been studied in mice and rats. Some animals exposed to 160 to 400 parts per million (ppm) methyl bromide, by inhalation, six hours a day, five days per week, for up to six weeks show degeneration of the seminiferous tubules in the testes.[170,171] Mice are more susceptible than rats to this effect. Another study in rats exposed to 200 ppm methyl bromide six hours a day for just five days failed to show any toxicity to testes or sperm but did show a marked decrease in testosterone levels.[172] However, plasma testosterone levels returned to normal with cessation of exposure. In a two-generation reproduction study of rats whose diets contained up to 500 ppm methyl bromide, no adverse effects were noted in reproductive success or tissue examination of parents or offspring.[173] Methyl bromide is listed in California as a known reproductive hazard.

Metam Sodium (Sodium N-methyldithiocarbamate)

Methylisothiocyanate (MITC) is the major breakdown product of metam sodium in organisms and in the environment. In animal tests, symptoms of acute toxicity of MITC include vomiting, diarrhea, weakness, and skin and eye irritation.[174] Short periods of inhalation of large amounts lead to convulsions and death. Smaller exposures over longer periods of time cause toxicity to the intestine, liver, kidneys, and ovaries. In rats, metam sodium causes adverse reproductive and developmental effects at doses of 10 mg/kg/day given on days 6 to 15 of pregnancy.[174] Some fetal deaths occur at this dose, with an increase in birth defects in surviving offspring at higher levels of exposure. Spinal column defects, brain swelling with hydrocephalus, umbilical hernias, and delayed skeletal development occur in rats and rabbits.

Metam sodium is heavily used in California agriculture. In 1991 a tank car carrying nineteen thousand gallons of metam sodium derailed, spilling its contents into the Sacramento River in northern California. Local residents were exposed mainly to MITC, formed when the pesticide mixed with water. It was subsequently discovered that officials at the EPA had received reports of the reproductive health effects of metam sodium in 1987 but had failed to review the information.[175] Consequently, local health officials were poorly equipped to inform and advise residents adequately.

An early population survey identified eight pregnant women who may have been exposed to MITC. Two women in the first trimester of pregnancy elected to have abortions. A follow-up investigation attempted to determine whether there was an increased incidence of spontaneous abortions (SAB) after the spill.[176] However, the study was limited by an inability to confirm one-third of the reported SABs in a follow-up interview. When all reported SABs (confirmed and unconfirmed) were included in the analysis, more occurred in women exposed to the chemical in the first trimester of pregnancy than in those unexposed. However, the small population made it impossible to achieve statistical significance with anything less than a 73 percent SAB rate.

Using only confirmed SABs, there was no increased incidence in those exposed.

This follow-up investigation demonstrates the often limited capacity of epidemiological studies to provide definitive answers after releases of toxic chemicals to the environment. Less than a two- to threefold increase of the outcome of concern in the exposed population is not likely to achieve statistical significance except when the study population is large.

Endocrine Disruptors

6

Hormones are chemical messengers that circulate in the blood and regulate many critical biological functions through intricate signaling mechanisms. In addition to the sex hormones estrogen, progesterone, and testosterone, there are others, such as thyroid hormone, insulin, melatonin, and cortisone. Endocrine disruptors (EDs) are chemicals that mimic or block hormones or otherwise interfere with normal hormone activity, often at extremely small doses. Evidence for endocrine disruption comes from studies of animals, humans, and laboratory cell cultures. Chemicals released into the environment have dramatically affected the reproductive success and development of wildlife by interfering with sex hormones. Humans are intentionally or inadvertently exposed to EDs in the workplace, home, and community and during medical care. Evidence of adverse health effects is overwhelming in some instances but only suggestive in others.

As early as the 1930s, studies in laboratory animals demonstrated estrogenic properties of a number of synthetic chemicals. Among them was bisphenol-A, now widely used in some plastics, resins, and dental sealants.[1] Estrogen-like effects of the pesticide DDT in chickens were reported in 1950.[2] In 1962, Rachel Carson's *Silent Spring* alerted the world to the harmful effects of pesticides on wildlife reproduction. She described a cascade of events resulting in contamination of the food chain, decline of egg survival, and destruction of populations of songbirds. Although unrecognized as hormone disruption at the time, that mechanism of toxicity for some chemicals later became clear. In the 1970s, scientists began to discuss hormone interference as a risk associated with widespread environmental contaminants of other types. In 1996, publication of the book *Our Stolen Future* was instrumental in bringing concerns about the effects of chemicals on hormone function to the general public, contributing to a broader debate about the health and environmental effects.[3]

Early discussions focused on the estrogenic effects of environmental contaminants; recent research has extended concerns to antiestrogens, androgens or antiandrogens, and some that interfere with prolactin, thyroid hormone, cortisone, and others.[4] The reproductive and developmental success of birds, fish, reptiles, and other wildlife species has been impaired where they have been sufficiently exposed to endocrine-disrupting chemicals.[5-7] Abnormalities include indeterminate sex, feminization of male animals, inability to reproduce successfully, and birth defects.

As more detailed understanding of the biological effects of EDs emerges, investigators have begun to study the potential relationship between exposure to these chemicals and a series of alarming human health observations. The incidence of breast, prostate, and testicular cancer has increased in this country and other parts of the world during the past several decades.[8] Between 1962 and 1981 there was a doubling of the frequency of undescended testicles in England and Wales.[9,10] The rate of hypospadias, an abnormality of the penis, doubled in the United States during the 1970s and 1980s.[11] There is increasing agreement that sperm counts in some regions of the world have fallen substantially and, in some individuals, approach levels that predict infertility.[12] The federal government reports that more than 2 million couples are involuntarily childless.[13] And a tragic twenty-year human experiment with a synthetic estrogen, diethylstilbestrol (DES), begins to explain how fetal exposures may result in serious health effects years later.

Nevertheless, there is considerable controversy over the degree to which humans are threatened. Some argue that there is no persuasive evidence of health effects at current environmental exposure levels in the general population. They focus on the lack of a rigorously proven, causal link between chemical exposures and human health observations. But this troubling issue is not easily dismissed by a prove-it-to-me response. Throughout the world, humans and wildlife are exposed to chemicals that, under certain circumstances, clearly alter hormone levels and function, sometimes with disastrous results. Often, however, the long-term effects of those changes at the individual or population level are unknown and difficult to predict. Better understanding depends on further research. Consequently, as with other public health and environmental concerns, how or whether to respond in the face of cause-and-effect uncertainty emerges as a more general policy question.

Endocrine disruption has gained the attention of lawmakers. In 1996 Congress passed the Food Quality Protection Act (FQPA) and amended the Safe Drinking Water Act, including in each statute a requirement that the EPA develop a screening and testing program for the estrogenic effects of food use pesticides and drinking water contaminants. The laws also allow the EPA administrator to consider other hormone-disrupting properties. In response, the EPA convened the multi-stakeholder Endocrine Disrupter Screening and Testing Advisory Committee (EDSTAC) to develop recommendations for the screening and testing program.

Diethylstilbestrol (DES)

From 1950 to 1971 diethylstilbestrol (DES), a synthetic estrogen with a chemical structure considerably different from naturally occurring estrogen, was used in an attempt to prevent spontaneous abortions in women. An estimated 5 to 10 million Americans were exposed to DES during pregnancy (DES mothers) or while in the uterus (DES daughters or sons).[14]

No harmful effects of DES exposure were suspected until 1970, when a rare form of vaginal cancer was reported in six young women, ages 14 to 21, who had been exposed to DES while in the uterus.[15] Previously this disease had occurred almost exclusively in older women. It is now known to be caused in younger women by exposure of the developing fetus to DES. The risk for developing vaginal cancer from birth to age thirty-four is estimated to be 1 in 1,000 to 1 in 10,000 for women exposed in the uterus, accounting for thousands of cases in the United States alone.

Later studies demonstrated that DES daughters often have abnormalities of their reproductive organs, reduced fertility, and unfavorable pregnancy outcomes, including ectopic pregnancies, miscarriages, and premature birth, as well as immune system disorders. DES sons are more likely to have small and undescended testicles, abnormal semen, and hypospadias.[16] DES mothers have a breast cancer risk about 35 percent greater than those not exposed.[17]

Animal studies in mice and monkeys show that prenatal DES exposure may result in masculinization of parts of the female brain and feminization in males.[18] Several studies in humans suggest similar results.[19]

Some DES daughters and sons are only in their mid-twenties. Many do not know that they were exposed. Their health status requires careful attention. There is no definite evidence for adverse health effects in the offspring of those who themselves were exposed to DES while in the uterus (DES grandchildren). However, since many are still young, it is too early to draw final conclusions. The issue is not resolved.

DES illustrates that an estrogenic chemical can cause reproductive and developmental abnormalities, immune system malfunction, and cancer decades later in some people exposed during fetal development.

Mechanisms of Action

The body produces many different hormones, each with its own receptor on the surface or inside of cells. To exert its effect, a hormone attaches to a receptor much like a key fits into a lock. Under normal circumstances, this attachment initiates a cascade of events that result in a biochemical reaction or chemical production in the cell. Endocrine disruptors may interfere in several different ways.

Hormones generally fall into three categories depending on their chemical structure: steroids, polypeptides, and amino acids. Sex hormones from the ovaries and testes and cortisone from the adrenal glands are examples of steroids. Thyroid hormone is a polypeptide. The receptors for these types of hormones are on the inside of cells. Hormone-receptor complexes are transported to the nucleus, where they attach to DNA and trigger genetic activity resulting in various gene products. Some hormones and neurotransmitters, such as those from the hypothalamus, are simple amino acids or peptides and attach to a receptor on the surface of cells. In turn, a series of "second messengers" initiates a cascade of events inside the cell, resulting in biochemical changes.

For some hormones, such as human chorionic gonadotropin, as few as 0.5 to 5 percent of the receptors in a cell must be occupied for full activation of response. For others, higher levels of receptor occupancy are needed.[20]

Endocrine disruptors may interfere with hormone function in a variety of ways:

1. An ED may mimic or block a naturally occurring hormone. If a chemical is similar enough to the natural hormone, it may occupy the binding site on the receptor and trigger the same sequence of events as the natural hormone—a hormone mimic causing a hormone-like effect. In some instances, it may occupy the receptor but not be similar enough to initiate a biochemical response. By attaching, it effectively blocks the receptor from occupancy by the natural hormone, acting as a hormone antagonist.

2. Hormones are transported through the circulation largely attached to carrier proteins. Most naturally occurring estrogen and testosterone, for example, is bound to sex-hormone-binding globulin (SHBG), confining the hormones to the bloodstream and limiting the amount of free hormone available to cells. The hormone-SHBG combination also appears to be able to influence cellular activity from the cell surface under certain conditions.[21] The degree to which a hormone binds to a receptor depends in part on the concentration of carrier proteins like SHBG. EDs may alter the levels of these carrier proteins or interfere with hormone attachment. Some pyrethroid insecticides, for example, displace testosterone from SHBG, and phytoestrogens from plants stimulate SHBG production.[22,23]

Thyroid hormone exists in two forms, T3 and T4, and is bound to several different proteins, including thyroid-hormone-binding globulin, transthyretin, and albumin. Thyroid hormone enters the developing fetal brain as T4 bound to transthyretin. The T4 is then converted to T3, essential for normal brain development. Some polychlorinated biphenyls (PCBs) and other industrial chemicals displace T4 from transthyretin, potentially disrupting thyroid hormone delivery to the developing brain.[24]

3. Exposures to EDs may interfere with hormone production, and fetal or infant exposures permanently alter baseline levels of hormones in some circumstances. For example, in rats, dithiocarbamate fungicides suppress signaling required for the ovulatory surge of luteinizing hormone (LH) from the pituitary. As a result, hormone production by the ovary is disrupted.[25] Rats exposed to small amounts of dioxin at critical times of pregnancy give birth to male offspring with permanently lowered testosterone levels.[26]

4. The number and kinds of hormone receptors normally fluctuate, subject to various hormonal or chemical influences. Some EDs inappropriately increase or decrease the number of hormone receptors in various organs of the body. Estrogen and estrogen mimics, for example, readily induce the formation of estrogen and progesterone receptors. In fact, the induction of progesterone receptors in the uterus is often used in laboratory tests as a measure of the potency of an estrogenic agent.

The Health Effects of Endocrine Disruptors

Hormones, circulating in extremely low concentrations, are essential to normal reproduction and are critical components of the signaling mechanisms that orchestrate development. In general, developing organisms are more susceptible to the effects of EDs than adults. Fetal or infant exposures may cause a range of health effects, including abnormalities of reproduction, growth, and development; impaired function of the immune and nervous systems; and cancer. This diversity emphasizes the fundamental nature of the processes potentially affected by these chemicals. Furthermore, effects of EDs may not be apparent for years or only in future generations, complicating attempts to study any link to early exposures in humans or wildlife. Functional abnormalities are not always easy to identify and may be difficult to attribute to exposure during pregnancy or infancy.

Universal exposures often make it difficult or impossible to identify unexposed comparison populations. Some ED chemicals have already accumulated in humans, domestic animals, and wildlife at levels that are near or above those that cause biological effects.[27] Additional exposures, though small, may be of great importance. These features argue for the redesign of toxicity studies used to determine the safety of chemical exposures to humans and wildlife.

Wildlife Health Effects
There is substantial evidence that sufficient exposure to some chemicals in commerce and in the environment disrupts the endocrine systems of a variety of invertebrates, reptiles, birds, fish, and mammals. Some examples illustrate the diversity of health effects.

- Various types of snails exposed to environmental levels of tributyl tin, an antifouling additive used in marine paint on ships, develop a condition called imposex in which affected female snails have irreversibly superimposed male sex characteristics.[28]
- Hermaphroditic fish are found in rivers below sewage treatment plants in Great Britain and the United States. Vitellogenin, a protein normally synthesized by female fish in response to estrogen, is utilized as an egg yolk protein to nourish the developing fish. Male fish have vitellogenin levels similar to egg-bearing females in some rivers.[29] Laboratory tests show that nonylphenol, an alkylphenol used in detergents and surfactants and found in effluent, behaves as an estrogen mimic and induces vitellogenin formation and testicular inhibition in male trout.[30] However, it is not entirely clear which chemical or combination of chemicals in the sewage effluent mixture is responsible for the observations in river fish. Some investigators believe that estrogens from the urine of women taking birth control pills also contribute.
- Alligators and red-eared turtles in Lake Apopka in Florida are demasculinized after exposure to a mixture of chemical contaminants including the pesticide dicofol. There are no normal male turtles in Lake Apopka. All hatchlings have either normal-appearing ovaries or are intersex.[31]
- Gulls breeding in the Puget Sound and Great Lakes regions show evidence of eggshell thinning and reproductive tract abnormalities, with feminization of male embryos. In some instances, populations have declined, and sex ratios are skewed.[32] These areas are contaminated with mixtures of DDT, PCBs, and polycyclic aromatic hydrocarbons, each of which may cause the observed effects. Birds from these areas and from locations far more remote from industrial activity show elevated tissue levels of contaminants.
- Great Lakes gulls and terns, as well as some western gulls, have, within the past several decades, shown supernormal egg clutches and female-female pairing.[33] Gulls in these colonies also show excessive chick mortality, birth defects, and skewed sex ratios, with an excess of females. These effects correlate with levels of persistent organic pollutants like PCBs and DDT.
- Seal populations have markedly declined in portions of the Wadden Sea in the Netherlands. Fish from the area of decline are contaminated with higher levels of PCBs and pesticides than those from other areas.

Captive seals fed fish exclusively from the contaminated area were less able to reproduce and had altered estrogen levels and thyroid function compared to seals fed less contaminated fish over a two-year period.[34–35]

Dicofol, DDT, and Deformed Alligators in Lake Apopka, Florida

In the 1970s, alligator populations in the United States were rapidly declining, leading to their protection as an endangered species. As a result of this protection, alligator populations made a strong recovery through the 1980s, with one major exception: alligators in Florida's Lake Apopka, which have suffered a severe population decrease that continues to the present.

Apopka is one of Florida's largest freshwater lakes—and one of its most polluted. Agricultural activities in the area and a nearby sewage treatment plant are part of the problem, but the greatest threat, at least for the alligators, may be due to a major pesticide spill at the adjacent Tower Chemical Company, now no longer in business. In 1980, the pesticide dicofol, contaminated with DDT and its metabolites, spilled into Lake Apopka. A dramatic decline in the number of juvenile alligators immediately followed.[36] Although the water in Lake Apopka no longer has detectable pesticide residues, these chemicals have made their way into lake sediments and into the food chain.

Scientists have compared the alligators in Lake Apopka with those in another relatively pristine lake.[37] They have found that Lake Apopka alligators have severely reduced survival rates; their eggs are three times as likely not to hatch as eggs from the cleaner lake; and they are forty times more likely to die soon after birth and four times more likely to die as juveniles than are the alligators from the unpolluted lake.

Even more worrisome, the alligators that do survive may not be able to reproduce normally. They appear superficially normal, but closer examination has revealed some strange, and troubling, abnormalities. One of the most striking features is that a significant proportion of the alligators demonstrate a condition known as intersex, in which the external reproductive organs have a combination of male and female features. All of the Lake Apopka females examined had abnormal ovaries, and most of the males had abnormal testes and small penises. Almost none of the alligators from the clean lake had similar abnormalities. Hormone testing found that the Lake Apopka males had elevated levels of a form of estrogen, and the females had elevated levels of testosterone.

Both dicofol and DDT are known to act as estrogens in the body. DDE, a metabolite of DDT acts as an antiandrogen. The evidence points strongly to these chemicals as the source of the abnormalities and subsequent population decline of the Lake Apopka alligators. Although DDT was essentially banned in the United States in 1972, dicofol is still used today in the United States and abroad.

Human Health Effects

There is little disagreement that wildlife have suffered reproductive and developmental abnormalities as a result of exposure to EDs and that DES is an important example of an endocrine-disrupting chemical in humans. There is less agreement about the importance to human health of exposure to "weaker" EDs. But the increasing incidence of endocrine-related cancers, genital abnormalities, and an apparent decline in sperm counts remain unexplained. Scientists from various disciplines are increasingly concerned that environmental contaminants are the common thread tying these conditions together.

Carcinogenesis

There is no doubt that DES caused the unusual vaginal cancers seen in some young women exposed to the drug as fetuses. Some investigators suspect that exposures to EDs may also contribute to the development of breast, prostate, and testicular cancer. In each case there are fragments of inconclusive evidence to support that concern. The mechanisms by which toxicants may foster development of each of these malignancies and the nature and timing of the relevant exposure(s) are matters of considerable debate and research interest.

One hypothesis consistent with current understanding of carcinogenesis proposes that hormone levels, environmental exposures at critical times in development, and genetic susceptibility interact to create the conditions for development of cancer. According to this view, precancerous changes resulting from early molecular, biochemical, and cellular events are transformed, sometimes much later, into recognizable cancer.

Breast Cancer Breast cancer incidence has increased in the United States over several decades. Today, one in eight or nine women will develop this cancer in her lifetime, resulting in over forty-four thousand breast cancer–related deaths annually.[38] The causes of breast cancer are not well understood but may include early biochemical and cellular events that increase susceptibility or cause changes that are later transformed into full-blown malignancy.[39]

The breast in males and females is quite similar until the prepubertal period, when female breast development begins. This is a time of rapid cell proliferation and differentiation, dependent on interactions of estrogen, progesterone, prolactin, and growth hormone.[40] Estrogen and prolactin levels at this time regulate the number of estrogen receptors in breast tissue. Hormonal effects on the breast are complex and vary with age, stage of cellular differentiation, and presence or absence of hormone receptors.

There is considerable evidence that the total lifetime exposure to estrogen influences the likelihood of developing breast cancer. High serum or urine levels of estrogen, early onset of menstruation, delayed menopause, and delayed first-child bearing are all risk factors for breast cancer.[41,42] Although environmental exposures may affect breast cancer risk, radiation exposure and alcohol are the only well-established links.[43] However, a considerable amount of interest and research is focused on other environmental contaminants as possible contributors, including organochlorine compounds, solvents, metals, and polycyclic aromatic hydrocarbons, which are products of combustion spread widely throughout the environment.[44-48] Breast milk contains a large number of these contaminants in complex mixtures, and some studies show that breast feeding reduces the risk of developing breast cancer in premenopausal women.[49,50] If true, risk reduction could be attributable to low estrogen levels during the period of breast feeding, decreasing chemical concentrations by elimination in breast milk, or some combination of the two.

A variety of chemicals mimic, block, or influence the levels of estrogen, progesterone, and prolactin. Whether breast cancer in adults may be initiated by fetal, prepubertal, or young adult exposures to hormonally active chemicals is unknown, but if it is, the timing of the exposure may be as critical as the nature of the chemical. Since studies of women with breast cancer are rarely able to determine the timing and magnitude of exposures with accuracy, this important question remains difficult to answer. Studies that do not account for important time windows of vulnerability may miss causative relationships if they exist.

There are two parallel pathways by which excess estrogen or estrogenic environmental contaminants may increase breast cancer risk. The first is through increased proliferation of estrogen–responsive cells, and the second is through direct DNA damage by estrogen metabolites.[51] Several studies suggest that breast cancer is related to tissue levels of organochlorines, like DDT, its by-product DDE, or PCBs.[52-55] In one study, for example, investigators compared PCB and DDE levels in stored blood specimens from 58 women who developed breast cancer with levels in the blood of women who were healthy. They found that DDE levels were significantly higher in women with breast cancer.[52] Another study of 150 women with breast cancer, with equal representation of Caucasians, African Americans, and Asians, showed no correlation with DDE or PCB blood levels. However, when just the Caucasian and African American women were included in the analysis, there was an increased risk of breast cancer among the women with the highest levels of DDE.[56] Several other studies show no relationship between organochlorine levels in breast tissue or blood and the risk of breast cancer, and the matter is unresolved.[57–59] If there is some relationship between chemical exposures and breast cancer risk, it may be that DDE or PCBs are only relatively crude markers for a more relevant exposure, explaining the discrepancy in study results.

There is also considerable debate about the role of estrogen metabolites as a contributor to breast cancer risk.[60] Various chemicals, including atrazine and organochlorine pesticides, alter the metabolism of estrogen, in some cases leading to an excess of a metabolite that itself is strongly estrogenic. Other metabolites may directly damage DNA, causing abnormal cells to proliferate and increasing breast cancer risk.[39]

Prostate Cancer   Prostate cancer is a common disease of older men, found frequently in those who die of other causes. Deaths from prostate cancer have increased over the past thirty years, suggesting that the disease has increased in frequency more than can be explained by better screening alone. In the United States, prostate cancer is responsible for about forty-thousand deaths per year.[61] It is rare in men of Asian origin and more common in African American males than Caucasians. Its natural history is variable, as some tumors behave much more aggressively than others despite treatment.

Factors that contribute to the development of prostate cancer are not well understood. However, there are suggestions that both naturally occurring estrogens and synthetic estrogenic toxicants may play a role. As with breast cancer, the evidence follows two parallel pathways, one of which emphasizes the cell-proliferative function of estrogenic agents and the other the cell-damaging effects of estrogen metabolites.

First, studies in mice show that estrogenic exposures during fetal life increase prostate weight in adult animals.[62] This has been demonstrated with estrogen, DES, bisphenol-A, and octylphenol. Also in mice, estrogenic exposures during the first three days of life initiate cellular changes in the adult resembling those associated with prostate cancer.[63] The abnormal cells have features, such as enlarged nuclei and abnormal organization, that are often identified as precancerous in other tissues. Moreover, when compared with unexposed mice, male mice exposed to DES only as fetuses also exhibit greater expression of an estrogen-responsive gene (c-fos—one of the genes responsible for cell division) when given estrogen after birth. There are estrogen-responsive sites in the prostate in dogs, monkeys, and humans as well.[64–66] These observations demonstrate the capacity of estrogenic agents to increase cell proliferation and cell division in the prostate, at least in part by altering gene expression.

A parallel line of reasoning holds that the products of estrogen metabolism may be significant. Estrogen can be transformed into metabolites (e.g., 4-hydroxy estradiol) that are sources of free radicals, short-lived fragments that can damage cellular proteins and DNA.[67,68] Although there are mechanisms that are constantly at work identifying and repairing damaged DNA, these mechanisms may fail, due to either rapid cell division, which overloads repair capacity, or reduced repair capacity associated with aging, and cancer may result. Moreover, as men age, estrogen levels rise relative to testosterone. This may be an important factor in the later development of prostate cancer.

In an autopsy study of 152 males ten to forty-nine years old who died from other, unrelated causes, detailed microscopic examination of their prostate glands revealed cancer in 34 percent of all men between ages 40 and 49, and 27 percent of men ages 30 to 39. In addition, cellular changes that may progress to cancer or, alternatively, be evidence of susceptibility to cancer were found in 9 percent of the age 20 to 29 group.[69] These results show that unrecognized prostate cancer sometimes begins quite early in life and is a disease of men much younger than previously thought.

Whether fetal exposure to estrogenic substances contributes to susceptibility to later development of prostate cancer in humans remains unclear, but the question obviously deserves further study. DES sons have not shown an increased incidence of prostate cancer, but sufficient time may not have passed for an increased risk to become apparent.

Testicular Cancer   The incidence of testicular cancer has increased dramatically and is now two to four times more common in industrialized countries than it was fifty years ago. However, it is still a relatively uncommon disease with an overall annual incidence of about 4 or 5 per 100,000 men. Testicular cancer is sometimes seen in infants but has its peak incidence in young adult men.[8] It is the most common malignancy in men twenty-five to thirty-five years old. Caucasians are more than twice as likely to develop this cancer as African Americans. It may arise from any of the cell types found in the testes, but more than 90 percent of cases develop from germ cells (immature cells that will develop into sperm).

In a recent review of the possible role of sex hormones in the development of testicular cancer, the authors conclude that despite uncertain mechanisms, cancerous changes of immature sperm cells "take place most probably during early fetal life. In this phase of development, germ cells are vulnerable to the influence of maternal hormones and other environmental agents."[8] The young cancer cells probably remain dormant until puberty, when hormonal changes stimulate their growth.

Several pieces of epidemiological and laboratory evidence support this conclusion. Testicular cancer is more likely in those with undescended testicles, a condition seen in DES sons. The fetuses and newborn of mice exposed to estrogen during pregnancy have testicular and germ cell abnormalities that look like precursors to cancer.[70] First-born male children have an increased risk of testicular cancer, and first pregnancies are associated with higher estrogen levels than subsequent pregnancies.[71,72]

Evidence linking in utero DES exposure with later development of testicular cancer is conflicting, with some studies finding a strong association and others finding none.[73,74] This discrepancy may result from two study-related problems. It is often difficult to determine the timing and amount of DES used in pregnancies years before a study, making exposure assessment problematic. Moreover, although the incidence of testicular cancer has increased, it is still relatively uncommon, and studies of small numbers of DES-exposed males are statistically unlikely to identify cases of cancer. It is not likely to be productive to concentrate exclusively on DES sons to help resolve the role of estrogenic substances in the development of testicular cancer.

Considering the laboratory and epidemiologic evidence as a whole, most investigators agree that a link to estrogenic compounds must be examined closely.[75]

Falling Sperm Counts, Undescended Testicles, and Hypospadias
In recent years there has been considerable controversy about sperm count trends in the general population. In a review and analysis of sixty-one papers published in the medical literature between 1938 and 1991, the authors concluded that there had been a substantial decline in semen quality over the past fifty years.[76] They excluded all studies of men from infertility clinics in order to avoid possible bias. Furthermore, they excluded any studies in which sperm counts had been done by methods not available during earlier years. Their analysis of the world's literature showed a 42 percent decline in sperm count, from 113 million sperm/cc of semen to 66 million sperm/cc. This report sparked intense debate, including disagreement over the appropriateness of the authors' statistical methods.[77,78] A reanalysis of the same data, using several different statistical techniques, confirmed the original conclusions but showed that the decline was seen in the United States and Europe but not in non-Western countries, although the data for the latter are limited.[79] Even within the United States, there appears to be considerable geographical variation.[80] Future research will need to account for this variable and search for explanations.

A review of twenty years of sperm bank data from a single laboratory in Paris showed that in 1,351 healthy men who were fathers, there was a yearly decline in the average sperm count and motile, normal-appearing sperm for donors of a given age.[81] The sperm count declined by 2.1 percent per year over the twenty-year period. This analysis took into account the period of sexual abstinence of the donors, a variable that influences sperm counts in individuals, and has the advantage of using data from a single laboratory.

An analysis of data from 577 semen donors collected over eleven years in a laboratory in Scotland showed that a later year of birth was associated with a lower number of sperm and lower number of motile sperm in the ejaculate. When men born in the 1970s were compared with men born in the 1950s, the total number of motile sperm was reduced by 25 percent.[82]

Other studies have not confirmed a decline.[83,84] However, different statistical techniques were used in the analyses, making comparisons difficult. For example, the Paris and Scottish studies compared the sperm counts of donors of the same age, while those finding no decline aggregated all donors for a given year and corrected for an average age. This latter technique runs the risk of missing a decline within each age group from year to year, which may be a more sensitive indicator of changes than an average across all age groups.

Any study of sperm donors presents significant challenges, since there is considerable daily variation in sperm numbers, even in the same individual. An autopsy study of Finnish men was designed to circumvent this limitation. Investigators compared the quality of sperm production in 1991 versus 1981 by postmortem microscopic examination of the testes of 528 men, showing that there had been a decline in the percentage of men with normal, healthy sperm production from 56 percent in 1981 to 27 percent in 1991.[85] There were as well a decrease in the average weight of the testes, a decrease in the size of the seminiferous tubules, and an increase in the amount of fibrous tissue. The investigators accounted for differences in age, weight, and history of smoking, alcohol, and drug use. Their results support the conclusion that there has been a significant decline in the quality of human semen in the past several decades.

Along with a decline in sperm counts, there also appears to have been a significant increase in hypospadias and undescended testicles over the past few decades.[86] There was a doubling of the frequency of undescended testicles in England and Wales from 1962 to 1981, and similar increases were reported in Sweden and Hungary.[87] A doubling of hypospadias rates in the United States in the 1970s and 1980s has also been reported.[11] There is now considerable concern that falling sperm counts, increasing incidence of undescended testicles, hypospadias, and testicular cancer may be linked to fetal exposures to endocrine-disrupting chemicals.

Behavioral and Learning Abnormalities

Fetal and neonatal exposures to some environmental agents also adversely affect neurological and intellectual development. For example, it has been known for some time that lead and mercury are neurological toxicants, although not through endocrine-disrupting mechanisms at likely levels of human exposure. More recent studies suggest that some behavioral and learning abnormalities, as well as general impairment of intellectual function, may result from endocrine disruption.

Intellectual development in children is impaired after fetal exposure to PCBs.[88] One theory holds that this is explained by PCBs' interference with thyroid hormone function during critical periods of fetal brain development. Thyroxine, a form of thyroid hormone necessary for normal brain development, is decreased after exposure to PCBs and dioxin.[89,90] In addition, some PCBs compete for binding to the thyroid receptor or thyroid transport proteins, and thyroxine must be attached to transthyretin in order to enter the fetal brain. PCBs may also increase the metabolism of thyroid hormone. Any of these mechanisms may interfere with brain development.

Reported thyroid-disrupting properties of other industrial chemicals raise concerns that they might also adversely affect normal development.[91-93] Among the challenges facing investigators is the recognition of adverse effects due to minor changes in thyroid status during fetal development, since neurological and behavioral effects are often difficult to measure.

This sparse but growing body of evidence of the neurological effects of some endocrine-disrupting chemicals furthers concern about their contribution to learning disabilities and behavioral abnormalities in the general population, but complete understanding is still a distant goal. Some studies show that social behavior may be influenced by prenatal chemical exposures; others show that social deprivation at birth can lead to exaggerated responses to chemical exposures later in life.[94] The complex interaction of early social factors, genetics, lifelong metabolic pathways, stress hormone levels, and chemical exposures requires considerable additional research for understanding.

The Debate

There are those who believe that concerns about environmental hormone-disrupting chemicals are much ado about nothing.[95] Although we know that the function of various hormones may be altered by some chemicals under certain circumstances, much of the general debate centers around the importance of low-dose exposures. Those who believe that these substances are unlikely to be important at current exposure levels often refer to the following:

1. The low potency of man-made chemicals, when compared to naturally occurring hormones, makes their role minor and insignificant. (DES is an exception. It is generally agreed to be "strong.")
2. Exposures to estrogenic chemicals naturally present in foods (phytoestrogens) are much larger than exposures to synthetic man-made chemicals with estrogenic action.
3. Feedback loops are resilient and easily able to adjust for minor fluctuations in hormone levels.
4. Man-made estrogenic and antiestrogenic chemicals tend to balance each other.

These observations are misleading, for several reasons. First, regarding argument 1, the strength of receptor binding of a synthetic chemical with that of a natural hormone is important but tells nothing about other factors that may contribute to hormone disruption—for example, alterations in hormone metabolism; distribution, storage, or bioaccumulation of a synthetic chemical; effects on carrier proteins, like SHBG and albumin; and interaction with the hypothalamic-pituitary-gonadal axis.

Regarding argument 2, although food contains naturally occurring phytoestrogens, some actually behave as estrogen antagonists in the presence of naturally occurring estrogens. It is too simplistic to conclude that a comparison of the receptor-binding capacities of two chemicals in a test tube will predict how each will behave in an intact organism. Moreover, man-made chemicals with endocrine-disrupting potential may be metabolized, stored, or protein bound in the body in very different ways from naturally occurring substances to which humans and wildlife have been exposed throughout their evolution. Comparisons of synthetic with naturally occurring chemicals must be made carefully.

Argument 3, that adult feedback loops may be resilient, ignores evidence of the exquisite sensitivity of the developing organism to minor hormone fluctuations at critical times in fetal life. The thresholds and sensitivities of adult feedback loops are set during fetal development and may be permanently altered by small exposures. For example, dioxin has developmental effects at levels well below those causing adult toxicity. 17-alpha estradiol (a close relative of 17-beta estradiol, the naturally occurring form of estrogen) is relatively inactive in adults but causes tumors in mice when given to newborns.[96]

Finally, argument 4, that man-made hormone mimics and antagonists will balance each other out, is not based on any evidence. There are specific chemicals that interact competitively with hormone receptors, but given multiple mechanisms of hormone disruption, to postulate a net effect of zero is purely speculative.

Conclusions

Humans and wildlife are exposed to a large number of naturally occurring and man-made chemicals capable of mimicking, blocking, or otherwise interfering with the endocrine system. They interact in complex ways with each other, food constituents, naturally occurring hormones, receptors, and carrier proteins and may disturb a wide range of reproductive and developmental events.

These chemicals are found in air, soil, food, water, and human and wildlife flesh throughout the world, in plastics, food wraps, cosmetics, baby bottles, detergents, and pesticides. Only about three thousand man-made organic compounds out of an estimated sixty thousand in drinking and waste water and sewage sludge have been identified.[97] A random screen of twenty of these chemicals showed nine of them to interact with the estrogen receptor.[98] We appear likely to be exposed to large numbers of unidentified and unstudied chemicals, many with endocrine-disrupting activity.

Animal, laboratory, and epidemiological studies demonstrate adverse health effects resulting from exposure to endocrine-disrupting chemicals and clarify the importance of timing as well as magnitude of dose. Small exposures during critical windows of vulnerability may cause lifelong changes in reproductive function and development. Although estrogenic or estrogen-antagonistic effects of some of these chemicals have been known for years, interference with the function of other hormones including androgens, thyroid hormone, insulin, cortisone, and neurotransmitters, is now apparent.

There are documented worldwide increases in a number of diseases or conditions of the reproductive system in infants, children, and adults

---

that may be linked to early exposures to hormonally active chemicals. For some, the connection is clear. For others, there is limited evidence of environmental cause. Several hormone-related cancers have also increased in frequency in recent decades. Biologically plausible hypotheses suggest ways in which these cancers may be related to exposure to endocrine-disrupting chemicals during periods of susceptibility. Consistent findings of delayed psychomotor development of children exposed to PCBs in the uterus come from several sources. Possible relationships between neurobehavioral disorders and chemical exposures are incompletely understood and are under investigation.

When entire populations of humans and wildlife are exposed, the consequences of a population-wide effect, even if subtle or difficult to detect in individuals, may be profound. For example, there are important social and economic consequences of small population-wide shifts in behavior patterns, learning capacity, and sperm counts. Very little population-wide change is needed to increase the need for special education or demand for fertility services markedly. It will take many years to acquire more thorough understanding of the importance of widespread, low-dose, multichemical exposures through fetal life, infancy, and growth and development. Meanwhile, the world's populations of humans and wildlife participate in the ongoing experiment.

Endocrine Disruptor Profiles

Many different and widely distributed man-made chemicals have the potential to interfere with normal hormone action. There are also naturally occurring substances produced by plants and fungi that have estrogenic or antiestrogenic effects. Total exposure to combinations of chemicals from all sources will influence their biological effects and cannot be predicted by simply adding doses and responses. The following sections review some of the chemicals known to have endocrine-disrupting activity.

Dioxin

Reproductive effects: Interferes with the production or activity of enzymes, hormones, other growth factors; adversely affects reproduction, growth, and development through a variety of mechanisms

Dioxins, among the better-known and -studied EDs, are a family of related compounds differing in the number and position of chlorine atoms on the basic underlying structure. The toxicity of each member of the family varies considerably and is usually described relative to the most toxic. Together, they demonstrate several different mechanisms of hormone-disrupting action and have diverse biological effects.

Dioxin results from heating mixtures of chlorine and organic compounds in industrial processes, such as the bleaching of paper pulp, production of some pesticides, or during incineration of chlorine-containing materials. Because many consumer products contain chlorinated organic compounds (e.g., polyvinyl chloride), municipal, medical, and hazardous waste incinerators are leading dioxin sources. It is not easily broken down in the environment, accumulating in soils and sediments and biomagnifying as it passes up the food chain. Dioxin bioaccumulates in fat tissue with an estimated half-life in humans of approximately seven years.

There may be significant regional variations depending on local industrial activity, but dioxin is widely spread around the globe. Beef, pork, fish, shellfish, and animal and human milk are the major sources of human exposure. Because breast milk has a high fat content, nursing infants are actually exposed to higher daily amounts of dietary dioxin than most adults and may receive more than 10 percent of their anticipated lifetime exposure during this particularly vulnerable period of mental and physical development.[99]

Dioxin has been under critical review by the EPA since 1991, when the American Paper Institute and the Chlorine Institute campaigned to convince regulators that dioxin was not nearly as dangerous as previously thought.[100] Their claims were based on a recount of tumors in a fourteen-year-old industry-sponsored rat study. Pressure from industry and environmentalists has been intense, revealing the highly political nature of the interpretation of scientific findings, as well as the regulatory response intended to flow from them.

The extensive six-year EPA review documents a wide range of health effects that result from exposure to dioxin, some of which occur at extremely low exposure levels, and provides important information about dioxin sources. Although there is some variation with geographical location and diet, many people have dioxin levels at or near those known to cause harmful effects in animal studies.[101]

Animal Studies

In animal studies, dioxin has a wide range of health effects, which differ among the fetus, newborn, and adult. Some are apparent only with large doses, but cancer, immune system toxicity, and reproductive and developmental effects occur at low levels of exposure (table 6.1). Dioxin causes the liver to produce metabolic enzymes at exposures of 1–10 picogram (pg) per kilogram daily, a level similar to average daily adult human exposures. (A picogram is one-trillionth of a gram.) These enzymes alter the metabolism of hormones and other endogenous or exogenous chemicals. Enzyme induction occurs at levels that also cause immune system toxicity in mice and reproductive effects in rats.[23] In rats, thyroid tumors occur at doses as low as 1,400 pg/kg/day.[102]

There is considerable variability in the toxicity of dioxin among adults of different animal species but much less among fetuses and infants, particularly with respect to the sensitivity of offspring to developmental effects. For example, adult hamsters are several thousand times more resistant to dioxin toxicity than adult guinea pigs.[103] But the hamster fetus is only ten times more resistant to dioxin than the guinea pig fetus. Similarly, early life stages of fish and birds are more sensitive to dioxin toxicity than adults.[104,105] From these data, one might suspect that dioxin toxicity in human fetuses would be similar to that in fetuses of other species, even if human adults were relatively resistant.

Sufficient exposure to dioxin during pregnancy causes prenatal mortality in the monkey, guinea pig, rabbit, rat, hamster, and mouse. The response is dose related, and there is a species difference. Monkeys and guinea pigs are the most sensitive, followed by rabbits, rats, hamsters, and mice, which are the most resistant. In these species, the maternal dose necessary to cause prenatal mortality ranges from 1 to 500 µg/kg (cumulative dose). The timing of maternal exposure is just as important as the magnitude of the dose, often demonstrating a window of vulnerability. In the guinea pig, for example, prenatal death is caused by a single dose of 1.5 µg/kg on day 14 of pregnancy, whereas later in pregnancy, larger amounts are needed.[106]

Similarly, a single low maternal dose of dioxin at a critical time in pregnancy may cause permanent developmental effects in male offspring, including altered sexual differentiation of the brain.[107] On day 15 of a typical twenty-one-day pregnancy in rats, most organs are formed, but

Table 6.1   Reproductive and developmental toxicity of dioxin, animal studies

Health effect	Species
Decreased fertility, litter size; offspring are sensitive to even lower doses as they reproduce (a second-generation effect)	Rats (0.1 µg/kg per day)[109]
Inability to carry pregnancies to term, suppressed estrogen levels	Monkey (50 ppt dietary dioxin)[110,111]
Endometriosis*	Monkey (5–25 ppt dietary dioxin)[112]
Decreased testis weight, sperm production, fertility	Adult rat (65 µg/kg) Mouse (100 µg/kg) (doses sufficient to reduce feed intake and/or body weight)[113–115]
Lower testosterone levels	Rat (15 µg/kg)[116]
Embryo mortality	Rainbow trout embryo (LD_{50}, 0.4 µg/kg egg weight) Juvenile rainbow trout (LD_{50}, 10 µg/kg body weight) Fertilized lake trout eggs (LD_{50}, 65 pg/g egg weight) Chicken embryo (LD_{50}, 0.25 µg/kg egg weight)[117]
Congenital heart defects	Chicken (eggs treated with 1 picomol/egg)[118]
Reduced size of the thymus gland and altered blood counts, altered immune system	Most laboratory animals at a range of doses depending on species[101]
Cleft palate formation and enlarged kidneys	Mouse: doses that are not maternally toxic (e.g., 1–4 µg/kg on day 6–15 pregnancy Other species as well but at higher doses
Learning disability (impaired object learning, not spatial learning)	Monkey: 5–25 ppt in maternal diet[120]

Note: The LD_{50} is the concentration of dioxin that will kill (lethal dose) 50 percent of those exposed.
A microgram (µg) is one-millionth of 1 gram. A picogram (pg) is one-trillionth of 1 gram. ppt = parts per thousand.
* Three of seven female monkeys exposed to 5 ppt dietary dioxin (43 percent) and five of seven animals exposed to 25 ppt dietary dioxin (71 percent) had moderate to severe endometriosis after four years of exposure followed by 10 years of no exposure. The frequency of disease in a control group was 33 percent.

the hypothalamic-pituitary-gonadal (HPG) axis is just beginning to function. The critical period of sexual differentiation of the brain extends from late fetal life through the first week of postnatal life. A single low maternal dose of dioxin (0.16 µg/kg) on that day of pregnancy reduces male testosterone levels, delays descent of the testicles, decreases anogenital distance (making it more female-like), and reduces prostate weight and sperm production in offspring.[26] It also demasculinizes their sexual behavior in the months that follow. A single maternal dose of just 0.064 µg/kg on day 15 of pregnancy causes a 43 percent reduction in sperm production in male offspring.

Dioxin does not attach to the estrogen receptor, yet it causes both estrogenic and antiestrogenic activity in different tissues of the body. Both dioxin and PCBs attach to another intracellular receptor, called the Ah-receptor, whose function is not otherwise fully understood. (Unlike dioxin, some forms of PCBs also attach to the estrogen receptor.) The occupied Ah-receptor is transported into the nucleus of a cell, where it attaches to DNA, influencing the activity of genes, which regulate chemical production. By this mechanism, dioxin indirectly influences estrogen activity. Its antiestrogenic effects, which seem to predominate, may result from causing the cells to produce an enzyme that metabolizes the body's normal estrogen or decreasing the number of estrogen receptors available for normally occurring estrogen.[108,101]

Epidemiological Studies

In the Ranch Hands study, reproductive histories of men who sprayed Agent Orange in Vietnam from 1962 to 1971 were examined beginning in 1978 in an attempt to see if exposure to dioxin might have had adverse effects in their children.[121] Agent Orange is a mixture of two herbicides, almost always contaminated with dioxin. Dioxin in the blood of participants was measured years after exposure, and an attempt was made to estimate earlier levels from those results. An increase in all nervous system defects in offspring was found. However, increases in spina bifida and cleft palates were too few to allow formal statistical analysis. One finding that is difficult to explain was an increased risk for spontaneous abortion, all birth defects, and specific developmental delays in the low- but not the high-dioxin exposure group.

Another study of Vietnam veterans found that opportunity for Agent Orange exposure was associated with an increased risk of spinal cord abnormalities (spina bifida) and cleft palates in offspring.[122] The National Academy of Sciences has concluded that there is limited but suggestive evidence of a relationship between paternal Agent Orange exposure and spina bifida in offspring.

In a study of 248 chemical production workers in New Jersey and Missouri, investigators found that workers with higher dioxin levels had higher amounts of luteinizing hormone and follicle-stimulating hormone and lower amounts of testosterone than a control group from the neighborhood.[123] These results must be interpreted with caution since it was a cross-sectional study (all measurements of dioxin, testosterone, and gonadotropins were done on the same blood specimen, making it difficult to determine cause-and-effect relationships), but the results are consistent with the effects of dioxin in animal studies.

In 1977, an industrial accident in Seveso, Italy, released large amounts of dioxin, contaminating the environment and exposing local residents. From 1977 to 1984, there was a marked increase in the female-to-male birth sex ratio among those most heavily exposed.[124] Almost twice as many girls as boys were born during those years. Over the next ten years, the ratio began to return to normal. The mechanism by which dioxin may have this effect on sex determination is unclear. In this same population, there was no increase in the rate of birth defects, as determined from a birth defects registry, when compared to an unexposed population.[125] However, in this study, the number of children of mothers with the highest likelihood of exposure was too small to assess specific categories of birth defects. Other limitations include possible exposure misclassification and unrecognized spontaneous abortions that may have resulted from fetal malformations. Children of exposed women have not been examined for subtle structural or functional developmental deficits.[126]

In Times Beach, Missouri, an area contaminated with dioxin-containing oil that had been spread on roads for dust control, there was no apparent increased risk of fetal deaths or low-birth-weight babies.[127] There was, however, a two- to threefold increase in risk of nervous system defects and undescended testicles, though this was not statistically significant. Because of the small sample size, a sixfold increase in risk would have been necessary in order to achieve statistical significance.

Investigators in the Netherlands found that higher dioxin levels in breast milk correlate with lower thyroid hormone levels in breast-feeding infants.[90] This finding is particularly important since the correlation appears at current levels of ambient dioxin exposure. Moreover, in preterm and low-birth-weight babies, decreased thyroid hormone in the first weeks of life is associated with increased risk of neurological disorders, including the need for special education by age nine.[128] Although the thyroid hormone levels in the Netherlands study were still in the normal range, it is possible that the observed changes will influence infant development, a subject that will require further research.

Polychlorinated Biphenyls (PCBs)

Reproductive effects: Adverse reproductive effects in many different species; may mimic estrogens and interfere with thyroid hormone function; are associated with decreased birth weight and delayed brain development in humans.

From 1929 to 1977, PCBs were manufactured and widely used in the United States in electrical transformers and capacitors, hydraulic fluids, plasticizers, and adhesives. They were banned from most uses in the United States because of environmental persistence, bioaccumulation, and toxicity; however, they remain widely spread in the environment, and because of bioaccumulation, human and wildlife consumption of food contaminated with even small amounts of PCBs inevitably leads to gradual increases in total body stores. Ninety-four percent of fish collected nationwide show PCB residues at an average concentration of 0.53 parts per million (ppm).[129] In marine mammals, amounts may be thirty thousand to sixty thousand times higher.[130] Inuit mothers in the Arctic have the highest known levels of PCBs in their milk as a result of a diet rich in marine mammal fat.[131]

PCBs and dioxin are related families of structurally similar chemicals. Each may have a different number of attached chlorine atoms, the number and position of which largely determine molecular shape and toxicity. Like dioxin, many PCBs attach to the Ah-receptor and have similar toxic effects. PCBs, however, also behave differently from dioxin. Some are capable of binding competitively to thyroid hormone carrier proteins, interfering with the transport of thyroid hormone, which is

---

Table 6.2 Reproductive and developmental toxicity of PCBs, animal studies[138]

Health effect	Species
Reproductive toxicity	
Reduced fertility[139]	Male rat exposed during lactation
Failure to conceive and abortion[140]	Monkey
Reduced progesterone levels[141]	Monkey
Estrogenic activity (stimulate uterine growth)[142]	Rat
Prolonged estrus cycle[140]	Monkey
Developmental toxicity[143,144]	
Prolonged gestation[145]	Rat, mouse
Low birth weight; reduced litters and infant survival[146]	Monkey, rat
Decreased thyroid function[147]	Rat fetus
Birth defect	Mouse (cleft palate—like dioxin)
Altered sexual differentiation[136]	Turtle
Reduced visual discrimination, increased activity level[134]	Rat
Increased locomotor activity[148]	Rat, monkey, mouse
Learning difficulties[149]	Rat, mouse, monkey

Note: Doses are not included since different PCB mixtures were used in the various studies, making it difficult to correlate doses with health effects easily across studies.

essential for normal growth and development.[24] Also unlike dioxin, some forms of PCBs occupy the estrogen receptor, causing an estrogenic or antiestrogenic effect. In some instances, estrogen-receptor binding is facilitated by metabolic alteration (hydroxylation) of one portion of the PCB molecule so that it more closely resembles a portion of an estrogen molecule. However, this metabolic transformation is not always necessary for estrogen-receptor binding.[132]

Animal Studies

The reproductive and developmental health effects of PCBs have been studied in a variety of animal species (table 6.2). Some of the reproductive effects occur after exposures that are considerably higher than any currently likely for humans in the United States, although wildlife are at much greater risk because of their specialized diets. Reduced fertility, spontaneous abortions, and reduced litter and infant survival have been observed in the wild and in laboratory studies at mg/kg or μg/kg doses depending on the PCB mixture used. Developmental effects often occur after much smaller fetal or neonatal exposures and also depend on the particular PCB mixture studied.

Of particular concern is the apparent neurotoxicity of some PCBs, which cause reduced learning capacity and altered behavior after low levels of exposure during the period of brain development. In rats, PCB-126 causes reduced litter size and infant survival, along with delayed neuromuscular development, after maternal dosing at 10 μg/kg on every second day from days 9 to 19 during pregnancy.[133] Maternal postpartum body weight was also slightly decreased compared to controls. The same PCB mixture administered in the same fashion at 2 μg/kg did not affect maternal bodyweight or the physical development of the offspring but caused poorer visual discrimination and increased activity levels in pups after weaning.[134] These results underscore the importance of examining for subtle neurological effects at low doses that do not cause maternal toxicity.

Monkeys fed from birth to age twenty weeks with a PCB mixture representative of the PCBs typically found in human breast milk showed significantly impaired learning and performance skills when tested at three years of age.[135] The affected monkeys had blood PCB levels at a low 2 to 3 parts per billion (ppb), similar to levels in the general human population.

Studies of the estrogenic influence of two types of PCBs on sexual differentiation in turtles demonstrate a synergistic interaction.[136,137] The sex of turtles, like many other reptiles, is determined by the incubating temperature of the fertilized egg. For most turtles, low temperatures produce males, and higher temperatures produce females. PCBs with estrogenic activity, applied to turtle eggs, can cause female development in eggs incubated at male-producing temperatures. Certain PCBs synergize with minor alterations in temperature to cause more dramatic sex reversals than would be predicted by simply adding the PCB effect with the temperature change effect. The same phenomenon occurs with small amounts of PCBs in combination.

Epidemiological Studies

Since PCBs have been banned in the United States and many other parts of the world, there is little opportunity to study their toxic effects in the occupational setting, where exposures might be expected to be high.

However, a pre-ban study of mothers potentially exposed to PCBs in an electrical capacitor manufacturing plant showed a small but significant decrease in the birth weight of infants.[150]

In the late 1970s, accidental human exposure to PCBs in Japan and Taiwan resulted from consumption of PCB-contaminated rice oil. Since then investigators have monitored the people exposed, their pregnancies, and their offspring.[151] The immune system of those exposed was affected so that they were more susceptible to infection and had decreased antibody levels. There were increases in prenatal deaths, retarded fetal growth, and infant mortality. Delayed brain development and behavioral abnormalities in the children exposed as fetuses persist years after the incident. They score lower on developmental testing, and their intellectual development lags behind that of their peers. According to teachers, they are hyperactive and exhibit more behavioral problems than unexposed children.[152] Some believe that the toxic responses were due not to PCBs but to polychlorinated dibenzofurans (PCDFs), other toxic chemicals that contaminated the PCB industrial fluid.[153]

One group of 212 children exposed to ambient levels of PCBs in the uterus or through breast milk has been followed in Michigan. In most cases, their PCB exposure increased with the amount of Lake Michigan fish that their mothers consumed before and during pregnancy. Those who were most highly exposed to PCBs as fetuses showed delayed or reduced psychomotor development and poorer performance on a visual recognition memory test.[154] When the data were analyzed to include only prenatal exposure (no exposure through breast milk), deficits in physical growth, memory, and attention persisted. The investigators have reported results of neurological and intellectual testing of these children at eleven years of age. They found that prenatal PCB exposure was associated with lower IQ scores after controlling for other factors such as socioeconomic status.[88] The most highly exposed children were more than three times as likely to perform poorly on IQ tests and tests designed to measure their attention span. They were more than twice as likely to be at least two years behind in word comprehension in reading.

Another group of children, being followed in North Carolina, shows similar results.[155] PCB exposures were determined by measuring maternal PCB levels at birth and in maternal milk. Children with higher transplacental exposure to PCBs consistently scored lower at six and twelve months of age on a psychomotor development test than children with lower exposures. In a New York study of several hundred newborn children whose mothers ate varying amounts of PCB-contaminated fish from Lake Ontario, those in the higher-exposure group showed abnormal reflexes and startle responses when compared to those with less exposure.[156] In the Netherlands, investigators found that higher levels of PCBs in breast milk were correlated with lower levels of thyroid hormone in mothers and higher TSH levels in nursing infants.[90] The subjects in this study were exposed to PCBs at ambient environmental levels.

It appears that despite a twenty-year ban on U.S. production, PCB exposures at current ambient environmental levels impair intellectual and motor development of children. The environmental persistence of these chemicals and their tendency to bioaccumulate ensure continued exposure for years to come.

Alkylphenols

Reproductive effects: decreased testicular size, reduced sperm counts, and feminization of males in some animal studies

Alkylphenols are industrial chemicals used in detergents, paints, pesticides, plastics, food wraps, and many other consumer products. Hundreds of thousands of tons of these chemicals are produced annually. Much ends up in sewage treatment works and is discharged to surface water.[157] Some alkylphenols accumulate in sewage sludge, and others remain dissolved in water. They may contaminate drinking water and food, leaching from plastics used in food processing and wrapping.[158,159] Some members of this family of chemicals are estrogenic, although their affinity for the estrogen receptor is substantially less than that of estrogen.

In a laboratory in which estrogen-sensitive breast tumor cells were being studied, investigators discovered that the plastic (polystyrene) used to make test tubes for routine laboratory procedures contained a substance that behaved like estrogen. They identified it as nonylphenol, a member of this family of chemicals, extracted it from the test tube plastic, and demonstrated its ability to cause estrogen-sensitive cells to grow in both tissue culture and in the uterus of rats.[160] Other laboratory studies confirm estrogen-like properties of these chemicals in fish, bird, and mammalian cells.[157] Male fish raised in water near sewage outflows contaminated with alkylphenols are feminized. They produce a female protein, vitellogenin, found in egg yolks. Some have genitals of both sexes.[161] Whether these abnormalities in river fish should be attributed entirely to alkylphenols or to estrogen from human urine is still a matter of debate. There is no information about the effect of alkylphenols on humans.

Bisphenol-A

Reproductive effects: Estrogenic effects in animal studies at exposures near current human exposure levels

Bisphenol-A is a major component of polycarbonate plastics, epoxy resins, and flame retardants. More than a billion pounds of bisphenol-A are produced annually in the United States, Europe, Japan, Taiwan, and Korea.[162] Polycarbonate plastics are among the largest and fastest-growing markets. Epoxy resins made of bisphenol-A are used to coat the inside of food cans, as dental sealants, and in a variety of dental, surgical, and prosthetic devices. Laboratory tests show that bisphenol-A and related chemicals leach out of polycarbonate containers or the epoxy coating on the inside of food cans, particularly when the container is heated in order to sterilize the contents.[163,164] These same chemicals are found in saliva after dental treatment with sealants, sometimes years after the original application.[165]

Bisphenol-A and related chemicals attach to the estrogen receptor, exerting estrogenic effects.[163,166] Bisphenol-A stimulates the growth of estrogen-responsive breast cancer cells in cell cultures, although it binds about two thousand times less avidly to the estrogen receptor than does estrogen in those studies.[165,167] When fed to rats, bisphenol-A also behaves like estrogen and stimulates prolactin production, but here it is only one hundred to five hundred times less active than estrogen—ten times more potent than would have been predicted from the cell culture studies.[168] Cell culture experiments that compare estrogen-receptor binding of bisphenol-A and octylphenol show that when carrier proteins are absent, octylphenol is one thousand times more potent than bisphenol-A; but when the proteins are added, bisphenol-A is one hundred times more potent than octylphenol.[167] These observations demonstrate the importance of considering protein binding and looking for a variety of biological effects before drawing conclusions about the hormone-disrupting potency of synthetic chemicals.

Previous research has shown that small increases in serum estrogen levels during mouse fetal life are related to enlargement of the prostate in adulthood. In one study, investigators fed two different dose groups of pregnant mice 2 and 20 µg bisphenol-A/kg on days 11 to 17 of gestation. Each of these doses resulted in significantly enlarged prostates in adult male offspring.[167] The larger of the two exposures also resulted in reduced sperm production.[169] These doses are near the estimated ranges of human exposure to this chemical, raising questions about the relative safety of the various uses of bisphenol-A.[170,171]

There have been no studies of the effects in humans exposed to bisphenol-A.

Phthalates

Reproductive effects: Reproductive and developmental toxicity at a variety of exposure levels; are testicular and ovarian toxicants and have estrogen-like activity in some cases

Phthalates, the most abundant man-made chemicals in the environment, are used in construction, automotive, medical, and household products; clothing; toys; and packaging.[98] Over 1 billion pounds of 25 different phthalate compounds are produced annually in the United States.[172] In their largest single application, they serve as plasticizers for polyvinylchloride (PVC). Like alkylphenols, phthalates may leach out of packaging material into food. Plastic wraps, beverage containers, and the lining of metal cans all may contain phthalates. Phthalates volatilize during their manufacture and use and disperse atmospherically. The two most abundant, di-2-ethyl-hexyl phthalate (DEHP) and di-n-butyl-phthalate (DBP), are found in soil; in fresh, estuarine, and ocean water; and in a variety of fish, including deep sea jellyfish from more than three thousand feet below the surface of the Atlantic.[172] All phthalates tend to accumulate in fat tissue, though some may be broken down and excreted from the body. They are easily absorbed through the skin.

Some phthalates attach to the estrogen receptor and in laboratory tests behave as weak estrogens.[173] However, they vary considerably in potency. In descending order of estrogenicity, as measured by receptor binding in test tube experiments, they are butyl benzyl phthalate (BBP), dibutyl phthalate (DBP), diisobutyl phthalate (DIBP), diethyl phthalate (DEP), and diisononyl phthalate (DINP). DEHP showed no estrogenic activity in this study.

In animal studies, DEHP reduces fertility and testis weight more readily than DBP.[174] Phthalates are therefore likely to be toxic to the testes through some mechanism other than estrogenicity. Research showing that phthalates, or breakdown products, interfere with the function of follicle-stimulating hormone (FSH) may better explain testicular toxicity, since FSH is required for normal Sertoli cell maintenance in the testes.[175] Developing animals are much more susceptible to this effect than adults. Interference with FSH function might also account for altered estrogen levels and ovulation in rats exposed to DEHP.[176] At larger doses in rats (maternal diet 2 percent BBP), BBP is toxic to the fetus, causing spontaneous abortions and birth defects.[177,178] In multiple-generation studies, the effects of BBP on the second generation are greater than the first.[179] Virtually nothing is known about the chronic effects of long-term low-dose human or wildlife exposure.

The largest source of human exposure to phthalates is likely to be from food. Estimates of average dietary intake of all phthalates range from 0.1 to 1.6 mg per person daily (0.002–0.03 mg/kg per day for a 130- pound person).[180] The average intake from infant formulas is larger, estimated at 0.13 mg/kg body weight per day for a newborn. If the usual uncertainty factors for extrapolating risks from animals to humans were applied to the animal data showing adverse effects on the male reproductive system, this level of exposure is several-fold larger than what would be considered a safe dose.

DEHP also leaches from the plastic of medical equipment and is found in the blood or tissues of people who have undergone blood transfusions or kidney dialysis.[181] Little is known about the metabolism, storage, and excretion of phthalates in humans. Because of the widespread presence of phthalates in water and sewage effluent, where concentrations range from nanograms to milligrams per liter, effects on fish and wildlife are also a concern.[182]

Pesticides

Organochlorines: Interfere with normal estrogen function; in animals: males exposed to some organochlorines in fetal life may be feminized, and females have altered estrus cycles and hormone levels

Dicofol, pentachlorophenol, dinoseb, and bromoxynil: Interfere with thyroid function

Pyrethrins and vinclozolin: some have antiandrogen activity

A number of pesticides belonging to several classes have endocrine-disrupting properties (see table 6.3; also see chapter 5).

Organochlorines

Some organochlorines have been banned from use in the United States because of environmental persistence and endocrine-disrupting properties. For example, kepone caused low sperm counts and sterility in exposed workers.[183] DDT was banned because of harmful effects on wildlife reproduction. Others, including endosulfan, methoxychlor, dicofol, and lindane, are still in use. Laboratory animal and wildlife studies demonstrate a range of toxic effects, some of which are due to interference with normal endocrine function. They have other mechanisms of toxicity as well.

DDT and its metabolic by-product, DDE, are weakly estrogenic. However, although DDE binding to the estrogen receptor is limited, it strongly blocks the binding of testosterone to the androgen receptor. It is an androgen antagonist.[184]

Methoxychlor is used as an insecticide on a variety of fruits and vegetables. Its metabolic breakdown product is estrogenic in birds and mammals and interferes with sexual development, reproduction, and behavior.[185–187] In a series of experiments designed to study the behavioral effects of prenatal exposure to estrogenic chemicals, investigators fed pregnant mice DDT, methoxychlor, or diethylstilbestrol on days 11 to 17 of pregnancy.[188] The behavior of male offspring in territory new to them was examined. Urine marking of territorial boundaries influences social and reproductive behavior in other mice. It elicits aggressive behavior in other males and attracts females. Male mice exposed to each of the estrogenic agents as fetuses demonstrated significantly increased urine-marking behavior as adults.

Table 6.3 Pesticides—Endocrine-disrupting effects

Pesticide	Species and health effect
Methoxychlor	Estrogenic Female rat: accelerates vaginal opening, abnormal estrus cycles, inhibits corpus luteum function, blocks implantation of fertilized eggs Male rat: suppresses Leydig cell function, increases prolactin levels, retards growth, increases aggressive behavior Rabbit: loss of litter, birth defects
Endosulfan	Estrogenic Rat: shrinkage of testicles Stimulates growth of estrogen-sensitive breast tumor cells in cell culture
Lindane	Antiestrogenic Rabbit: reduced ovulation rate
Dicofol	Estrogenic Birds: eggshell thinning, reduced hatching, abnormal testes development, aberrant submissive behavior, high infertility of offspring Displaces thyroxine from transthyretin
Atrazine	Alters estrogen and testosterone metabolism Interferes with hypothalamic-pituitary-gonadal axis
Pentachlorophenol	Displaces thyroxine from transthyretin Lowers thyroxine levels
Dinoseb	Displaces thyroxine from transthyretin
Bromoxynil	Displaces thyroxine from transthyretin
Dithiocarbamates	Lower thyroid hormone levels, increased TSH levels
Vinclozolin	Antiandrogen Male rat with prenatal and early postnatal exposure: reduced anogenital distance, nipple development, abnormal penises with hypospadias
Cypermethrin	Antiandrogen Male rat with prenatal and early postnatal exposure: decreased anogenital distance, decreased prostate weight

Some growers of fruits and vegetables use endosulfan on the food supply. Although much less potent than estrogen, endosulfan binds to the estrogen receptor and, in cell cultures, stimulates the growth of estrogen-sensitive breast cancer cells.[189]

Lindane is used as an insecticide on trees and is available by prescription to treat the human body louse (scabies). It is easily absorbed through the skin. Lindane accumulates in the ovarian follicles, fallopian tubes, and uterus of test animals. Most investigators conclude that lindane has antiestrogenic properties.[190,191]

Dicofol is of the same chemical family as DDT, and commercial preparations are contaminated with DDT to varying degrees. Dicofol has estrogenic activity in birds, where it causes feminization of male embryos, abnormal submissive behavior in male offspring, and impaired reproductive success.[192] Dicofol also strongly competes for the thyroxine binding site of the thyroid hormone carrier protein, transthyretin.[193]

Another organochlorine, pentachlorophenol (PCP), also has thyroid-disrupting effects. PCP has been widely used for many years as a wood preservative, although its use has been somewhat restricted since 1984. However, population monitoring studies in 1994 concluded that an estimated 64 percent of the U.S. population had PCP residues in their urine.[194] PCP is also a potent competitor for human transthyretin, binding to the protein twice as readily as the naturally occurring hormone, thyroxine.[195] In rats, PCP also lowers thyroid hormone levels significantly.[196] One study showed that PCP directly reduced uptake of thyroxine into the brain.[197] The thyroxine-transthyretin complex is essential for transport of thyroid hormone into the fetal brain, where it is required for normal development. Observations of the thyroid-disrupting effects of dicofol and PCP raise concerns about their effect on the developing brain in humans. Two other nonorganochlorine pesticides in widespread commercial use, dinoseb and bromoxynil, have similar effects.[198]

Fungicide

Vinclozolin is a fungicide used on fruits, vegetables, ornamental plants, and grass. It is degraded in soil or in plants into two by-products, also detected in rats treated with the fungicide.[199] Test tube experiments and studies in rats show that the by-products of vinclozolin metabolism bind to androgen receptors and effectively block testosterone, causing feminization of male rats and other birth defects.[200,201] However, in the absence of testosterone, vinclozolin by-products actually behave as androgens rather than antiandrogens, demonstrating that hormone effects of synthetic chemicals depend in part on the presence or absence of natural hormones.[202]

Pyrethroids
Pyrethrin and synthetic pyrethroid insecticides are heavily used in home and agricultural pesticide products. Studies of fluvalinate, permethrin, and resmethrin in cell cultures demonstrate that they bind to the androgen receptor in competition with testosterone.[22] When cypermethrin was administered by injection to pregnant rats during the last seven days of gestation and to male offspring for the first thirty days of life, there was a significant decrease in anogenital distance but no change in sperm counts.[203] These findings are consistent with an antiandrogenic effect. Some pyrethrins and pyrethroids also displace testosterone from sex-hormone binding globulin.

Triazine Herbicides
The triazine herbicides atrazine, simazine, and cyanazine, are heavily used in large agricultural areas in the United States and are under special review by the EPA. Atrazine contaminates major groundwater aquifers used as drinking water in many parts of the country. Among toxicologic concerns are the endocrine-disrupting properties of this widespread contaminant. Depending on the experimental design of animal studies, atrazine may have either estrogenic or antiestrogenic effects.[204] It also causes breast cancer in one strain of rats.

A series of laboratory and animal experiments show that atrazine does not seem to exert hormonal effects through binding to the estrogen receptor.[204] However, rats given 17 mg/kg of atrazine daily during pregnancy give birth to male rats with significantly fewer testosterone receptors in their prostates than a control group. Female offspring have significantly altered enzyme activity in their pituitary glands. Moreover, when this dose of atrazine is administered throughout pregnancy and lactation, the male offspring also have permanently altered enzyme levels in their pituitary glands, resulting in decreased ability to convert testosterone to dihydrotestosterone, its active form. Much larger doses are necessary to have a similar effect in adult male rats.[205]

Atrazine is also reported to alter the metabolism of naturally occurring estrogen, resulting in a metabolite that is even more highly estrogenic.[206] Atrazine appears to disrupt hypothalamic-pituitary regulation of ovarian function and interferes with the biochemical conversion of testosterone and its interaction with the testosterone receptor in the prostate.[207–209]

Dithiocarbamate Fungicides

Exposure to ethylene thiourea, a breakdown product of dithiocarbamate fungicides, causes a decrease in thyroid hormone (T4) levels and a corresponding increase in thyroid-stimulating hormone (TSH) in both rats and mice.[210] The degree of thyroid hormone decline will determine whether an exposed organism suffers ill effects solely from the hormone deficiency. But the resultant constant stimulation of the thyroid by TSH is thought to be the cause of an increase in thyroid cancers in exposed animals. A study of dithiocarbamate appliers and landowners in Mexico where the pesticide was used showed elevated TSH levels but no decrease in thyroid hormone levels.[211]

Phytoestrogens

Many grasses, grains, soybeans, vegetables, nuts, berries, and some fungi contain naturally occurring chemicals with estrogenic or antiestrogenic properties. Known as phytoestrogens, their hormone-like properties have been recognized for years. Sheep that eat a particular kind of clover rich in phytoestrogens develop menstrual irregularities and infertility. Studies in laboratory animals confirm that disruption of the estrus cycle, pituitary hormone levels, puberty onset, male sexual behavior, and changes in the size of portions of the hypothalmus of the brain may result from prenatal or postnatal exposure to these compounds.[212–217] Phytoestrogens can affect uterine growth and implantation of fertilized eggs, inhibit ovulation, or cause eggs to degenerate.

The compounds often behave differently under varying circumstances. Some are estrogenic; others are estrogen antagonists. Whether a particular phytoestrogen is estrogenic or an estrogen antagonist depends in part on levels of naturally occurring estrogens. In the presence of estrogens, a phytoestrogen often behaves as an estrogen antagonist by blocking the estrogen receptor from occupancy by the naturally occurring hormone.[218]

Phytoestrogens also stimulate production of sex-hormone binding globulin in the liver of premenopausal women, decreasing the availability of free estrogen to cells. The combination of estrogen antagonism and increased binding globulin may explain how phytoestrogens seem to have a mildly protective effect against breast and uterine cancer among premenopausal women whose diets are rich in these substances.

Human Exposures to Reproductive Toxicants

7

Exposure to chemicals can be hard to quantify, yet it is a critical piece of the toxics puzzle. It is also a particularly difficult piece, because there are so many points along the path from chemical production to a target tissue where relevant information may be gathered, and each point along the path provides only a limited snapshot of information.

The Environmental Protection Agency defines exposure as contact between the outer boundary of the human body (e.g., skin, nose, throat) and a pollutant or pollutant mixture.[1] It is difficult to measure a person's total daily exposure to a chemical at the point of entry into the body, and even harder to measure the total exposure to the complex mixtures that people encounter every day. Despite the technical difficulties and expense, some studies have attempted this type of exposure assessment for some known or suspected reproductive toxicants. There are also ways of estimating human exposure using surrogates, such as chemical production, use, and release. These data are poor estimates of potential human exposure, but sometimes they are the only available information on a particular chemical.

In order to result in exposure, a chemical must be in indoor or outdoor air; in water that is used for drinking, swimming, or fishing; or in food or dust—all sources that can be sampled and measured. Once people have been exposed, some chemicals can be measured in blood, urine, fat, or breast milk. This biomonitoring information lets us know that human exposure has occurred but often gives little information about the route of exposure or the original source. Figure 7.1 outlines

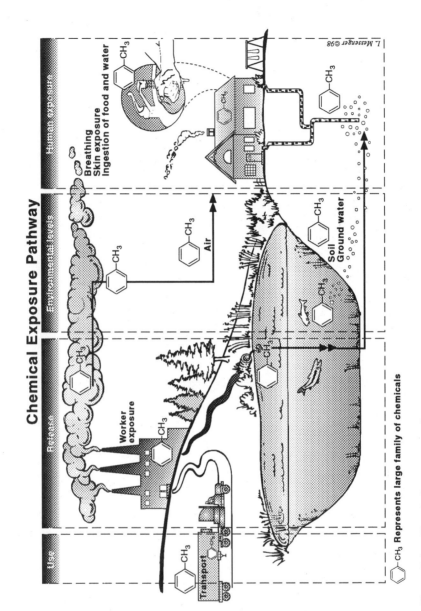

Figure 7.1   Chemical exposure pathway

Key:

Symbol	Represents	Measurement/Monitoring Mechanism
PRODUCTION AND USE	Chemical production, sales, and use in industry and agriculture.	SOC, TURA, PUR, NCFAP
RELEASES	Chemical releases to air, water, and land.	PUR, TRI
ENVIRONMENTAL LEVELS	Levels of chemicals in the environment—air, water, food.	TDS, USGS, NAQET, FCA, PDP, FDA Residue Testing, NOPES, USDA Meat Inspection, NHEXAS, TEAM
BIOMONITORING	Levels of chemicals in humans through breathing, drinking, eating, and skin absorption. Blood, urine, exhaled air, fat tissue, breast milk.	NHEXAS, TEAM, NHANES, NHATS, Breast milk monitoring

Monitoring databases. See chapter text for more details.

SOC—Synthetic Organic Chemicals: U.S. Production and Sales, International Trade Commission. TURA—Toxic Use Reduction Act. In Massachusetts and New Jersey; limited in a few other states. Large facilities report use to state. PUR—Pesticide Use Reporting. Only in California and limited in New York. Pesticide applicators report use to state. NCFAP—National Center for Food and Agricultural Policy, Pesticide Use in U.S. Crop Production. National pesticide use estimates. TRI—Toxics Release Inventory. Large facilities report chemical releases to EPA inventory. USGS—U.S. Geological Survey's National Water Quality Assessment. Pesticide and solvent residues in water. NAQET—National Air Quality and Emissions Trends. Solvents and metals in ambient air. TDS—Total Diet Study. Pesticide and chemical residues in prepared food through "market basket" samples. PDP—Pesticide Data Program. Pesticide residues in foods measured by the U.S. Department of Agriculture. FDA Residue Testing—Pesticide residues on food—whole and unwashed. FCA—Federal and state fish consumption advisories on chemical levels in fish. NOPES—Non-Occupational Pesticide Exposure Study. Environmental levels of pesticides. USDA Meat Inspection—Pesticide residues in meat and poultry. NHEXAS—National Human Exposure Assessment Survey. Environmental levels in air, water, food; human levels in urine, hair, blood. TEAM—Total Exposure Assessment Methodology. Environmental levels, human exhaled air. NHANES—National Health and Nutrition Examination Survey. Levels in blood and urine. NHATS—National Human Adipose Tissue Study. Levels in human fatty tissues.

the steps on the pathway between chemical production and the presence of a chemical in the human body and summarizes some data sources that can provide information relevant to exposure. Lack of evidence of exposure may be due to difficulties in measurement, failure to look for the chemical at all, or a true absence of exposure.

Exposure monitoring has many purposes. It can identify whether people are exposed to certain chemicals and by what route most exposure occurs. Quantification of exposure may allow comparisons with levels known or suspected to cause health effects. Good measurement of exposure is important for epidemiological research because exposed and comparison groups need to be clearly identified and separated. Finally, assessments over time allow us to identify trends, set reduction goals, and assess the efficacy of exposure-reduction efforts.

Most studies with information relevant to exposure to reproductive toxicants are local or regional, assess groups that may not be representative of the general population, or are fairly small. Some were done in other countries and may not reflect conditions in the United States. In this chapter, we focus on a few publicly available, large, systematic databases or studies relevant to potential exposure to the general population of the United States. We selected them to span the spectrum from chemical production and use through emissions into environmental media, to levels in human tissues.

Table 7.1 presents an overview of the major data sources used in this chapter. Each individual source of information provides data on some chemicals and some routes of potential exposure. Together they complement the health effects data presented in earlier chapters. The chapter reviews each of these data sources, explaining what they are, their strengths and weaknesses, and the information they provide. Then we mine the information from these sources to outline what we know about human exposures to the known and suspected reproductive toxicants previously discussed in this book.

Data Sources Relevant to Human Exposure

There are numerous sources of data that can be useful for estimating potential human exposures to chemical toxicants. The data sources, however, are not generally available in one place. Some of the informa-

Table 7.1   Data sources with implications for human exposure

Study	Metals	Solvents	Pesticides	Endocrine Disruptors★
Exposure surrogates				
Production and sales data (SOC report)		Production, sales	Production, sales	Production, sales
Pesticide use data (NCFAP, U.S. EPA)			Use	
Toxics Release Inventory (TRI)	Release	Release	Release	Release
Environmental Measures				
Air quality and emissions trends	Environmental level	Environmental level		
U.S. Geological Survey water testing		Environmental level	Environmental level	
Total Diet Study (TDS), Pesticide Data Program (PDP)	Intake		Intake	Intake
Fish advisories	Intake		Intake	Intake
USDA meat inspection	Intake		Intake	Intake
Drinking water monitoring	Intake	Intake	Intake	
Non-Occupational Pesticide Exposure Study (NOPES)			Intake	
Total Exposure Assessment Methodology (TEAM)		Intake		
Biomonitoring				
National Human Exposure Assessment Survey (NHEXAS)	Intake/ absorbed dose	Intake/ absorbed dose	Intake/ absorbed dose	
National Health and Nutrition Examination Survey (NHANES)	Absorbed dose	Absorbed dose	Absorbed dose	
National Human Adipose Tissue Study (NHATS)			Absorbed dose	Absorbed dose
Breast milk monitoring (various studies)	Target tissue/ fetal intake	Target tissue/ fetal intake	Target tissue/ fetal intake	Target tissue/fetal intake

★ Excludes pesticides. Includes only the following subset of endocrine-disrupting chemicals: dibenzo-dioxins, dibenzofurans, PCBs, selected phthalates, octylphenol, nonylphenol, and bisphenol-A.

tion is updated annually, while some is collected only once or intermittently. Much more information is available beyond the sources profiled here, yet these are the major nationwide data sources. The data sources are presented in order along the exposure chain from information about chemical production volumes to information about levels in human tissues.

SOC Report

From 1917 until 1995, the U.S. International Trade Commission (ITC) issued an annual report, *Synthetic Organic Chemicals: United States Production and Sales* (known as the SOC Report),[2] with information on pharmaceuticals, pesticides, plastics, and other synthetic organic chemicals. Raw materials, including metals such as lead, mercury, and cadmium, are not included in the report, nor are imported chemicals. The data for the report are compiled from responses to questionnaires sent to chemical manufacturers.

Although chemical companies are required by law to complete the questionnaire, the results are not independently verified by the ITC. Small companies are often not included, import data are omitted, and there are reporting thresholds, so the numbers in the report are likely to be underestimates of true production and sales. In addition, due to trade secrecy considerations, statistics for chemicals are reported only when there are more than three U.S. producers of that chemical. Yet even when more than three companies produce the chemical, the statistics are not reported if the results are provided to the ITC "in confidence." Due to these limitations, there are many gaps in the reporting scheme. Nevertheless, despite its many weaknesses, the SOC Report has been a useful source of trend data for chemical production over time and for identifying large-production-volume chemicals.

The last annual SOC Report used 1994 data and was published in 1995, and now this seventy-eight-year rich data source has been eliminated. The House Committee on Ways and Means requested in October 1995 that the ITC terminate publication of the SOC Reports and said that users should "identify and develop alternative sources for this information."[3] Unfortunately, there are no other sources of this information readily accessible to the general public. The loss of the SOC Report is a significant step backward in terms of the public's right to know.

National Pesticide Use Estimates

The National Center for Food and Agricultural Policy (NCFAP) issues a national report estimating total use of herbicides, insecticides, fungicides, and fumigants in the United States.[4] This information is available publicly for purchase and includes state-specific data, national summary reports, and a database with information specific to pounds of active ingredient per crop and state. Data are compiled from the California pesticide use reports, the U.S. Department of Agriculture's (USDA's) Cooperative Extension Service reports for individual states and selected crops, surveys by the National Agricultural Statistics Service of pesticide use, and USDA's pesticide benefits assessments. The most recent update covers the period from 1990 to 1993.

The report makes some extrapolations and estimates in order to come up with rough totals of pounds of pesticides used in agriculture and number of acres treated. It reports on two hundred pesticide active ingredients. The NCFAP receives extensive funding from the pesticide industry and solicits industry review of the database prior to publication. It does not have access to industry proprietary data on sales or use.

U.S. EPA also makes rough estimates of pesticide sales and usage annually in *Pesticide Industry Sales and Usage*.[5] The level of detail it provides is fairly crude, but this information complements the NCFAP report by also providing some estimates of household use of pesticides.

These national reports of pesticide use are the best that are currently widely available, but they are rather difficult to check for accuracy, and the many assumptions used to extrapolate the data to the entire United States leave quite a lot of room for error. In addition, unlike the California pesticide use reports, these data do not provide detailed information about the location of pesticide use. This information does provide a rough estimate of use and some comparative data for different pesticides, and it can potentially be used to follow trends of use.

Toxics Release Inventory

The Emergency Planning and Community Right-to-Know Act (EP-CRA) requires companies with ten or more employees, which process or manufacture certain chemicals in excess of reporting thresholds, to report releases of these chemicals to land, air, and water on the Toxics Release Inventory (TRI).[6] Originally the TRI covered a list of slightly over three hundred chemicals. In 1994, the EPA expanded the list, and starting in 1997, 643 chemicals are reportable under the TRI. Another expansion of TRI has also recently occurred. Originally only manufacturing industries were covered in the law; in 1997, EPA added seven new industry sectors, including bulk petroleum storage, hazardous waste treatment, mining, and electric utilities. Releases will be reported by these new industry sectors starting in 1999.

TRI provides only indirect information about what chemicals may be stored and used on-site. In addition, since the reports of releases come from the companies, they are difficult or impossible to verify for accuracy, and some companies fail to file reports at all. Poor enforcement makes it likely that the data collected under TRI underestimate true emissions. TRI does not contain chemical use data, which are collected in certain states, including Massachusetts and New Jersey. Use data can be helpful for estimating potential worker exposures, exposures during accidents, chemicals shipped out in product, and chemicals brought in by road. Nonetheless, the TRI is an invaluable tool for gathering information about chemical releases from a particular company, in a particular neighborhood, or by an entire industry. Some of the TRI data, particularly the stack and fugitive air emissions data, may indirectly reflect exposures to nearby populations. Although TRI data are far from an ideal measure of human exposure, they are a surrogate exposure measure that should not be ignored.

National Air Quality and Emissions Trends

The EPA reports annually on air pollution trends in the United States. The focus is on the six pollutants regulated under the Clean Air Act: ozone, particulates, sulfur dioxide, carbon monoxide, nitrogen dioxide, and lead. These are called the *criteria pollutants*. Lead is therefore closely tracked. EPA also monitors some of the solvents discussed in this book in certain high-ozone regions because they are ozone precursors. The Aerometric Information Retrieval System (AIRS) database contains much more detail than the summary report and is accessible to the public over the Internet.

Although the data on lead are useful, there is very limited or no information about outdoor ambient air levels of most of the reproductive toxicants discussed in this book. The information we do have indicates that outdoor air contributes only a small fraction of total exposure to most reproductive toxicants in most areas. Toxicants in air can be important if they settle to the ground and concentrate in food or water, accumulate in dust where children play, or are concentrated in certain hot spots—local areas with increased exposure due to, for example, a local polluting industry. Information on lead, for example, indicates that there are major hot spots around the country that could be targeted for exposure-reduction efforts, and certain other chemicals probably follow this pattern of uneven distribution. Although pesticide drift is a major concern in some areas, ambient air monitoring for pesticides is virtually nonexistent.

Water Monitoring Programs

The U.S. Geological Survey (USGS) monitors water quality for pesticide residues. The Pesticide National Synthesis Project of the National Water Quality Assessment (NAWQA) analyzed about five thousand samples of water for a panel of eighty-five pesticides and pesticide metabolites between 1991 and 1995. Samples were collected from urban areas, agricultural areas, and major aquifers and from groundwater and surface water. Not all of the water analyzed is used for drinking, but the analysis does provide some information about the extent of pesticide runoff and the potential for drinking water contamination.

In addition to the pesticide analyses, the NAWQA program is planning a national assessment of volatile organic chemicals (VOCs), many of which are organic solvents, in major aquifer systems and rivers. No data are yet available, but the plan is to analyze samples for over fifty different VOCs in groundwater, rivers, and aquifers that serve as public drinking water supplies. Unfortunately, only five of the solvents that are discussed in this book will be included in the testing protocol. Water monitoring is also required under the Safe Drinking Water Act and under the Clean Water Act. These data are available to the public and can be very helpful locally.

These data are interesting and helpful as far as they go, but numerous pesticides and solvents that are known to be water soluble or have previously been reported in water are not included in the monitoring programs. Certain endocrine-disrupting chemicals such as some phthalates, bisphenol A, and alkylphenol polyethoxylates have also been reported in water and should be assessed systematically.

Total Diet Study/Pesticide Data Program

The Food and Drug Administration (FDA) has been conducting yearly surveys of a host of contaminants in the U.S. food supply since the early 1960s. The Total Diet Studies (TDS) are done each year in the Northeast, South, West, and North-Central states. Food selections are based on an FDA food consumption survey. These studies are market basket samples: researchers collect foods at local supermarkets in three cities—up to a total of 261 food items—prepare the foods for eating, and blend together samples in amounts proportional to those normally eaten by individuals in various age and sex groups. These blended samples are then analyzed for residues of pesticides, toxic metals, some radioactive materials, and some other hazardous chemicals such as polychlorinated biphenyls (PCBs). Tap water samples are also collected in the regions of study.

In addition, the FDA carries out an ongoing regulatory program that tests for pesticide residues in samples of foods that are imported or shipped in interstate commerce. Approximately ten thousand samples are tested each year, taken from less than 1 percent of foods shipped interstate or imported. Grains, fruits, vegetables, milk, dairy products, eggs, fish, shellfish, baby food, and some processed foods are tested for residues of 345 pesticides under this surveillance monitoring system. These food samples are analyzed whole and unwashed.

Finally, the USDA tests for pesticide residues in foods as part of the Pesticide Data Program (PDP) instituted in 1991.[7] This program provides pesticide residue data gathered in ten states and collects data on food consumption. Food is collected primarily in supermarkets and distribution centers, washed and prepared for consumption, and analyzed for the presence and concentration of more than one hundred pesticides. About six thousand individual food samples are collected, representing only a small subsample of food types. For example, in 1998, only five kinds of fresh fruit or vegetables, three types of canned or frozen commodities, three juices, two kinds of grain, milk, and corn syrup were collected and analyzed.

The TDS, PDP, and residue testing programs provide useful estimates of the actual human intake of this group of toxic substances in food and, to a limited extent, in water. These studies measure residues at the point of human exposure and are quite useful for determining how much of these chemicals people are consuming. If a particular chemical is detected in great excess in a particular sample, it may be possible to trace the sample back to the point of origin.

Fish and Wildlife Advisories

The U.S. EPA maintains a list of fish and wildlife consumption advisories.[8] Advisories rely on surveillance of the levels of certain contaminants in fish and wildlife species, so there can be major variation in the surveillance and advisories of different states. Advisories may identify highly polluted areas or may simply identify areas where there is good environmental monitoring. In addition, only a small subset of known pollutants (mercury, PCBs, and certain pesticides) are generally monitored. Other contaminants in fish and wildlife go untested and unreported. The advisories may be issued by states, tribal governments, or the federal government and may cover one water body where contaminated fish have been detected or may be statewide. Fish advisories may also include recommendations to limit consumption, to avoid fish if pregnant, or to avoid consumption altogether. Wildlife advisories cover certain species of animals that may be hunted. The EPA's Listing of Fish and Wildlife Advisories (LFWA) is available on the Internet and is updated regularly. Information is readily available by state, contaminant, or type of advisory, and some information on time trends is available.

This information is critical to those who fish or hunt and eat their catch. Because fish and meat can accumulate certain environmental toxicants, they may be a major route of exposure to a subset of the population.

USDA Meat Inspection

The USDA's Food Safety and Inspection Service regularly tests meat and poultry for residues of organochlorine and organophosphate pesticides and veterinary drug residues.[9] Although this monitoring is not carried out with the same level of sophistication as the FDA's TDS, it does provide some information about the particular contribution of meat to peoples' chemical intake.

The meat testing program is particularly important because certain residues accumulate in meat rather than in fruits and vegetables. This is due to bioaccumulation, which occurs when food animals eat plant material and concentrate certain chemical residues in their bodies. Many of the worst chemical hazards tend to concentrate in fat, and meat products contain more fat than do fruits and vegetables. Unfortunately, the USDA measures only a small subset of chemicals routinely. The only metal measured is arsenic, because it is used as a veterinary drug. Many other important environmental toxicants are overlooked.

NOPES Study

The EPA performed the Non-Occupational Pesticide Exposure Study (NOPES) in the late 1980s in an effort to assess total daily exposures to pesticides. The NOPES was done throughout a year in Jacksonville, Florida, and in the Springfield–Chicopee, Massachusetts, area. Jacksonville was chosen to represent a region of predicted high pesticide usage, and Springfield was chosen as representative of a low-use area. A total of 261 people over the age of sixteen who had no occupational exposure to pesticides were chosen using a sampling technique designed to draw participants with a range of predicted levels of pesticide use. The participants were questioned about their pesticide use, and then twenty-four-hour personal air monitoring for twenty-eight pesticides, stationary air monitoring inside and outside the home, and an activity log for the monitoring period were undertaken. Drinking water was also analyzed, and for some subjects, skin exposure was estimated. Subjects wore cotton gloves while spraying pesticides, and the levels were measured on the gloves.[10]

The NOPES study attempted to quantify the relative contributions of various pathways to people's total daily pesticide exposure. This attempt was only partially successful, and children's exposure was not quantified at all. Yet due to activities such as crawling on the floor and hand-to-mouth activities, children's exposure is likely to greatly exceed adult exposure.

TEAM Studies

In the 1980s, the EPA carried out a major investigation of exposures in its Total Exposure Assessment Methodology (TEAM) study. Its purpose was to measure peoples' total daily exposures to roughly two dozen

VOCs. Some, such as benzene, toluene, and xylene, are due primarily to exposure to gasoline. Others, such as chloroform, are found in drinking water, and people are exposed to perchlorethylene from freshly dry-cleaned clothes. The TEAM study was a complicated, multiphase project, and it yielded crucial information about daily exposures to organic solvents. The main study selected six hundred people in four states (New Jersey, North Carolina, North Dakota, and California) who were picked in such a way as to represent the broader population in the states. Selected individuals were studied intensively for twenty-four hours with simultaneous personal exposure monitoring, outdoor air sampling near the home, and drinking water sampling. Finally, the study subjects answered a questionnaire and contributed exhaled breath samples at the end of the sampling period.[11]

The TEAM study was a difficult, expensive, and painstakingly careful piece of exposure assessment because it was focused on chemicals that are not persistent in the environment or in the human body. VOCs quickly evaporate from water, dissipate in air, and break down and are eliminated from the human body. The short-lived nature of VOCs is not necessarily reassuring from a health effects standpoint, however. These chemicals are ubiquitous, and people are exposed on a daily basis, so repetitive dosing may make the exposure important over time.

The TEAM studies allow fairly good estimates of overall daily intake of organic solvents and provide some information about the major sources of exposure to these chemicals. Since the exhaled breath measurements are a form of biological monitoring, this study spanned the two exposure categories of measuring environmental levels of these chemicals and biomonitoring. Nevertheless, these studies may have underestimated the total exposure to VOCs because skin absorption was not considered in the environmental exposure monitoring.

National Health and Nutrition Examination Survey

The National Health and Nutrition Examination Survey (NHANES) is a periodic national health study carried out primarily by the Centers for Disease Control. There have been seven surveys to date, and another is scheduled to begin in 1999. The NHANES surveys are designed to estimate the prevalence and trends of selected diseases and risk factors.[12] The small amount of environmental exposure assessment in these surveys is essentially an add-on.

The first survey that assessed some environmental exposures was NHANES II (actually the fifth NHANES-type survey), conducted in 1976–1980. This survey included blood lead levels and some limited blood testing for pesticide residues. Hispanic HANES (1982–1984) also contained blood lead monitoring and some pesticide monitoring. NHANES III (1988–1994) measured only blood lead and urine cadmium levels in the main study. A special study on a subset of volunteers ages twenty to fifty-nine in NHANES III involved blood monitoring of thirty-two VOCs and urine monitoring of twelve pesticides or pesticide metabolites.

The methodology used in NHANES III involved selection of a random sample of approximately forty thousand people over two months of age from eighty-one counties within the United States.[13] These people were asked to complete an interview and underwent a physical examination in a mobile examination center, with blood and urine samples collected for metabolic, nutritional, and toxicologic analyses.

The major weakness of NHANES is that so few environmental toxicants are sampled in the surveys. However, the data that are available are of excellent quality and allow for some extrapolation to the rest of the population, as well as determination of some time trends. NHANES IV, which will begin data collection in 1999, will include a special biomonitoring effort for an array of pesticides and endocrine disruptors.

National Human Adipose Tissue Study

The NHATS tissue survey was done annually by the EPA from the early 1970s to 1990, when it was discontinued, in part due to data quality problems. EPA collected fatty tissue samples from hospital pathologists around the country in order to characterize levels of exposure in the U.S. population. Tissue samples came from surgical specimens and autopsied cadavers. Samples from cadavers came only from individuals who died suddenly from an accident or a cardiac arrest and were not known to be occupationally exposed to toxic chemicals. Information collected on the source of the specimen was limited to the age group of the subject, sex, race (white and nonwhite), and the census region from which the sample came. The survey tested the samples for over one hundred compounds, including some pesticides, PCBs, dioxins, dibenzofurans, VOCs, and some other chemicals of interest, including several phthalates.

The NHATS survey had numerous quality control problems[14] and collected no information about the possible origin of the detected residue levels. In a review of the NHATS program, the National Academy of Sciences (NAS) concluded that the program was fundamentally flawed and should be phased out and replaced with an improved tissue monitoring effort. The major criticisms of the program included insufficient sample size, nonrandom sample selection, exclusion of certain segments of the population (particularly those living in rural areas), sampling errors, and poor communication of the results to the public.[15] As a result, the NHATS study has been discontinued, yet despite the clear recommendation of the NAS committee, no other program has been put into place to collect this important information. This is highly unfortunate because tissue sampling is a useful way of collecting exposure information. In addition, we are no longer able to follow time trends of residues in human fatty tissue because no current data are available.

National Human Exposure Assessment Survey

NHEXAS is a government-funded exposure assessment that uses both a TEAM approach and biomonitoring. Phase I of NHEXAS is ongoing in the north-central states,[16] in Arizona,[17] and in the Baltimore metropolitan area. Approximately a thousand people, including children, will be sampled in this phase. Phase I measurements include lead, arsenic, cadmium, manganese, and other metals, a range of VOCs, and a small number of pesticides.

Exposure measurements are ascertained by means of questionnaires, activity diaries; monitoring of personal air, indoor air, and outdoor air; sampling of drinking water and food; collection of soil and house dust; and urine, hair, and blood samples. Sampling is done over a week or repeatedly during a year on the same people to assess the consistency of exposures over time. Measured exposure levels can then be correlated with suspected risk factors.

One of the central questions that Phase I will answer is how well current exposure models predict actual exposures. If this phase reveals that current exposure models do not accurately predict actual exposures, and simple changes are not sufficient to bring the models into alignment with reality, then the EPA plans to undertake Phase II, a nationwide study of over five thousand people using similar exposure and biomonitoring methods.

Unfortunately, exposures to many common pesticides and most known or suspected endocrine-disrupting substances will not be measured. The VOC measurements essentially replicate the TEAM study with only minor improvements, and we already know a lot about human exposures to lead. Therefore, despite a great deal of time, money, and effort, we will end up only with good exposure assessment data on a handful of chemicals, most of them already fairly well understood. Although time trend data are useful, people are exposed to hundreds or thousands of different chemicals every day, and it may not be the best use of resources to focus again on the same small subset of chemicals.

Breast Milk Monitoring

Numerous smaller studies have evaluated levels of persistent organic chemicals, particularly some pesticides and endocrine disruptors, in breast milk. The National Human Milk Monitoring Program studies, done by the EPA in the 1970s to evaluate the extent of human milk contamination with organochlorine pesticides and PCBs, are the most recent nationwide studies. The most recent data available are from the second survey, done from 1976 to 1978, of 1,842 milk samples from women living in urban and rural areas. Participation was voluntary, and information about the mothers included geographic region, age, dietary information, race, and number of children nursed. Since then, there has been some information from smaller studies, which may allow us to extrapolate some trends into a more recent period, but no recent systematic nationwide study on pollutants in breast milk.

Although organic solvents are lipophilic (fat soluble) and tend to appear in breast milk,[18] they are much more short-lived than some pesticides and PCBs and are rarely measured. Yet these solvents may be quite important toxicologically.[19] For example, a published case report concerning a nursing infant with jaundice found that the mother was environmentally exposed to perchloroethylene.[20] Her breast milk contained

---

significant quantities of this liver toxicant, which likely caused the infant jaundice.

Should I Breastfeed?

Human milk carries a load of synthetic chemicals. Many concentrate in breast fat because they are lipophilic. About 20 percent of the mother's total body burden of lipophilic chemicals is transferred to her child during six months of breast feeding. Breast feeding infants may get a dose of dioxin, per kilogram of their body weight, within the first six months of life that exceeds health-based limits,[21] and may receive five times the allowable daily intake of PCBs for a full-grown adult. Cow's milk with levels of PCBs this high would be too contaminated for sale in the United States.[22]

The first child gets a larger dose of contaminants than later children. Older women tend to have higher levels, as do black women and cigarette smokers. Women with higher levels of DDE in their milk breast-feed for significantly less time and stop because of difficulty producing milk, indicating that DDE may interfere with lactation.[23] No health effects in children have been specifically linked to exposure during breast feeding. PCBs adversely affect neurological development in infants, but these effects appear more likely to result from prenatal exposure rather than breast feeding.

Human breast milk is extremely important for optimal child health. It contains antibodies, white blood cells, and proteins that guard against infection.[24] Studies comparing breast and bottle-fed infants show that breast milk may help protect against infections of the intestine, middle ear, respiratory tract, and urinary tract and from meningitis, and may improve the efficacy of vaccines.[25-27] Other benefits may include protection against allergies,[28] a decreased risk of inflammatory bowel disease,[29] and lower risk of juvenile diabetes.[30] Breast feeding also may confer an advantage in neurological development.[31,32] However, no studies have been done comparing the intelligence of bottle-fed infants with those who consumed highly PCB-contaminated breast milk.

The unfortunate result of this conflicting information is that women who breast-feed may be fraught with anxiety, and perhaps even guilt. Although it is clearly still advisable for most women to breast-feed, current levels of contamination in breast milk are not acceptable and point to the importance of addressing the problem of chemical exposures. The levels of chemicals in breast milk will provide a measure of our progress toward cleaning up our environment.

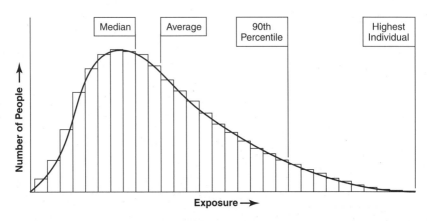

Figure 7.2   Typical population distribution of exposures

Data Relevant to Human Exposure to Reproductive Toxicants

Most people in the United States today are regularly exposed to low levels of many chemical toxicants. High-level exposures, however, generally affect a smaller number of people who are disproportionately exposed (see figure 7.2). The resulting skewed distribution of exposures means that reporting only average exposure levels may obscure the high-end exposures. There is often a specific group of individuals who bear a disproportionate share of the exposure (and of the risks) compared with the rest of the population. The most exposed population varies for different chemicals. In some cases, individual choices such as a consumer product use pattern or dietary habits can lead to high exposures; in other cases, occupation, proximity to pollutant sources, or other factors may be responsible. An understanding of the distribution of exposure for a particular chemical allows health effects studies and disease prevention efforts to be better targeted.

This section examines the same chemicals discussed in previous chapters and summarizes data relevant to human exposure ranging from chemical production through levels in human tissues. It is not possible to follow a particular batch of a specific chemical from production through residence in human tissue, so there are many missing links in the exposure chain.

Dose levels relevant to human exposure are impossible to present consistently and in some cases are difficult to access publicly. For example, chemical releases may be presented in pounds, air levels in parts per million (ppm), and blood levels as micrograms per deciliter (μg/dl). Comparisons among these units are difficult or impossible. Reporting also ranges from presenting averages, the percentage of samples with detectable levels, to the percentage in violation of a regulatory standard. Without a full reanalysis of the raw data, it is impossible to present the information in a consistent way and thus derive a full picture of potential exposures to certain chemicals. In this chapter, we note specific levels only if they are in a generally familiar metric (such as pounds), if the levels measured can be put into context for readers to compare with potential health impacts, or if they illustrate a point relating to exposure distributions in the population. Otherwise we present the data roughly as percentage of samples with detectable levels or other general estimates.

Metals

Production and Release of Metals

No production, sales, or use information is readily available for metals because metals are raw materials and are mined rather than produced synthetically. Use data are not available for the entire United States and will not be presented in this section. All of the metals discussed in this book are reportable on the TRI (see table 7.2).

Environmental Concentrations of Metals

Lead is the only metal listed as a criteria pollutant under the Clean Air Act, so information is available on the levels of lead in outdoor air that is not available for other metals. Lead levels in outdoor air average 0.01 micrograms per cubic meter ($μg/m^3$) in rural areas and 0.06 $μg/m^3$ in urban areas. These levels represent nearly an 80 percent decrease over the decade and a 97 percent decrease in the past twenty years, largely due to virtual elimination of lead from gasoline. About 60 percent of the lead in outdoor air comes from industrial processes, and the rest comes from transportation and fuel combustion. There are a few notable hot spots around the country located near metal smelters in Pennsylvania, Missouri, Nebraska, Montana, Illinois, Tennessee, Florida, and Ohio.[33]

Table 7.2   Reproductive toxic metals released into the U.S. environment in one year (in pounds)

Metal	Air	Surface water	Groundwater	Land	Off-site disposal	Total releases
Lead	1,805,420	62,419	794	14,979,456	23,220,634	40,068,723
Mercury	17,097	541	9	537	25,884	44,068
Cadmium	44,664	4,624	82	553,447	938,552	1,541,369
Arsenic	154,918	4,468	61,280	1,849,786	1,243,489	3,313,941
Manganese	8,963,136	2,018,602	17,696	50,189,866	40,570,018	101,759,318

Source: 1996 Toxics Release Inventory Data.[6]

Mercury, although not a criteria air pollutant, is nonetheless known to be significantly concentrated in the northeastern United States, particularly in New England. This is thought to be due to prevailing weather patterns that bring mercury emitted from coal-fired power plants in the Midwest into the northeastern states.[34]

Approximately 20 percent of human exposure to lead is from drinking water, primarily from lead pipes in older water systems and from lead solder in household plumbing. In addition, brass fixtures such as faucets and water meters may contain leachable lead. Private wells may be contaminated by lead and other metals from nearby toxic waste sites. The regulatory action level for lead in drinking water is 15 parts per billion (ppb). EPA estimates that in 1995 over 69 million people were served drinking water that exceeded 15 ppb, and over 26 million people were supplied with water that contained more than 30 ppb of lead. About 2 million people in the United States have been receiving water with lead levels over 130 ppb, nearly ten times the EPA-designated action level.[35]

Mercury levels in drinking water in the United States are generally low, averaging from 0.0003 to 0.025 µg/liter of water. Nevertheless, it is important to remember that mercury and other metals can accumulate in the aquatic food chain, so despite the low levels of mercury in the water itself, the levels in fish can be quite high. Arsenic can also be high in certain water systems, including parts of the northern Midwest due to naturally occurring sources and in mining regions due to contamination from mine tailings.

Human Intake of Metals

Estimated lead intake from food varies substantially by age, with averages ranging from 1.82 to 4.17 μg/day. When reported in terms of dose per kilogram of body weight, the greatest lead exposures from food are to infants under age two.[36] In women of childbearing age, the FDA estimates that 43 percent of their average total daily lead intake derives from food; dust provides about 31 percent, water 22 percent, and air 4 percent. In children, about 16 percent of the total daily lead intake is estimated to come from food, with about 75 percent from dust and smaller fractions from water, soil, and air.[37] These proportions may vary widely among individual children. The FDA's TDS has documented success in reducing foodborne lead exposures over the past thirty years. The proportion of food cans that were seam-sealed with lead solder has decreased from over 90 percent in 1979 to near zero today. In parallel with this intervention, dietary lead intake has dropped more than 80 percent over the past two decades. The dietary lead intake for children under age two has dropped from a peak of about 45 μg per day in 1979 to less than 2 μg per day today. The dietary lead intake for women of childbearing age has also dropped, from over 35 μg/day in 1982 to less than 4 μg/day today.[38]

Much less information is available about dietary levels of other metals. Cadmium intake in the United States is approximately 10 μg/day among adults and somewhat higher among infants. The major sources of cadmium in the diet are shellfish, cereal, grains, and potatoes.[39] Overall, however, the levels of cadmium in the U.S. diet are far below any that have been associated with health effects.

Dietary levels of mercury among adults are approximately 3.5 μg/day in the United States, although people who eat a lot of fish may have a dietary intake that is much greater than average. Children obtain two to three times more mercury exposure per unit of body weight than adults. Mercury levels in shark, swordfish, and tuna steaks average 1 μg/g of fish, a significant exposure for a pregnant woman or a child. EPA estimates that 1 to 3 percent of reproductive-age women in the United States are overexposed to mercury through fish consumption.[34] Freshwater fish can be significantly contaminated with mercury. Mercury advisories increased 98 percent from 1993 to 1997, and forty states issued

1,782 mercury advisories in 1997. An advisory means that high concentrations of mercury have been found in the local fish and the level detected may pose a risk to members of the public who consume the fish. All of the New England states, the District of Columbia, New York, New Jersey, Florida, Indiana, and Michigan currently have mercury advisories on all water bodies.[8] Seventy-eight percent of all fish advisories in the United States are issued because of mercury contamination.

Manganese levels are not routinely measured in food, although they are not anticipated to be excessively high. Nevertheless, now that manganese is being added to gasoline, it will be useful to know if manganese levels are increasing over time.

Meat is routinely tested for arsenic because arsenic compounds are given as a drug to prevent bacterial gastrointestinal infections in food production animals. Arsenic is detectable in up to one-third of the pigs, half of the turkeys, and two-thirds of the chickens tested by the U.S. Department of Agriculture for arsenic residues. Less than 1 percent of the animals tested had arsenic levels high enough to be in violation of a legal standard.

Absorbed Dose of Metals

From the late 1970s to the late 1980s, the average blood lead level in the U.S. population fell from 12.8 µg/dl to 2.8 µg/dl, and the prevalence of blood lead levels over the CDC level of concern of 10 µg/dl declined from 77.8 percent to 4.4 percent. This remarkable decline precisely parallels the removal of lead from gasoline.[40] Data from the early 1990s show that the decline is continuing, and the average blood lead level has fallen to 2.3 µg/dl. Still, almost 1 million children under age five continue to have blood lead levels over 10 µg/dl (see table 7.3).[41,42] This skewed distribution of blood lead levels, with a relatively small number of children bearing a disproportionate share of the exposure, is an example of the uneven distribution of exposures, and risks, across the population.

Cadmium was also measured in the urine of 960 people in the late 1980s. Cadmium levels in urine were found to increase with age in both men and women, with levels in cigarette smokers higher than in nonsmokers. Four individuals had measured urine cadmium levels above 20 µg/liter for unclear reasons and may represent a disproportionately exposed group.[43]

Table 7.3 Average blood lead values over time

Dates of testing	Average blood lead (μg/dl)	% with blood lead over 10 μg/dl
1976–1980	12.8	77.8
1988–1991	2.8	4.4
1991–1994	2.3	4.4

Source: National Health and Nutrition Examination Surveys.

Table 7.4 Levels of metals detected in breast milk

Metal	Median (μg/liter)	Range (μg/liter)
Arsenic	0.3	0.1–0.8
Cadmium	0.1	0.1–3.8
Lead	5.0	0.0–41.1
Manganese	18.0	7.0–102.0
Mercury	2.7	0.6–257.1

Breast Milk Levels

All of the metals discussed in this book have been detected in breast milk. Levels of lead, manganese, and mercury vary widely, with some very high levels detected. Results of a World Health Organization study on trace elements in breast milk are summarized in table 7.4.[44] These levels, which reflect both maternal absorbed dose of these metals and infant exposure, illustrate the large ranges of exposure across the population. Among women who eat a lot of fish, for example, levels of mercury in breast milk may exceed levels in unexposed women by 100-fold, and significant infant intake of mercury has been documented during breast feeding.[45]

Summary of Data on Metals

Lead and mercury have been extensively studied, and biological monitoring for blood lead has become inexpensive, easy, and standardized. Although lead levels have declined significantly in the United States in the past two decades, lead is still being released in large amounts from manufacturing facilities. There is a definable portion of the population still receiving major exposures to lead and to mercury. Exposure monitoring allows us to identify exposure sources and to design interventions to lower exposure. Unfortunately, data of the quality that exist on lead and mercury are not available for most other chemicals, including other metals.

Solvents

Organic solvents, most of which are also known as VOCs, are widely used in industry, hobbies, and consumer products, so human exposure is expected. These chemicals are volatile and may be present in air, can be absorbed through the skin, and are also often present in water. They can be hard to measure because they move rapidly through the environment, are short-lived in any individual environmental medium, and are quickly excreted from the body. In addition, there are numerous organic solvents, only some of which are measured in any analytic study. Tables 7.5 and 7.6 summarize information about the production and release of solvents into the environment.

Environmental Concentrations of Solvents

Twenty photochemical assessment monitoring stations (PAMS) located around the country measure ambient air concentrations of about fifty-six VOCs.[33] The monitoring is designed to measure levels of ozone precursors, because VOCs interact with sunlight to create smog and ground-level ozone. Of all the VOCs monitored, toluene, xylene, and formaldehyde were among the most abundantly detected in the environment. Overall levels of VOCs, and particularly of benzene, have been declining since the mid-1990s in the twenty monitored areas. These monitored areas of high ozone pollution are likely hot spots for outdoor air VOCs in the country. Several studies that have compared rural and urban measurements of VOCs have found enormous differences. Maximum urban air levels are as high as 2,500 parts per trillion (ppt) for trichloroethylene and 57,000 ppt for methylene chloride. Urban averages are in the hundreds to thousands of ppt. Rural areas, in contrast, have outdoor air levels in the 10 to 50 ppt range.[46] This difference is almost completely attributable to vehicle traffic.

Table 7.5 Solvent production information, 1994 (in pounds)

Solvent	Production
Benzene	14,189,646,000
Toluene	5,778,021,000
Xylene	14,123,817,000
Styrene	10,886,689,000
Trichloroethylene	*
Perchloroethylene	246,408,800
Formaldehyde	8,146,840,000
Phenol	6,836,652,000
N-methyl-pyrrolidone	78,291,000
Methylene Chloride	402,497,000
Chloroform†	478,504,000
Glycol ethers‡	*
Epichlorohydrin	*

Source: Synthetic Organic Chemicals: Production and sales (SOC Report), 1995.[2]
* Data not reported because there are fewer than three U.S. producers or data reported in confidence to ITC.
† Sales quantity listed; production quantity not available.
‡ Refers to the ethylene-based glycol ethers and acetates discussed in this book.

Table 7.6 Reproductive toxic solvents released into the U.S. environment in one year (in pounds)

Solvent	Air	Surface water	Groundwater	Land	Off-site releases	Total releases
Benzene	8,119,471	27,376	312,766	76,157	65,750	8,601,520
Toluene	125,382,228	68,697	329,275	557,160	1,022,535	127,359,895
Xylene	87,729,509	43,517	183,980	330,008	508,478	88,795,492
Styrene	41,929,161	12,864	228,317	266,690	3,251,349	45,688,381
Trichloroethylene	21,272,166	541	1,291	23,140	76,327	21,373,465
Perchloroethylene	7,861,170	1,311	13,436	30,442	22,071	7,928,430
Formaldehyde	11,419,200	320,003	9,403,275	114,406	329,509	21,586,393
Phenol	9,552,502	72,555	2,045,370	159,059	1,016,261	12,845,747
N-methyl-pyrrolidone	3,090,538	52,339	2,907,704	66,949	550,926	6,668,456
Methylene chloride	53,420,465	10,060	749,507	4,957	116,409	54,301,398
Chloroform	9,321,418	340,396	45,387	32,709	38,868	9,778,778
Glycol ethers	40,171,792	143,511	99,208	58,625	653,180	41,126,316
Epichlorohydrin	331,024	20,735	0	2,205	4,137	358,101

Source: Toxics Release Inventory, 1996.

Human Intake

Only some of the solvents discussed in this book were studied as part of the TEAM monitoring. In virtually every case, indoor air levels exceeded outdoor air levels by at least a factor of two, and personal exposures generally exceeded both outdoor and indoor air levels. At least three VOCs that are known or suspected reproductive toxicants—benzene, xylene, and 1,1,1-trichloroethane—were measured in almost 100 percent of personal air monitoring samples and were considered to be "ubiquitous." Chloroform, trichloroethylene, and styrene were detected in up to 79 to 92 percent of samples. Methylene chloride and toluene were also reported as prevalent in a subset of the TEAM study in eight homes in the Los Angeles area. These two solvents were not studied elsewhere.[47] The maximum measured exposure generally exceeded the average exposure by 10- to 1,000-fold, with average exposures to most VOCs in the 10 to 100 $\mu g/m^3$ range and maximum exposures in the range of 100 to 100,000 $\mu g/m^3$.

Some homes were found to have persistently high indoor air levels of certain VOCs for unidentified reasons. In other cases, numerous factors were identified as increasing measured concentrations of selected VOCs:

Smoking (benzene, styrene, xylene)
Occupational exposures (perchloroethylene, xylene, others)
Visiting a dry cleaner (perchloroethylene)
Driving in traffic or going to a gas station (benzene)
Painting or removing paint (benzene)
Showering, swimming, washing dishes and clothes (chloroform)

Some of these activities raised peoples' breath residues of these chemicals by up to 10- or even 100-fold compared to periods with no particular exposures.[48] Smokers had six to ten times the level of benzene and styrene in their exhaled breath compared to nonsmokers.[49] Because solvents are excreted (like alcohol) through exhaled breath, there is evidence that workers exposed to solvents can expose their families by exhaling solvent vapors in the home.[50] Laboratory experiments on low-dose exposure to toluene also indicate a gender difference in exposure, with biomonitored levels (urine and exhaled breath) of toluene consistently higher in women than in men, even when body weight, height, and air level of toluene are controlled. In addition, there is a large variation among individuals in the amount of toluene absorbed, not explainable by exercise intensity, height, weight, or any other factors.[51]

VOCs were also measured in drinking water as part of the TEAM study. Chloroform, a by-product of water chlorination, was detected in all samples measured, and trichloroethylene, perchloroethylene, and 1,1,1-trichloroethane were found in about half of the tap water tested. Benzene was found in up to 25 percent of drinking water samples. Detected levels were generally quite low.

Small spin-off studies from the main TEAM study provide information about the relative contribution of inhalation versus skin absorption for VOCs in water. Volunteers who showered wearing rubber suits had exhaled breath levels of chloroform half as great as those who showered without rubber suits, indicating that half of the exposure to chloroform in the shower comes from absorption through the skin.[52] The total dose of chloroform and other solvents received from showering in contaminated water is approximately equal to that from drinking 2 liters of the water.

Swimming: An Unexpected Dose of Chloroform

In order to avoid microbial contamination, public swimming pools are often highly chlorinated. The chemical reactions between the chlorine and unfiltered organic material in the water create chloroform and a host of other chlorinated VOCs.

Swimmers at indoor pools have been shown to absorb chloroform. The factors influencing the absorbed dose include the number of swimmers in the water (more swimmers mean more splashing and higher air levels of chloroform), the intensity of the workout (faster breathing rate means more intake of chloroform), the time spent swimming, and the concentrations of chloroform in water and air.[53]

When some swimmers were given scuba tanks supplying pure air, biological monitoring showed that skin absorption accounts for approximately 25 percent of the chloroform dose, and inhalation accounts for 75 percent.[54] The fact that inhalation is a more important route of exposure from swimming rather than showering is likely due to the faster breathing rate during exercise. Swimming is excellent exercise, but pool areas should be well ventilated, and techniques that keep chloroform levels low (good filtration, avoidance of overchlorination) should be used.

Absorbed Dose

A subsample of six hundred volunteers was recruited from the third NHANES for sampling of a panel of thirty-two VOCs in blood. Eight of the VOCs sampled were suspected reproductive or developmental toxicants; all were detected in varying proportions of the blood samples.[55] Toluene and xylene were detected in 100 percent of the people sampled, and benzene, styrene, perchloroethylene, and 1,1,1-trichloroethane were detected in over 75 percent of the samples. Chloroform was detectable in over half; trichloroethylene and methylene chloride were much less commonly detected (around 10 percent).

One observation from the NHANES III study was that exposures to VOCs were not normally distributed across the population. For example, half of the population sampled had concentrations of less than 0.063 part per billion (ppb) of perchloroethylene in their blood, yet the top 5 percent of the population had exposures more than tenfold higher, at 0.62 ppb or more.[56] Further analysis is necessary to reveal the specific exposures that lead to the higher levels in the top 5 percent, so interventions could be designed to reduce the exposures. In addition, information about the distribution of exposures throughout the population can be used as a reference level for comparison with particular subpopulations or even particular individuals suspected of being heavily exposed.[57]

NHATS measured seventeen VOCs in human fat, of which seven are discussed in this book. There was no association between age and levels of VOCs in fat, probably because these chemicals do not accumulate significantly over time. The VOCs of interest were not concentrated in any particular geographic region, implying that exposures to these chemicals are distributed without regard for geography.[58]

Numerous organic solvents have been detected in human breast milk, including benzene, chloroform, methylene chloride, styrene, perchloroethylene, toluene, trichloroethylene, 1,1,1-trichloroethane, and xylene.[19] Breast milk levels are frequently higher than blood levels of these compounds because solvents are lipophilic and because breast tissue does not clear solvents as quickly as they are cleared from the blood.[18] Perchloroethylene, in particular, is known to concentrate in breast milk.[59]

Solvent Summary

Unlike metals, solvents are volatile and move rapidly through various media. They evaporate quickly from water and break down in the environment and in the human body. As a result, they are more difficult to measure than more persistent substances. Despite these difficulties, some excellent work has been done quantifying human exposures. Subpopulations of people are exposed to much higher levels—ten or one hundred times greater than the average. These data reveal that there is plenty of room for exposure reduction to VOCs.

Pesticides

Numerous organochlorine pesticides are banned in the United States, yet their residues are persistent in the environment and are still widely found in human tissues. Other pesticides are still heavily used and found commonly in water, food, indoor air, and house dust. Human monitoring for nonorganochlorine pesticides has been limited, but indicates high exposure potential.

Production and Release of Pesticides

The TRI covers only pesticide emissions from manufacturing facilities, not from agricultural uses (see table 7.7). Agricultural uses are estimated in pesticide use report data (table 7.8). The TRI numbers look quite low compared to the estimates of commercial agricultural pesticide use, but they add to the overall information about pesticide emissions throughout the life cycle of the products and can be relevant to communities located near a facility that releases these substances.

The EPA estimates that about 20 percent of the total U.S. use of pesticides occurs in nonagricultural sectors, particularly in the home. 2,4-D is the most common pesticide used by consumers in the home and garden: 7 to 9 million pounds per year were applied annually in 1994–1995. Dicamba is used by home owners at a rate of 3 to 5 million pounds per year, diazinon and chlorpyrifos at a rate of 2 to 4 million pounds per year, and carbaryl at a rate of 1 to 3 million pounds annually.

Environmental Concentrations

Groundwater and surface water has been tested in many areas of the United States as part of the U.S. Geological Survey's water testing.[60] In the period from 1992 to 1996, seventy-six out of the eighty-three

---

Table 7.7 Pesticides reported as released into the environment, manufacturing only (in pounds)

Pesticide	Air	Surface water	Groundwater	Land	Off-site releases	Total
Organophosphates						
Diazinon	15,587	21	0	0	1,000	16,608
Chlorpyrifosmethyl	2,010	0	0	3,653	0	5,663
Acephate	1,505	0	0	0	1,400	2,905
Tetrachlorvinphos	365	5	0	0	2,030	2,400
Organochlorines						
Lindane	510	5	0	250	276	1,041
Fungicides						
Thiram	3,634	40	0	2,000	94,436	100,110
Herbicides						
Cyanazine	1,915	421	0	0	2,695	5,031
Simazine	4,591	93	0	0	54,457	59,141
Atrazine	27,011	1,326	1	614,353	188,963	831,654
2,4-D	5,989	832	0	255	6,017	13,093
Paraquat	1,000	0	0	0	5	1,005
Dicamba	1,059	132	59,200	0	0	60,391
Bromoxynil	15	0	0	0	1,388	1,403
Fumigants						
Methyl bromide	2,299,843	7	303	6	0	2,300,159
Metam sodium	3,449	4	0	2	15,937	19,392
Ethylene oxide	789,439	4,474	22,200	551	1,048	817,712

Source: Toxics Release Inventory, 1996.
Note: The following pesticides are listed on TRI but reported releases totaled less than 1,000 pounds in 1996: benomyl, dicofol, dimethoate, fenvalerate, maneb, methoxychlor, nabam, parathion, propargite, resmethrin, and zineb.

pesticides and pesticide degradates tested for in water were detected at least once. More than 95 percent of water samples collected from streams contained at least one pesticide, compared with about half the samples collected from groundwater wells. The most commonly detected pesticides were the herbicides atrazine, simazine, and metolachlor. Atrazine was detected in over 78 percent of all water sources tested. Insecticides, particularly diazinon and chlorpyrifos, were more often found in streams in urban areas than in agricultural areas.

Detections of pesticides were less frequent in groundwater than in surface water. Only the triazine herbicides were detected with a significant frequency. Atrazine was detected in slightly over 31 percent of groundwater samples and simazine in over 12 percent.

Several states have undertaken drinking water testing programs. Again, the triazine herbicides are by far the most common drinking water

Table 7.8 U.S. pesticide use estimates, agriculture only (in thousands of pounds)

Pesticide	NCFAP estimate[4]	EPA estimate[5]
Organophosphates		
Parathion	5,962	4,000–7,000
Diazinon	1,266	★
Chlorpyrifos	14,765	9,000–13,000
Acephate	3,390	★
Dimethoate	2,619	★
Propargite	3,628	★
Organochlorines		
Lindane	61	★
Endosulfan	1,797	★
Methoxychlor	89	★
Dicofol	1,392	★
Pyrethroids		
Cypermethrin	228	★
Fenvalerate	332	★
Resmethrin	★	★
Fungicides		
Benomyl	1,198	★
Maneb	3,525	★
Mancozeb	8,062	6,000–9,000
Thiram	199	★
Vinclozolin	135	★
Iprodione	874	★
Herbicides		
Cyanazine	32,190	24,000–29,000
Simazine	3,978	★
Atrazine	72,315	68,000–73,000
Dicamba	9,064	6,000–10,000
2,4-D	41,938	31,000–36,000
Fumigants		
Ethylene dibromide	★	★
Metam sodium	29,095	49,000–54,000
Ethylene oxide	★	★
Methyl bromide	44,197	39,000–46,000

Source: Pesticide Use in U.S. Crop Production: National Summary Report, 1995.
★ Data unavailable.

Table 7.9   Pesticide detections by food group, 1997

Product category	Percentage of samples with detectable residue	Violating tolerance (%)
Domestic		
Grains	40.6	0
Milk, dairy, eggs	3.0	0
Fish, shellfish	32.0	0
Fruit	54.7	1.2
Vegetables	28.5	2.4
Imported		
Grains	13.0	0.9
Milk, dairy, eggs	10.6	0
Fish, shellfish	6.3	0
Fruit	38.2	1.2
Vegetables	34.9	2.1

Source: FDA Pesticide Residue Monitoring Program, 1997.

contaminants. The state of Ohio tested 144 water systems and detected at least one triazine herbicide in each of these drinking water systems. Some water systems contained up to five different herbicides. In Ohio, atrazine was detected in 94 percent of drinking water systems tested, simazine in 70 percent, and cyanazine in 76 percent tested.[61] Similar results have been found in other midwestern states, and even by the pesticide industry in its own monitoring studies.[62]

There are drinking water standards for numerous pesticides, and individual water systems do test regularly for these contaminants. Testing results are available to individuals and organizations concerned about their own drinking water supply.

Human Intake

FDA and USDA pesticide monitoring shows frequent detection of pesticides on foods, but at low levels that rarely violate the regulatory tolerances. USDA found that overall, about 72 percent of fruits and vegetables tested had detectable pesticide residues. The majority of samples contained residues of more than one pesticide, up to a record of fourteen residues on one sample of peaches.[7] Ninety-two different pesticides were found in the over ten thousand samples of food analyzed by the FDA in 1996 (see table 7.9). The FDA detected residues on both domestic and imported produce, with violations slightly more frequent on imports.

Table 7.10 Most commonly contaminated foods

Commodity	Samples with detectable residue (%)	Number of different pesticides detected
Apples	98	39
Peaches	96	27
Wheat	91	16
Spinach	88	39
Oranges	84	19
Grapes	80	29
Carrots	78	17
Sweet potatoes	67	20
Apple juice	66	14
Tomatoes	64	23

Source: USDA, 1996.
Note: Washed, prepared food tested.

Although it is somewhat reassuring that most residues are below the regulatory tolerance, this shows only that most of the crops are treated within the parameters of expected use, not that the actual levels on the crops are safe.

The potentially reproductive toxic pesticides most commonly detected in foods in the USDA Pesticide Data Program (1996) and in the FDA Total Diet Studies (1991–1996) were chlorpyrifos, chlorpyrifos-methyl, DDT/DDE, diazinon, endosulfan, iprodione, and malathion. DDT/DDE was detected in 25 percent of the foods tested by FDA even though it has been banned in the United States for decades. The others were detected in 10 to 20 percent of foods tested.[63] Baby foods also contain pesticide residues, with detections in over 10 percent of foods tested for malathion, chlorpyrifos, dimethoate, carbaryl, permethrin, endosulfan, and iprodione.[64] Dimethoate was the most common detection in baby foods, found in 32 percent of samples tested. The most commonly contaminated fruits and vegetables are listed in table 7.10. Detected residues do not mean that the levels exceeded regulatory tolerances, and in most instances they do not. However, this table does illustrate the prevalence of pesticide residues on a wide variety of domestic and imported foods.

Fish advisories occasionally occur for organochlorine pesticide residues. Despite the fact that most of the organochlorine pesticides have been banned, the total number of fish advisories for these pesticides has increased over the past few years. This increase may be due to better monitoring and increasing awareness rather than actual increases in contamination. Nearly one hundred fish advisories are in place around the country because of contamination with pesticides. The pesticides most commonly responsible for advisories are DDT and chlordane.[8]

Meat is also tested for pesticides, including organochlorines and organophosphates. The panel of chemicals evaluated includes chlorpyrifos, DDT, dieldrin, endosulfan, endrin, heptachlor, lindane, and methoxychlor. PCBs are included in the same screen. Results are not available by chemical but instead are reported for the entire class of chlorinated organics and organophosphates. Overall, 3 to 12 percent of meats tested by USDA contain detectable pesticide residues. The lowest prevalence of detections is in turkey, the highest in beef.

Multipathway Exposures

NOPES found that indoor air levels of pesticides were generally 10- to 100-fold higher than outdoor air levels. Five pesticides were found in the majority of homes sampled—chlorpyrifos, chlordane, propoxur, heptachlor, and orthophenylphenol—and most homes contained detectable residues of three to nine pesticides, with one home containing residues of twenty different pesticides. Pesticides that had been banned years ago were nevertheless detected in many homes. Several pesticides were detected frequently in house dust samples.[65] For infants, skin absorption from contact with dust may account for nearly 70 percent of their total exposure to pesticides.[66] Concentrations of pesticides in dust were significantly higher than in indoor air (see table 7.11).

There was little day-to-day variation in pesticide levels indoors. There was, however, significant variation by geographic region, with detection frequency and level generally greater in Florida than Massachusetts; pesticide levels were generally about tenfold higher in Florida, probably because the indoor pest burden is greater in warm, moist areas, leading to heavier pesticide use. The average concentration of pesticides was also more than tenfold higher in dust than in air.

The spraying events that were measured made it clear that skin absorption may be a major route of exposure for people who use

Table 7.11 Pesticides in indoor air and house dust

Pesticide	Detection frequency in air* (%)	Detection frequency in dust† (%)
Organophosphates, carbamates		
Malathion	4–21	NR
Diazinon	17–83	82
Chlorpyrifos	30–83	100
Carbaryl	NR	45
Organochlorines		
DDT‡	9–12	82
Methoxychlor	0–1	NR
Lindane	10–32	NR
Heptachlor‡	0–3	91
Dicofol	0–12	NR
Dieldrin‡	12–22	91
Triazines		
Atrazine	NR	18

* Ranges represent averages for Massachusetts and for Florida. Lower numbers represent Massachusetts, with the exception of DDT, DDE, and dicofol, where detection frequency and levels measured in Massachusetts were higher.
† Only eleven houses were sampled for residues in dust.
‡ Banned.
NR = not reported.

pesticides. In this study population, drinking water was a minor contributor to total exposure. Some pesticides were found most frequently in air, and others were mainly ingested in food. Pesticides to which people were mostly exposed through air included aldrin, chlordane, and heptachlor. Pesticides mainly consumed in food included DDT, dicofol, dieldrin, malathion, methoxychlor, and carbaryl. In the case of chlorpyrifos, diazinon, and lindane, the exposures came from both sources nearly equally. Dust was a major source of exposure to toddlers.

Absorbed Dose
In two NHANES, subsamples of volunteers had measurements of blood pesticide residues as part of an effort to estimate exposure of the general population to pesticides. In NHANES II (1976–1980), a sample of nearly six thousand people ages twelve to seventy-four from throughout the United States had serum sampled for approximately thirty-six

Table 7.12   Pesticide residues in human blood

Pesticide	Detection frequency (%)	Range (ppb)
DDE	99.5	0–378.6
DDT	35.7	0–57.8
Dieldrin	10.6	0–16.1
Nonachlor	7.1	0–17.0
Heptachlor	4.3	0–22.4
Lindane	0.2	0–6.2

Source: NHANES II survey, 1976–1980.

Table 7.13   Pesticides in human urine

Pesticide	Detection frequency, 1976–1980 (%)	Detection frequency, 1988–1994 (%)
Chlorpyrifos	7	82
Parathion	3	35
Carbaryl	2	16
Lindane	★	10–20
2,4-D	1	12
Pentachlorophenol†	79	64

Sources: NHANES II, 1976–1980; NHANES III, 1988–1994.
★ Data unavailable.
† In the later study, all 197 children tested in Arkansas had measurable PCP levels. The group had median, 95th percentile, and maximum concentrations of 14, 110, and 240 ppb, respectively.

pesticides and pesticide metabolites (table 7.12).[67] Some urine sampling was also done for a smaller panel of pesticide products (table 7.13).[68] In NHANES III (1988–1994), a subsample of over nine hundred volunteers from all regions of the country, aged twenty to fifty-nine, underwent sampling of pesticides in urine.[69,70] Because of the choice to measure in urine, only a dozen pesticides that are metabolized into water-soluble products and eliminated in urine could be measured.

Because NHANES III did not measure pesticide residues in blood, there is limited means for comparison over time. In the case of pesticide metabolites in urine, there are worrisome trends. With the exception of

Table 7.14 Human fat levels of certain organochlorine pesticides over time (in nanograms/g)

Pesticide	1982	1984	1986
DDT	189	123	177
DDE	1840	1150	2340
Nonachlor	109	105	130
Heptachlor	59.4	68.3	57.6

pentachlorophenol (PCP), all the other pesticides measured have been increasingly detected in human urine. Chlorpyrifos detections, in particular, have increased more than tenfold, from 7 percent around 1980 to 82 percent around 1990. Some of the increase in detection frequency is attributed to more sensitive laboratory methodology. Trend information on human exposures is critical to determining the risks to the population and in evaluating the efficacy of regulation.

Over the four-year period reported in the most recent publication from NHATS, the trends in organochlorine pesticide residues are quite flat despite the fact that these pesticides were all previously banned (table 7.14).[71] This persistence in human fat is not surprising, because exposure is still continuing through the food chain. Newer data are not available from the NHATS survey.

Fat-soluble pesticides accumulate in breast fat and end up in breast milk. Detectable levels of the DDT metabolite DDE are found in the breast milk of virtually all women in the United States today. During the course of lactation, concentrations of these compounds in the breast milk tend to decrease as they are transferred to the nursing infant (table 7.15).[72]

Numerous other organochlorine pesticides have been detected in breast milk in over 60 percent of women sampled, including dieldrin, heptachlor, and chlordane. In the case of all of these pesticides, the typical intake for a nursing infant would be likely to exceed the allowable daily intake set by the World Health Organization.[22]

Pesticide Summary

There are major gaps in the data available on pesticides. For example, there is considerable information about pesticide residues on food, but little or no information about pesticide concentrations in parks, schools,

Table 7.15　DDE levels in breast milk during lactation[72]

Duration of lactation	Detection frequency (%)	Median concentration (ppm)	Maximum concentration (ppm)
Birth	>99	2.4	25.4
Three months	99	2.1	23.4
Six months	99	1.9	22.5
One year	100	1.5	12.7
Eighteen months	100	1.3	11.9

offices, and other indoor structures. There is also a lack of information about levels of exposure through skin absorption and through home use, likely significant exposure sources for numerous people.

As expected from their pattern of widespread use, pesticides are found throughout the environment. Highly water-soluble pesticides such as the triazine herbicides are common contaminants of drinking water. Numerous pesticides are found on foods, albeit at low concentrations. Although farmworkers and farmers are clearly disproportionately exposed populations, even people who are not occupationally exposed to pesticides can have significant exposures. For children, house dust may be a major source of pesticide exposure. There is good evidence of seasonal and geographic variation in pesticide levels due to variable pest pressures and to crop cycles. Persistent, fat-soluble pesticides that were banned many years ago are still found in numerous places including house dust, human blood, fat, and breast milk.

Endocrine Disruptors

For the purposes of this section, we are including only a small subset of known or suspected endocrine disruptors. Many pesticides are endocrine disruptors but exposure information on these was previously addressed. The only chemicals discussed in this section are the dioxins, dibenzofurans, polychlorinated biphenyls (PCBs), bisphenol A, octylphenol, nonylphenol, diethylhexylphthalate (DEHP), benzylbutylphthalate (BBP), and dibutylphthalate (DBP). There is very little systematic information relevant to human exposure to these latter six chemicals. In the case of the dioxins, dibenzofurans, and PCBs, there is extensive information,

Table 7.16   U.S. production data on endocrine-disrupting chemicals, 1995

Chemical	Production (pounds per year)
Ether phenols*	688,303,000
Nonylphenol ethoxylate	620,712,000
Dibutyl phthalate	17,054,000

Source: Synthetic Organic Chemicals: production and sales, 1996.
* Includes bisphenol A ethoxylates and related compounds.

Table 7.17   Some endocrine disruptors released into the U.S. environment in one year (in pounds)

Chemical	Air	Surface water	Groundwater	Land	Off-site releases	Total
DBP	85,126	452	180,000	313	25,217	291,108
DEHP	464,429	274	0	70,311	1,762,843	2,297,857
PCBs*	255	0	0	9,205	51,086	60,546

Source: Toxics Release Inventory, 1996.
* PCBs were banned in the United States in 1977.

which indicates significant and widespread human exposures even though PCBs were banned two decades ago; dioxins and their close relatives, dibenzofurans, are unintentional by-products of industrial processes.

Production and Release of Endocrine Disruptors

Very little production information is available on endocrine-disrupting chemicals (table 7.16). Some, such as the dioxins and PCBs, are not intentionally produced; others are manufactured by only one or two companies, and production data are considered trade secrets.

TRI data (table 7.17) also illustrate important gaps in reporting. Bisphenol A, octylphenol, nonylphenol, and benzylbutylphthalate are not required to be reported on the TRI. Dioxins were recently added to the TRI, but their release will not be reported for several more years. The general estimates of dioxin emissions in the United States show that incinerators appear to be a major current dioxin source.

Environmental Concentrations of Endocrine Disruptors

Other than some pesticides and PCBs, endocrine disruptors are not routinely measured in any environmental media, including air, water, or food. Dioxins have been measured sporadically in air. Estimates of total dioxin/dibenzofuran concentration in air range from 0.4 to 5.6 pico-grams per cubic meter (pg/m³). In soil, dioxin/dibenzofuran concentrations are much higher, ranging from 100 to 12,000 pg/cubic centimeter, with significant regional variation.[73]

PCBs are measured in fish and meat. In the case of freshwater fish, consumption advisories for PCBs increased 84 percent from 1993 to 1997, in part due to improved monitoring. In all, thirty-five states have PCB advisories on at least one water body in the state, and the District of Columbia and Rhode Island have issued statewide advisories for PCBs. The Great Lakes states (Minnesota, Michigan, Wisconsin, New York and Indiana) have issued the majority of the fish advisories, although numerous other states are also affected.[8]

Human Intake

Dioxins and dibenzofurans are found in dairy, meat, and fish sold for human consumption. One study in New York State detected total dioxin/dibenzofuran residues on meat in the range of 0.8 to 61.8 ppt. Levels in fish were lower, at 0.4 to 3.4 ppt, and levels in dairy products were intermediate, at 0.9 to 19.0 ppt. From these results, the authors estimated a daily adult dose of dioxin toxic equivalents (TEQs) of 0.3 to 3.0 pg/kg.[21]

The FDA's TDS, which estimates total dietary exposure to PCBs in food, found that the average intake has decreased significantly since 1971, from 6.9 µg/day to 0.05 µg/day.[74] This significant decline in food residues has exceptions only for those who consume a lot of freshwater sport fish and for breast-fed infants.

Absorbed Dose

There is no information available about levels of phthalates, alkylphenols, or bisphenol A in the general population. The only available information on the chemicals covered in this section is on PCBs, dioxins, and dibenzofurans, which are persistent in human fat, blood, and breast milk. Levels found in human tissues are close to those that are known to cause health effects in animals. It appears that the concentrations of these chemicals in human tissues have peaked and are beginning to decline; however, this may be because they are spreading more evenly throughout the environment and the population.

The NHATS tested human fat samples for dioxins, dibenzofurans, and PCBs. Overall dioxin TEQs are slightly higher in people who live in the northeastern United States and significantly increase with increasing age. Nonwhites also tend toward having higher levels than whites, and women may have slightly higher average levels than men, although neither of these differences is statistically significant. Overall average TEQ fat burden in the United States from the 1987 survey was 27.9 pg/g. When compared with the 1982 NHATS, there was no significant change in dioxin levels in the population over that five-year period.[75] No more recent data are available from the NHATS survey. In the case of PCBs, NHATS showed an increasing trend in total PCB concentrations in fatty tissue in the period from 1982 to 1986. During this time, PCBs increased from an average of 407 nanograms per gram (ng/g) in 1982 to 508 ng/g in 1984 and 672 ng/g in 1986.[71] Geographically, PCB exposures are highest in the southeastern and south-central states, with high exposures also in the north-central and Great Lakes states.[58]

Older NHATS data tell an interesting story about PCBs. In 1972, about 85 percent of the population studied had detectable concentrations of PCBs in their fat, and nearly half of these had high levels, exceeding 1 ppm. By 1976, the year that PCBs were banned in the United States, 100 percent of the population sampled had detectable PCBs in their fat, but the percentage with levels over 1 ppm had declined to 30 percent. Since 1976, the percentage detectable has remained at 100 percent and the average PCB level has increased somewhat, but the number of people with residues at the high end of the distribution, over 1 ppm, has declined to less than 5 percent.[76] It appears that PCBs are distributing themselves more evenly throughout the population.

PCBs, dioxins, and polybrominated biphenyls have been reported in breast milk since 1966.[77,78] As with DDE, PCB levels in milk have been shown to decrease during the course of lactation as exposure is transferred to the fetus. When adjustment is made for duration of lactation, concentrations tend to increase with increasing maternal age.[72] In the late 1970s, PCBs were detectable in 30 percent of the human breast milk tested, with an average level of 87 ppb and a range from 50 to 409 ppb. This is in contrast with the FDA action level for cow's milk of 62.5 ppb.[22] More recent broad-scale testing data are not available. The average level of dioxins and dibenzofurans in breast milk in the United States is 16.7 pg/g, from World Health Organization studies in the late 1980s.[44]

Endocrine Disruptors Summary

Many of the suspected endocrine-disrupting chemicals that are still in use have minimal or no systematic data relevant to estimating human exposures. Numerous organochlorine pesticides and the PCBs, which have been banned, have extensive information on human exposures but no current use or release. The dioxins, which are not intentionally produced, have also been extensively monitored in human tissue. The data indicate that despite efforts to decrease human exposure, these chemicals remain widespread at worrisome levels in the population. Exposure to the nursing infant is particularly high due to the fact that these chemicals concentrate in breast milk.

The Relevance of Exposure Information

If we study only the health effects of chemicals, we are likely to go astray in setting priorities for public health. Very toxic chemicals may not be a significant problem if human exposures are virtually nonexistent, whereas chemicals of moderate toxicity may be very important if there is widespread exposure. In addition to the presence or absence of human exposures, it is important to assess the route of the exposure and the time when the exposure is likely to occur.

The exposure route may be important for chemicals that are detoxified in the liver. When a compound is ingested orally, it is transported directly to the liver and metabolized. When the same compound is inhaled or absorbed through the skin, it bypasses the liver and may deliver a greater dose directly to the fetus. In addition, if exposure is likely to occur during pregnancy or early infancy, the magnitude of the dose may not be as important as the timing of the exposure.

Unfortunately there is no centralized collection of exposure information, and pieces of information relevant to human exposures are collected in various places. Some of this information is readily available to the public; some is not or is difficult to access and use. Because it is difficult to measure exposure at the point of entry into the human body, we customarily use other pieces of information as surrogates of exposure: information on chemical use, emissions, concentrations in environmental media, and biomonitoring in human tissues. But chemicals behave in different ways in the environment, so some surrogates of exposure may be more useful than others. For example, some chemicals are volatile and short-lived, while others are persistent and bioaccumulate. Both may have important implications for human health, but the techniques for estimating exposure must be very different.

There are widespread exposures to many of the substances discussed in this book. Many of the substances are produced, sold, used, and released in large quantities into the environment. They reach people in food, water, indoor and outdoor air, and dust. Many of the chemicals have been detected in human tissues. The concentrations, with some exceptions, are generally low, but the exposures are to multiple chemicals, through multiple routes, on multiple occasions. The health implications of these widespread exposures are generally unclear. However, this information does give us direction about ways to limit exposures and means of comparison to see if exposure-reduction efforts are succeeding. For many of these chemicals, there are identifiable subpopulations with excessive exposures.

There are major data gaps in human exposure assessment. Toxic chemicals that are not listed on the TRI should be made subject to mandatory reporting. Drinking water monitoring should be more systematically available and should include important reproductive toxicants that currently are not monitored. Indoor exposures via air and dust are important and should be more systematically evaluated, especially because they likely contribute disproportionately to the exposures of infants and children. Nonagricultural pesticide use, particularly in schools, parks, and other public places, should be reported. Finally, ongoing human monitoring of an expanded panel of chemical toxicants is an important part of a public health infrastructure and should be given priority and support.

A Guide to Investigating Environmental Threats to Reproduction

III

The Regulation of Hazardous Chemicals and Your Right to Know

8

Regulatory laws to protect public health and the environment have evolved over several decades, shaped by scientific knowledge, economic and political power, and varying worldviews. The historical record documents failures to protect workers, the general public, and the environment from exposures to harmful and unstudied synthetic chemicals. These practices often continue today because of restrictive language in laws, inadequate implementation, narrow court decisions, corporate lobbying for loopholes, or outright deceit.

Our society uses fragmented analyses to derive public policy, separately addressing water, air, soil, pesticides, other industrial chemicals, cosmetics, and food additives. As we reactively divert public health resources from threat to threat, we often fail to consider the health of entire systems and neglect to develop a more integrated public health infrastructure. For example, early Superfund legislation was in direct response to unconscionable environmental contamination at Love Canal in New York. Thousands of tragic deaths and injuries in Bhopal, India, resulting from the negligent release of a poisonous industrial gas over a sleeping city, provided impetus for right-to-know legislation in the United States. Although these are extremely valuable pieces of legislation, we have yet to focus comprehensively on sustainable development, life cycle analyses, and pollution prevention. Instead, a bewildering mix of legislation and regulatory agencies has evolved. The following overview summarizes the distribution of regulatory responsibility among federal agencies and strengths and weaknesses of some of the major regulatory statutes.

Regulatory Responsibility

Environmental Protection Agency

The EPA has the authority and responsibility to regulate toxic air pollutants under the Clean Air Act (CAA), toxic water contaminants under the Clean Water Act (CWA) and the Safe Drinking Water Act (SDWA), pesticides under the Federal Insecticide, Fungicide, and Rodenticide Act (FIFRA), toxic chemicals generally under the Toxic Substances Control Act (TSCA), toxic wastes deposited in or on the ground under the Resource Conservation and Recovery Act (RCRA), and hazardous waste sites under the Comprehensive Environmental Response, Compensation, and Liability Act (CERCLA). The Pollution Prevention Act of 1990 requires the EPA to prioritize source reduction as a tool to prevent all types of pollution.

The factors that the EPA must consider in setting standards under each of these acts vary considerably. For example, the CWA and CAA impose technology-based standards dictated by the best available technology, while the SDWA imposes health-based standards to protect human health. FIFRA and TSCA require cost-benefit analyses of the impact of proposed standards in addition to consideration of health and environmental effects. TSCA specifies that the administrator will consider social as well as economic costs when making regulatory decisions.

Sometimes two or more of the laws overlap, and EPA may have discretion over which legislative authority it chooses to limit toxic discharges. In other situations, EPA may share the authority to regulate with one or more agencies. For example, both the Occupational Safety and Health Administration (OSHA) and EPA have a responsibility to regulate worker exposure to harmful chemicals.

Testing requirements also vary considerably among the applicable laws. Manufacturers of pesticides (regulated by EPA under FIFRA) and pharmaceuticals (regulated by FDA) are required to conduct testing that addresses the effectiveness and safety of their products prior to their review for registration purposes. By placing the burden squarely on the manufacturer to address the safety of the product prior to registration and marketing, the federal government has adopted a precautionary approach to the regulation of these consumer products.

In contrast, the majority of industrial chemicals, regulated under TSCA, undergo much less scrutiny. In the United States there is no minimal absolute requirement for premanufacturing or premarketing toxicity testing. Instead, the burden is placed on the EPA administrator to find a reason to believe that the chemical is unsafe or that a large number of people will be exposed before requiring further testing or proposing action to control exposures. This is distinctly different from the European Union, where premarketing notification of substances produced in excess of 1 ton per year must include testing results for acute and subacute toxicity, mutagenicity, and short-term carcinogenicity screening.[1]

Toxic Substances Control Act, 1976

The purposes of the TSCA are to encourage or, under specific circumstances, require industry to develop adequate data on the health and environmental effects of chemicals, and to regulate chemicals that pose unreasonable risk of injury to health and the environment and to take actions against imminent hazards. It does not seek to impede technological innovation unnecessarily. It thus gives the EPA authority to regulate industrial chemicals broadly, from outright bans to labeling requirements. There are now about seventy-five thousand chemicals in the TSCA inventory, and more than one thousand new chemicals are proposed for manufacture annually. Unlike other environmental laws of the 1970s, which focused on controlling emissions and cleaning up air, land, and water resources, TSCA was designed to allow pollution to be addressed at its source. The act does not cover chemicals regulated under other acts such as pesticides, food additives, drugs, cosmetics, alcohol, tobacco, and nuclear material. It does cover all other industrial chemicals, including biological and genetically engineered substances, and mixtures of chemicals in some instances. When the act was passed, the sixty-two thousand chemicals in commercial use were automatically placed on the TSCA inventory.

Provisions

The act requires the administrator to determine if "there is a reasonable basis to conclude that a chemical substance or mixture presents or will present an unreasonable risk of injury to health or the environment." If data or lack of data indicate that the substance may "present an unreasonable risk," "enter the environment in substantial quantities," or present a likelihood of "significant or substantial human exposure," then the administrator may require the manufacturer to test the substance. There are also provisions for taking action when the administrator determines that the chemical "presents an imminent and unreasonable risk of serious or widespread injury to health or the environment."

What is meant by "unreasonable risk" is never defined in the act, but the following factors must be considered: (1) the effects of a chemical on health and the environment, (2) the magnitude of human and environmental exposures, (3) the benefits of the chemical and the availability of substitutes for similar uses, and (4) the "reasonably ascertainable economic consequences of the rule after consideration of the effect on the national economy, small business, technological innovation, the environment, and public health." Clearly the law intends risk to the economic health of industry, as well as to the environment and humans, to be part of the "unreasonable risk" analysis.

New chemicals must be proposed to the agency by a premanufacturing notification (PMN), after which the agency has a ninety-day period to review the submission. This may be extended another ninety days for good cause. If the administrator takes no action on the PMN, the new chemical becomes part of the inventory and is treated as an old or existing chemical. The administrator must explain reasons for no action in the *Federal Register*. The PMN requires identification of the chemical structure (if known); anticipated by-products of manufacture, use, and disposal; the estimated amount to be produced or imported; and anticipated worker exposure and environmental releases. There is no requirement for toxicity data.

The act established the Interagency Testing Committee (ITC) to develop a priority list for testing chemicals already on the TSCA inventory. Prioritization is to be based on the quantity manufactured, the likelihood and degree of exposure or environmental releases, and the existence of data on health and environmental effects.

The act requires the administrator to respond to information indicating that "there may be a reasonable basis to conclude that a chemical substance or mixture presents or will present a significant risk of serious or widespread harm to human beings from cancer, gene mutations, or birth defects." The administrator is required, "within a 180 day period, . . . to initiate appropriate action . . . to prevent or reduce to a sufficient extent such risk or to publish in the Federal Register a finding that such risk is not unreasonable."

Chemical manufacturers that possess or obtain information supporting the conclusion that a chemical or mixture presents a substantial risk to health or the environment must immediately inform the EPA adminstrator of such information. The manufacturer does not have to conduct new studies, however.

Among the other provisions are these:

Claims of confidential business information   TSCA prohibits the administrator from publicly disclosing submitted information identified by the manufacturer as confidential except under specified circumstances.

Public redress   TSCA provides opportunities for citizens to bring lawsuits to enforce the provisions of TSCA and to petition the EPA to take action under TSCA.

Relationship to other laws   If EPA determines that an unreasonable risk may be prevented or reduced by action under another federal law not administered by EPA, the agency must refer that risk information to the other agency administering the other law.

Major Weaknesses of TSCA

TSCA's legal standards are so high that they have discouraged the EPA from using its authority to require premanufacturing toxicity testing of nearly all chemicals. According to the law, the EPA must have a reason to believe that a chemical poses an unreasonable risk of injury to human health or the environment or find substantial or significant exposure before requiring testing. In the complete absence of toxicity or exposure information, the burden is on the agency to identify a reason for generating the data. As a result, many of the chemicals to which workers and the public are regularly exposed have had no formal toxicity evaluation. Some are chemicals that have been in use for some time; others are newly proposed for commercial use and fail to trigger testing requirements for reasons that are often political, statutory, or bureaucratic rather than biological.[2]

A second weakness is that industry has no obligation to test new chemicals for toxicity before notifying the EPA of its intent to manufacture. Fewer than half of all premanufacturing notifications contain any toxicity data, and of the ones that do, most have information only about acute effects, not chronic effects.[3]

With more than one thousand chemicals proposed for manufacture annually, comprehensive agency evaluation of each is impossible. In the absence of toxicity data, virtually all of these easily clear the ninety-day review period and automatically become part of the TSCA inventory, acceptable for manufacture and use.

For existing chemicals, the legal requirement for promulgating a test rule requiring toxicity testing has been extraordinarily time-consuming and costly. Developing a test rule for a single chemical may take two years or more and cost up to $234,000.[3] As of 1997, EPA had issued test rules covering only 121 chemicals from the entire TSCA inventory.

On the basis of its experience with TSCA, the EPA has concluded that confidentiality claims have been extensively and inappropriately overused. At times toxicity data generated by industry have been withheld, based on claims of business necessity.[3]

History and Analysis
The history of implementation of TSCA reflects a fluctuating political landscape and is marked by outright disregard of TSCA requirements, legal challenges, and congressional oversight hearings in 1981 and 1988.[4] In general, TSCA implementation has amounted to a series of negotiations with industry. By 1997, only about six hundred of the seventy-five thousand chemicals on the TSCA inventory had undergone some degree of testing by voluntary agreement or enforceable testing consent agreement—not by rule making. This compares to test rules for just 121 chemicals over the same period, none of which has been issued in the past seven years, though several are under development. In other words, the administrator has rarely chosen or been able to exercise his or her formal rule-making authority. A negotiated testing plan has been more common, but even that has been developed for fewer than 1 percent of chemicals in the inventory.

There are several explanations for this woefully timid exercise of testing authority. First, the law requires the EPA to judge whether available data are sufficient to determine or predict the effects of the chemical on human health and the environment. In order to do this, agency personnel must conduct a thorough search and review of relevant scientific literature, an expensive and time-consuming process. In contrast, pesticide regulatory policy obligates the manufacturer to provide relevant information. For TSCA chemicals, therefore, EPA resource constraints foster minimal use of the testing authority.

Second, chemical manufacturers have challenged TSCA test rules in court. An immediate effect of each court challenge is to delay or stop test rules in progress. For example, in one instance, manufacturers argued that the administrator had not demonstrated that the substance enters the environment in substantial quantity with the potential for human exposure.[5] The court decision revolved around the meaning of "substantial" and ordered the EPA to develop standards for a definition. While EPA addressed the issues raised in the challenge, progress on every test rule using the exposure criteria as the basis for requiring testing was stalled for four years. On another occasion, a test rule requiring neurotoxicity testing of ten heavily used solvents was challenged in court, resulting in a negotiated settlement. The EPA was required to enter into consent agreements that reduced testing requirements for seven solvents, eliminated testing requirements for two, and postponed its decision on one.

Third, the agency has tended to be generally conservative with respect to issuing test rules; legal counsel has historically urged that test rule proposals be fully fleshed out and well documented, often in excess of what is required by the Administrative Procedures Act.[6] Moreover, inasmuch as the EPA operates within the executive branch of government, agency activity naturally reflects the political alignment of the administration in power. During the Reagan presidency, for example, when EPA administrator Anne Gorsuch and her staff were faced with evidence of the carcinogenicity of formaldehyde, they concluded that section 4(f) of TSCA should be reserved for a crash effort to remedy a very serious hazard to public health.[4] In fact, section 4(f) requires the administrator to take action when test data or any other information indicate that there may be a reasonable basis to conclude that a chemical presents a significant risk of cancer, gene mutations, or birth defects. The agency may also face criticism, additional hurdles, and funding restrictions if it is perceived by an industry-friendly Congress as pursuing test rules too aggressively.

As a result of these factors, a 1994 Government Accounting Office (GAO) report found that while TSCA had over seventy-two thousand chemicals in the inventory, only four new and five existing substances had been controlled by EPA for posing unreasonable health risks.[3]

Court challenges have not been the only means by which manufacturers have subverted a precautionary implementation of the law. In 1991, EPA initiated the Compliance Audit Program (CAP), a one-time voluntary program that encouraged companies to audit their files for information required by TSCA to be reported to the EPA. The program waived fines for companies that submitted studies they should have provided earlier. During the program, EPA received over ten thousand notices concerning health and environmental effects of chemicals. Chemical manufacturers had clearly been ignoring legal requirements for disclosing knowledge of toxic effects.

It is obvious that TSCA, as written, challenged, interpreted, and implemented, does not provide the agency with the authority to require adequate toxicity testing of chemicals prior to their manufacture and use. Agency officials and the ITC have reverted to making testing decisions based largely on computer modeling, which predicts toxicity, and not on actual biological experiments. This approach uses the concept of structure–activity relationships (SAR), which is premised on the assumption that chemicals with similar structures have similar toxic effects. Those who believe in SAR see it as a very useful tool for screening large numbers of compounds. Others point out that SAR has serious limitations and think that it is folly to base an important regulatory program on that technique. At its best, SAR analysis may predict toxic effects of chemicals very similar in structure to well-known toxicants but will be unable to predict toxicity unique to the compound of interest. In other words, it is a technique guaranteed to fail to predict surprises.

In 1993, the EPA compared results of its SAR analyses with actual toxicity tests on the same chemicals, performed and submitted as part of the registration process for chemicals in Europe. They showed that the validity of SAR predictions varied with the toxic effect or chemical property being compared. SAR-predicted physical characteristics of the chemical, such as vapor pressure or boiling point, were correct for only about half of the chemicals. This degree of inaccuracy has important implications, since an underestimated vapor pressure will result in under-estimation of inhalation exposures. SAR predictions were more accurate for biodegradation and acute toxicity in fish. The SAR approach was less accurate in predicting other forms of toxicity, primarily because of a lack of toxicity data on other chemicals and lack of understanding of how toxicity varies with structure.

It is unlikely that Congress intended TSCA to serve as a research tool to develop SAR technology over several decades, while failing to provide the testing authority to the EPA necessary for protecting human health and the environment. But that is exactly what has evolved. On July 13, 1994, Joseph Dear, assistant secretary of labor for occupational safety and health, gave his assessment of TSCA to the Senate Subcommittee on Toxic Substances and Research and Development:

The testing authority granted to the EPA has been a disappointment. There is insufficient information about the long-term or chronic effects of chemicals used on the job. We do not know enough about which chemicals at which concentrations cause cancer, heart disease, or pulmonary disease. We know even less about the combined effects when workers are exposed to numerous chemicals daily. A survey undertaken for the National Academy of Sciences found that fewer than 20% of industrial chemicals had been adequately evaluated for possible human toxicity. Though OSHA is a member of the ITC, OSHA has no authority to require chemical testing, so that ITC provides a vehicle for OSHA to obtain data which can be used to protect workers. The Consumer Product Safety Council also needs data to protect consumers. The EPA, through TSCA, has not met those needs.

Claims of confidentiality by chemical companies have sometimes included results of toxicity testing. Companies argue that they do not want their competitors to know that they are interested in a chemical to the extent that they would pay for a toxicity test. During a two-year period of reviewing confidential claims under the Confidential Business Information Review and Challenge Program, industry voluntarily amended and withdrew over six hundred confidentiality claims after challenges by EPA.[3]

Although there is some debate about whether the existing law could be made to work better, the evidence seems clear that TSCA as written is fundamentally flawed. The law has probably been successful at keeping out of commerce many acutely toxic chemicals to which large numbers of people are exposed. But delayed effects or effects from chronic exposure are simply unknown, and TSCA testing authority has not been effective at routinely obtaining that information. At every step of the process, the burden of proof is on the agency, with the result that putting a warning label on a chemical requires the same effort as an outright ban. Moreover, people are often uninformed of their exposures and are virtually never told of the lack of safety information. Until a law is written that requires toxicity evaluation before a chemical is put into commercial use, exposures to unsafe and untested substances will continue.

Federal Insecticide, Fungicide, and Rodenticide Act, 1947

The Federal Insecticide, Fungicide, and Rodenticide Act (FIFRA) was passed by Congress in 1947 as a labeling statute for pest control products, to be administered under the U.S. Department of Agriculture (USDA). Amendments were added subsequently authorizing the department to deny, suspend, or cancel registered products after an appeal process.

Its purposes are to evaluate risks to health and the environment posed by pesticides; classify and certify pesticides by specific use; regulate the use of harmful pesticides through labeling requirements, crop use restrictions, or outright bans; and enforce requirements through inspections, labeling, notices, and state regulation.

In 1972, the authority for pesticide registration was transferred to the EPA. Many pesticides that had been in use for years were grandfathered, eliminating any immediate requirement for testing the majority of products already in use. Registrations for others, found to be carcinogenic in animals, were cancelled. Court challenges led to the establishment of a special review process used when EPA contemplates major use restrictions or cancellation of a registered pesticide. The process includes rights of appeal and cost-benefit analyses.

In 1988, Congress passed legislation requiring more rapid reregistration of pesticides already in use. Products registered before 1984 must be reregistered in order to consider new scientific information and ensure compliance with new policies. The EPA publishes a priority list for reregistration, and manufacturers must fill in toxicity data gaps with additional testing where necessary. The EPA estimates that reregistration will not be complete until 2006. The most recent FIFRA amendments are included in the 1996 Food Quality Protection Act (FQPA), which requires that EPA review and reregister pesticides every fifteen years.

Major Provisions

Unlike TSCA, FIFRA requires extensive toxicological testing of individual pesticidal chemicals proposed for registration. Testing requirements vary according to proposed uses of each chemical. Health studies are based on acute, subchronic, chronic, and reproductive studies performed in two or more animal species. Testing in individual wildlife species, like bees, birds, and young fish, is used to estimate ecological risks. Evaluation of the final formulation, which may include several active and inert ingredients, is limited to testing acute exposures. The effects of chronic prolonged exposures to the final product are unknown.

Registration requirements now include animal testing for skin and eye irritation, organ damage, cancer, reproductive toxicity, teratogenicity, and mutagenicity (damage to genes that may lead to cancer or inheritable disorders). Tests for neurotoxicity and behavioral effects are not required on all pesticides but are considered case by case.

During the registration process, the EPA considers economic, social, and environmental costs and benefits of pesticide products, along with possible adverse health and ecological effects. If use of a particular pesticide may cause unreasonable adverse health or environmental effects, the EPA may deny registration, restrict use to certain crops, or require labeling, describing potential toxicity and recommended protective actions. Until recently, Congress authorized the EPA to consider costs, benefits, and agricultural practices when setting food pesticide residue tolerances under the Federal Food, Drug, and Cosmetics Act (FFDCA). However, the 1996 FQPA, which amended FIFRA, requires food pesticide tolerances to be health based, with particular attention to the health and eating habits of children. The administrator is also to consider whether a food pesticide residue protects consumers from adverse effects on health that pose a greater risk than the residue itself and whether the pesticide is necessary to avoid a disruption in production of an adequate, wholesome, and economic food supply. The latter provision provides an opportunity for cost-benefit assessment when setting food tolerances.

Major Weaknesses

FIFRA is used very narrowly to register pesticides for particular uses. It is not used, as it might be, to foster the use of nonchemical pesticidal alternatives. Studies show that commercial applicators often use the wrong pesticide, in the wrong amounts, inappropriately where pests are known to be resistant, and with inadequate knowledge of alternatives.[7] FIFRA addresses none of these very practical problems. It serves as a vehicle for chemical registration, with the authority to require labeling, warnings, and define uses of pesticides. It leaves enforcement to individual states, where the oversight of pesticide use is often inadequate.

Preregistration toxicity testing is generally confined to single active ingredients rather than final formulations, except for very limited acute toxicity testing. In addition to active ingredients, final formulations often contain solvents, surfactants, or other carriers that have their own toxicity and may boost the toxicity of active ingredients.

Required tests do not routinely include evaluation of many delayed or functional developmental effects. The effects of prolonged, low-dose exposures are usually unknown.

The pesticide registration process does not address the total pesticide exposure burden experienced by consumers and farmworkers.

Benefits assessments are an important part of the regulatory process. A GAO report found that quantitative estimates of pesticides' benefits are generally imprecise because of poor-quality or missing data.[8] There are few sources of reliable data on the quantity of pesticides used on food crops and on the effect that various pesticide alternatives would have on crop yields. The EPA often relies on pesticide manufacturers to provide cost-benefit data.

The effectiveness of warning labels relies entirely on users' having the opportunity and ability to read and understand them. Many farmworkers are excessively exposed to pesticides, in part because of lack of information or inability to read English.

Finally, ecological risk assessments continue to rely heavily on toxicity testing in a few surrogate species. Proposed EPA guidelines for ecological risk assessment recognize that this approach may seriously underestimate risks and that other factors must be considered.[9]

Food Quality Protection Act, 1996

In 1996, the U.S. Congress passed the Food Quality Protection Act (FQPA), which amended both FIFRA and the FFDCA. Its purpose was to establish health-based criteria for setting tolerances for pesticide residues on food.

Major Provisions

The EPA, according to this law, is required to establish a single, health-based standard for all pesticide residues in food, whether raw or processed. This modifies the previous law, which allowed tolerances based on normal agricultural use rather than health effects. To do this, it must consider a more thorough assessment of potential health risks when setting tolerances, including delayed neurotoxicity, effects on the endocrine system, and effects of in utero exposure, with special protections for potentially sensitive populations, such as infants and children. This will require a reassessment of approximately nine thousand existing food tolerances. Mention of the endocrine system is an important addition to the list of health effects that must be considered. The FQPA and the Safe Drinking Water Act require the EPA to develop a screening and testing program for chemicals that disrupt estrogen function. Each law gives the administrator the authority to extend that program to testing for disruption of other hormones as well.

The EPA must assess total exposure to pesticides with a common mechanism of toxicity including through the diet, in drinking water, and as a result of household pesticide use. It is also to assess health effects of exposure to multiple pesticides with a common mechanism of toxicity. Aggregate exposure and health effects data are to be used when considering pesticides for registration, reregistration, or particular uses.

The EPA is then to develop consumer information, to be displayed in grocery stores, on the risks and benefits of pesticides used in or on food, identifying pesticides on foods that are allowed to exceed health standards, and recommending ways in which consumers may reduce dietary exposure to pesticides while maintaining a healthy diet.

The EPA reviews food tolerances every ten years and reregisters pesticides every fifteen years.

Major Weaknesses

The FQPA addresses pesticide policy only insofar as it affects the safety of the food supply. It does not address pesticides that are not used on food nor does it require more comprehensive or improved alternatives assessments. It does not require examination of interactive and additive effects of dissimilar pesticides, inerts, and other chemicals to which humans and the environment are exposed.

The FQPA allows a "negligible" level of carcinogenic pesticides in food. Previously, under what was known as the Delaney clause, carcinogenic pesticides that concentrated in processed food were entirely prohibited.

By excluding occupational exposures from the aggregate exposure assessment, farmworkers will not enjoy added protection from this law and will continue to risk the health effects associated with multiple and cumulative pesticide exposures.

Finally, there is no comprehensive consumer right to know about pesticide residues on food.

Emergency Planning and Community Right-to-Know Act (EPCRA), 1986

The purposes of EPCRA are to establish requirements for federal, state, and local governments and industry regarding emergency planning and community right-to-know reporting on hazardous chemicals.

Major Provisions

EPCRA expanded the role of state and local governments and citizens in preparing for emergencies and managing chemical risk. EPCRA requires industry and federal, state, tribal, and local governments to establish local emergency planning committees to identify chemical hazards, plan for emergencies, provide for emergency notification, and convey public information. It also expands the rights of citizens to obtain information about the presence and release of hazardous chemicals in their communities, the latter through Toxics Release Inventory (TRI) reporting requirements.

EPCRA acknowledges the public right to information about toxic chemicals and thus requires the owners and operators of certain industries to report annually to the EPA and state their environmental releases (to land, air, and water) and off-site transfers of many industrial toxic chemicals. The data must also be made available to the public in a computerized TRI, the first publicly accessible, on-line computer database mandated by federal law.

The TRI provision was passed despite lack of support from the EPA. The new law would require more work for the EPA, including assembling huge amounts of data and making those data accessible directly to citizens. This new role of information provider was a divergence from EPA's historical command-and-control regulatory function. Environmental groups, labor organizations, and grass-roots activists campaigned for the law, recognizing the need for useful information on facilities' toxic waste generation. The tragic accident at a pesticide factory in Bhopal, India, in December 1984, in which thousands of nearby sleeping residents were injured or killed by a release of poisonous gas, provided context for the hotly contested congressional debate on the bill. Ultimately, the key TRI provision of the law was passed in Congress by a one-vote margin (212–211).

EPCRA requires facilities that annually process or manufacture any of more than six hundred chemicals in excess of twenty-five thousand pounds, or otherwise use in excess of ten thousand pounds, to report releases to air, land, and water to the TRI. Since 1986, the TRI has been heralded by industry and citizen activists as a successful model of legislation that recognizes citizens' right to know.

Citizens' groups have effectively used the data in negotiations with industry and government officials, resulting in numerous success stories, including the early phase-out of ozone-depleting chemical use by factories in California and Massachusetts, funding for air toxics monitors in Ohio, better regulation of toxic releases in Louisiana and North Carolina, the creation of an accident prevention plan in New Jersey, and the passage of toxics use-reduction laws in Massachusetts, New Jersey, and Oregon.[10] Even industry officials adamantly opposed to the law have found the annual data to be an opportunity for positive public relations—if their company has achieved measurable reductions in toxic environmental releases.

The current list of over six hundred chemicals that must be reported under TRI was most recently modified by EPA in 1994, when 286 chemicals were added. The addition of 152 of those chemicals to the list provoked a lawsuit by the Chemical Manufacturers Association (CMA), which argued that the federal agency had exceeded its authority in adding specifically those chemicals linked to chronic health effects such as birth defects and cancer. On May 1, 1996, a federal court sided with

EPA in determining that the agency had acted properly in expanding the list of chemicals.[11]

Major Weaknesses

EPCRA does not go far enough in that it exempts too many chemicals and industries (including small-quantity generators such as dry cleaners, collectively a major source of perchloroethylene releases).

Data collection and maintenance at the local and regional levels are often incomplete or incorrect. EPCRA relies on the polluter to estimate releases, and there is no consistency in how estimates of emissions are made. Moreover, EPCRA does not include chemical use data, and threshold reporting requirements for highly toxic or environmentally persistent chemicals are too high.

Many communities have not properly prepared emergency response plans, and there is no organized national database addressing the location of hazardous substances.

Safe Drinking Water Act, 1974, amended 1986, 1996

Under the Safe Drinking Water Act (SDWA), the EPA is required to establish standards or treatment techniques for contaminants that may have an adverse impact on human health. To achieve this, the EPA issues maximum contaminant levels (MCLs) for a series of chemicals.

Major Provisions

As amended in 1996, the SDWA requires the EPA to set MCLs using the best available, peer-reviewed science. In this work, it must consider costs and benefits when proposing MCLs, plus consider sensitive subpopulations such as infants, children, pregnant women, the elderly, and individuals with a history of serious illness.

The EPA is also required to conduct studies to understand the mechanisms by which chemicals cause adverse effects. It is also required to develop new approaches for understanding the effects of contaminant mixtures and establish a screening and testing program for drinking water contaminants that may disrupt estrogen function. The administrator may require screening and testing for interference with other hormones.

Beginning in 1998, operators of community water systems are to prepare an annual report to customers that identifies sources and levels of contaminants in drinking water for which maximum contaminant levels have been established. The EPA will compile information from local systems into a national occurrence database, required to be in place by August 1999.

The Occupational Safety and Health Administration

The Occupational Safety and Health Act (OSHAct) (1970) is intended to "assure so far as possible every working man and woman in the Nation safe and healthful working conditions."[12] Recognizing the lack of accurate and comprehensive data about the effects of chemical exposures, Congress provided for the establishment of standards for toxic chemicals in the workplace, based on the best available evidence.[12] However, employers have no obligation to conduct tests to make certain that operations will not present health risks.

Major Provisions

The employer, according to this legislation, has a general duty to furnish employment and a place of employment free from recognized hazards that are causing or likely to cause death or serious physical harm. To this end, the OSHAct created the Occupational Safety and Health Administration (OSHA), within the Department of Labor, to set occupational standards and conduct inspection of workplaces covered under the Act in order to ensure compliance with standards and the general duty obligation. OSHA is authorized to inspect places of business covered under the act to levy fines for violations of standards. Employers are required to keep records of workplace injuries and illnesses.

Any employee who believes that a violation of health or safety standard or of the general-duty obligation has occurred is authorized to request an OSHA inspection. The employee need not notify the employer of the request and may remain anonymous.

The act also established the National Institute for Occupational Safety and Health (NIOSH) for research and education.

Major Weaknesses

The burden is placed on OSHA to determine the presence of risk and attempt to define its magnitude before standards may be set. Moreover, since OSHA has no independent authority to require toxicity testing, it must rely on other sources of toxicological data to assess the safety of worker exposures.

Implementation

Since 1971, OSHA has promulgated fewer than thirty health standards.[13] It routinely takes up to ten years until a regulation becomes final. On one occasion OSHA updated exposure limits for 376 substances in one rule making, but the court found that to be unacceptable without substantial evidence in support of each change.

Supreme Court decisions interpreting the law have found that a protective standard is appropriate, but only when it is necessary to avoid a significant health risk. In 1980, the court overturned OSHA's benzene standard because the agency had not clearly stated what it considered a significant health risk to be. In an example of standard setting by the Court, Justice John Paul Stevens wrote:

Contrary to the government's contentions, imposing a burden on [OSHA] of demonstrating a significant risk of harm will not strip it of its ability to regulate carcinogens, nor will it require [OSHA] to wait for deaths to occur before taking any action. First, the requirement that "significant" risk be identified is not a mathematical straitjacket. It is [OSHA's] responsibility to determine . . . what it considers to be a significant risk. Some risks are plainly acceptable and others are plainly unacceptable. If, for example, the odds are one in a billion that a person will die from cancer by taking a drink of chlorinated water, the risk clearly could not be considered signficant. On the other hand, if the odds are one in a thousand that regular inhalation of gasoline vapors that are 2% benzene will be fatal, a reasonable person might well consider the risk significant and take appropriate steps to decrease or eliminate it. Although [OSHA] has no duty to calculate the exact probability of harm, it does have an obligation to find that a significant risk is present before it can characterize a place of employment as "unsafe."[14]

This example, embedded in a Court decision, has had the effect of establishing a one-in-a-thousand cancer risk as the workplace standard, less strict than the one-in-a-million standard considered "negligible" in other circumstances. Any attempt to set a more protective standard is likely to be challenged in the courts.

Once again, the burden is placed on the agency to determine the presence of risk and attempt to define its magnitude before standards may be set. Without independent authority to require toxicity testing, OSHA must rely on other sources of toxicological data to assess the risk of worker exposures. And since the EPA, through TSCA, has not been effective at obtaining toxicity testing data, workers are regularly being exposed to industrial chemicals with unknown health effects.

The Food and Drug Administration

The Food, Drug and Cosmetic Act (FDCA), enacted in 1938, provides the Food and Drug Administration (FDA) with the authority to regulate food for humans and animals, human and veterinary drugs, medical devices, and cosmetics. This is the statute under which the EPA sets tolerances for pesticide residues on food. Drugs and medical devices are also regulated by the FDA under this statute.

Major Provisions

Food The act prohibits the marketing of any food containing added substances that may make it injurious to health. The 1958 Food Additives Amendment requires that the additive be "reasonably certain to be safe" before marketing the product. However, this amendment exempted additives "generally recognized as being safe" (GRAS). In effect, the exemption sanctioned continued use of additives that had been in use prior to 1958 without requiring toxicity testing to demonstrate safety.

Three types of additives are subject to regulatory standards. First, pesticide residues on raw agricultural products are regulated by the EPA and are allowed if they meet tolerances established by the EPA. The FDA is responsible for monitoring these residues on fruits and vegetables, and the USDA is responsible for the meat and poultry supply. Monitoring programs and market basket surveys measure residues in foods prepared as they would be consumed. Critics of the monitoring program note that the FDA samples only a tiny fraction of the food supply and uses laboratory techniques that do not detect some important toxic compounds.[15]

Second, animal drug residues must be shown to be safe for humans. Carcinogen residues are permitted if they pose no more than a negligible risk. In other regulatory circumstances, this has generally been interpreted as a risk of no more than one excess cancer in a million consumers after a lifetime of exposure. However, it is not explicitly defined in this case.

Finally, a substance in contact with food is regulated by FDA if, when used as intended, it may reasonably be expected to become a component of that food. The FDA does not prohibit the use of carcinogenic food contact substances if the risk does not exceed the one in one million threshold.[16] As analytical techniques have become more sensitive, extraordinarily small amounts of chemicals that migrate from packaging are now detectable in food, so FDA has proposed to exempt food-contact substances from requirements when the potential for migration is trivial.

The 1938 act also authorizes establishment of tolerances for "added poisonous or deleterious substances" that cannot be avoided through good manufacturing practices. Here the agency considers price and availability of the food as well as the health effects.

Drugs   The law requires premarket approval of all new drugs, based on considerations of safety and efficacy. Traditionally, human clinical trials are conducted after extensive animal testing before a drug is released to the market.

Medical Devices   Amendments to the Food, Drug, and Cosmetic Act in 1976 provided new authority to the FDA to regulate the testing, marketing, and use of medical devices. The authority to regulate extends to substances that may be absorbed by a patient during product use.

Cosmetics   The FDA has the authority to deny the marketing of cosmetics that contain "a poisonous or deleterious substance which may render it injurious to health."[17] Except for color additives, which must be shown to be safe, the burden of proof is on the FDA to demonstrate violations of this provision. Neither cosmetic products nor cosmetic ingredients are reviewed or approved by the FDA before they are marketed. The FDA cannot require cosmetic manufacturers to do safety testing of their products before marketing and has no data collection or reporting requirements.

A large number of chemicals are used in thousands of cosmetic products—over five thousand in the fragrance industry alone. Only six chemicals are prohibited from use, and eight others have limited uses. Except for fragrances, which may be considered trade secrets, the ingredients of a cosmetic product must be listed on the label.

Very little testing is reported in publicly available literature, a matter of concern since many cosmetics applied to the skin, like other chemicals, are absorbed to some degree. Some cosmetic ingredients that have been tested have been detected in breast milk and adipose tissue.[18] Very few cosmetic ingredients have been adequately studied for reproductive, developmental, and endocrine-disrupting effects.

Consumer Product Safety Commission

The Consumer Product Safety Act created the Consumer Product Safety Commission (CPSC) in 1972, which is charged with protecting the public against unreasonable risks of injuries and deaths associated with consumer products. The CPSC must balance the likelihood of harm against the product's utility, cost, and availability.

Implementation

The commission has banned several chemicals from consumer products, including vinyl chloride as a propellant and TRIS, a fire retardant, both of which are carcinogens.[19] In 1983, it attempted to ban urea formaldehyde foam insulation from schools and residences because of carcinogenicity and acute irritant effects. The U.S. Court of Appeals for the Fifth Circuit set the ban aside, based on the commission's poor analysis of toxicological data pertinent to both the carcinogenicity and irritant arguments.[20] Since then, the CPSC has played virtually no role in addressing the risks of toxic chemicals in consumer products. Like OSHA, it has no authority to require toxicity testing of chemicals in consumer products, once again leaving that to the EPA through TSCA or FIFRA.

Conclusions

Regulatory laws intended to protect human health and the environment have been limited in their effectiveness. Narrow court rulings and industry pressure on regulators have resulted in the marketing of products that are not safe or those with unknown safety. Premanufacturing toxicity testing authority is granted only for pesticides, pharmaceuticals, and color additives to cosmetics. All other industrial chemicals may be produced and used without precautionary testing, as if manufacturers have that fundamental right. The burden, in most cases, is on the regulatory agency to show that a chemical poses unreasonable risks to human health or the environment before adequate testing or controls may be required.

Some recent legislation is beginning to acknowledge the public's right to information necessary for making choices protective of health and the environment. These efforts are still quite limited. however.

Special interest groups will continue to attempt to limit or deny the right to know to ordinary workers and citizens. Limited access to exposure and toxicity information is unjust, reflects a lack of due process, and contributes to adverse health effects and environmental harm. Regulations protective of health and the environment will require a new approach, including mandatory toxicity testing, toxicity studies of mixtures, integrated regulations that eliminate shifting pollution from one environmental medium to another, systems and life cycle analyses, and assessment of the need for new chemicals and the availability of alternatives.

Informed Consent and the Right to Know

Right-to-know laws are commonly described as "granting" citizens access to some specified information, implying that the public does not "have" a general right to information, but rather is "granted" the right by benevolent politicians who believe it is the right thing to do. Relying solely on the whims or conscience of legislators to grant access to information relevant to public and environmental health ignores economic and moral arguments for the right to know.

The economic argument is based on information flow as a necessity for free-market correcting factors to operate.[4] Potential harm or benefits that result from some industrial activity or use of a product, if concealed from the public, cannot be fully expressed in marketplace costs and choices. By this argument, whatever information is necessary to make risks and benefits fully apparent should be readily available.

Utilitarian and rights-based moral arguments may also be used to justify right-to-know laws. Since chemical use reporting requirements were established in Massachusetts in 1989, use and environmental releases of listed chemicals have substantially declined. Some people use this observation to justify the law; looking at the consequences, they argue that compulsory reporting requirements have had a beneficial effect. However persuasive, that argument is quite different from one based on a fundamental right to the information—that is, that people have a right to know irrespective of whether the law leads to an identifiable beneficial effect.

A rights-based analysis addresses two important questions: What is the nature and origin of the right? What information is it that we have a right to know?

There is a moral basis for right-to-know laws, similar to that of informed consent in the practice of medicine. It is based on the principles of personal autonomy and privacy. Informed participation in decisions affecting personal health, safety, and physical integrity is an essential component of personal autonomy. Exposing someone to a chemical, for example, without informing him or her is no different from exposing someone to a medical treatment without an explanation of risks and benefits. In a world where personal autonomy and privacy are respected, denying access to information necessary for informed decision making can be based only on the conclusion that individuals lack the capacity to make those decisions. Indeed, when proposed right-to-know laws are debated in a political setting, it is common for chemical manufacturers, product distributors, and some risk analysts to argue that the public will not know what to do with the information and that it will needlessly frighten them. Such paternalism, intended to restrict information flow, is disrespectful and, worse, violates the concept of informed consent.

There are both positive and negative rights; each confers distinct duties and obligations to individuals and, by extension, to institutions controlled by individuals. A negative right expresses what may not be done—for example, the obligation not to harm another individual. This is the ethical basis for workplace exposure regulations or smoking restrictions in public places and is used when we believe we can estimate the likelihood of harm.

Positive rights include the right to justice and fairness.[21] When access to information confers advantages or disadvantages, depending on how it is distributed, then justice and fairness must be considered. Justice and fairness require disclosure of information when the uninformed are thereby prevented from making informed decisions directly affecting their personal autonomy and health. If, for example, a toxic substance that confers otherwise desirable properties to a product is not clearly identified on the label, the manufacturer has a distinct advantage and stands to profit from consumer ignorance. This unequal distribution of information impedes free market correcting factors and is unfair and unjust inasmuch as it deprives the consumer of informed participation in decisions that may affect health. There is room here for legitimate confidential business interests, but not at the expense of personal autonomy.

The requirement for justice and fairness is an imperative beyond the right not to be harmed. It recognizes the positive right to information necessary to make informed decisions and holds that the employer or manufacturer has an obligation to respect the autonomy of the worker or community member by disclosure. It says that individuals have the right to remain free from harm *and* the right to know about exposures to themselves and the environment, potential exposures, and effects of exposures so that they may make informed choices. It follows that information that is inaccessible or incomprehensible does not satisfy the obligation.

Together, positive and negative rights help form the concept of due process, which is based on fundamental fairness. In the legal process, presumably the practical expression of our collective choices among moral principles, due process specifically requires that information be shared so as to avoid unfair advantages and disadvantages. This, then, also provides a constitutionally based argument for the right to know.

California's Proposition 65

No longer wishing to remain unknowingly exposed to cancer-causing or reproductive toxic chemicals, California voters passed Proposition 65, the Safe Drinking Water and Toxic Enforcement Act, in November 1986. It was passed through a direct voter initiative by a margin of two to one against massive industry opposition. Through its two central provisions, Prop 65 prohibits the discharge to drinking water sources of chemicals that are known to cause cancer or reproductive disorders and requires businesses to give clear and reasonable warnings whenever they knowingly and intentionally expose people to chemicals that are known to cause cancer or reproductive disorders.

Prop 65 is unique due to the incentives it uses to achieve its goals and enforce its provisions. For instance, rather than implementing new restrictions on the use of certain toxic chemicals in products, the law requires that the manufacturer place warning labels on products containing those chemicals. The result has been a move by industry away from the use of Proposition 65–listed chemicals in California. In addition, there has been a greater inclination on the part of businesses to embrace source reduction as a means of reducing pollution. In order to avoid issuing public warnings about the danger of their products, companies now tend to avoid the use of some toxic chemicals.

Industry response indicates that the success of Prop 65 reaches far beyond California. Some companies and industries have reformulated their products to remove cancer-causing or reproductive toxic agents—for example:

- Gillette reformulated its correction fluid product, Liquid Paper, by removing trichloroethylene, which is known to cause birth defects.
- Dow Chemical removed the carcinogen perchloroethylene from its spot remover, K2R.
- Industry ceased the use of lead in foil caps on wine bottles, and the drinking water faucet industry agreed to remove lead from all faucets.
- Old El Paso eliminated the use of lead-soldered cans.
- Sara Lee removed the carcinogen contained in its Kiwi waterproofing spray for shoes.

Opponents of Prop 65, including the food and drug industries, continue to argue that the law has created an industry-unfriendly environment in California that is harmful to the state's economy. They claim that consumers have been unnecessarily frightened by some products due to the chemical listing process and that this process does not distinguish between chemicals that pose a serious risk to the population and those that are less dangerous. Supporters of the law point to reduced toxic exposures and consumer right to know. They encourage other states to pass similar legislation; to date, industry opponents of the law have successfully prevented any other state from adopting such legislation. (In contrast, for example, Massachusetts law establishes a list of toxic chemicals and requires reporting their use.)

Material Safety Data Sheets

Material Safety Data Sheets (MSDS) are documents intended to address potential health hazards associated with exposure to chemical products. Requirements for MSDSs appear in several pieces of federal and Massachusetts legislation.

Under OSHA's Hazard Communication Standard (HCS), for example, chemical manufacturers and importers are required to obtain or develop an MSDS for each hazardous chemical they produce or import and provide these MSDSs to distributors and employers. And Title III of the Superfund Amendments and Reauthorization Act of 1986 (SARA) requires businesses covered by OSHA's HCS to submit MSDSs to local emergency planners and responders, subject from there to public disclosure. Since 1987, the HCS has applied to manufacturing and nonmanufacturing businesses. Trade secret information is protected from disclosure except in specific emergency and nonemergency situations described in the standard.

Each MSDS is supposed to contain the following information: the chemical and common name, subject to trade secret restrictions; physical and chemical properties of the substance; physical and health hazards; possible routes of exposure; any established exposure limits; handling precautions; control measures; emergency procedures; date of MSDS preparation; information for access to the manufacturer or importer; and whether the substance is listed as a carcinogen. Employers are permitted to rely on the information supplied by the manufacturer and are not required to address inadequate MSDS information. OSHA's HCS requires that employees be informed about the standard, the location of hazardous chemicals in the workplace, and the availability and location of MSDSs.

Although MSDSs are an important and legally required means for disseminating information to workers and the public about health hazards of chemical exposures, they are of little or no value when incomplete, uninformative, in error, or difficult to understand.

A 1989 study focusing on reproductive and developmental hazard warnings analyzed MSDSs for glycol ethers and lead on file with the Central Massachusetts Department of Environmental Protection.[22] Each substance is a reproductive and developmental toxin covered by both federal and state laws requiring disclosure of health hazards. Sixty-two percent of the documents made no reference to effects on the reproductive system and were completely uninformative; 41 percent mentioned or implied the reproductive target organ without specifying signs or symptoms; 28 percent referred only to developmental effects; 2 percent referred only to fertility effects; and 29 percent mentioned both fertility and developmental risks. All descriptions of fertility effects pertained only to male exposures, representing a gender bias.

In a 1993 study of one hundred unionized manufacturing workers in Maryland, investigators learned that only about two-thirds of the health and safety information presented on MSDSs is understood by those workers.[23] Participants attributed their difficulties in understanding to wordiness, technical language, or confusing layout of the documents. The investigators also described a previous report to OSHA in which MSDSs were found to be "accurate" or "partially accurate" with respect to health effects in only 37 percent of those sampled.

Twentieth-Century Seminal Events Relating to Toxic Chemicals

History provides important lessons about toxic chemical use and regulation. For example, some chemicals once heralded as solutions to serious environmental or public health problems have later been found to be toxic killers. In addition, it is clear that many major legislative acts regulating chemicals have been in reaction to severe chemical accidents or environmental tragedies. Following is a brief time line that identifies a few of the important discoveries, experiments, events, and disasters that have shaped our relationship to toxic chemicals.

1929   Polychlorinated biphenyls are first manufactured in the United States and are widely used in such products as electrical transformers and capacitors, hydraulic fluids, and adhesives. They are banned in the United States in 1977.

1938   The Food, Drug and Cosmetic Act is passed and provides the Food and Drug Administration with the authority to regulate food, drugs, medical devices, and cosmetics.

1945   World War II nerve gas research leads to the development of chemicals that are toxic to insects and to a postwar explosion of agricultural pesticides, including DDT and 2,4-D, into the commercial marketplace.[24]

1947   The Federal Insecticide, Fungicide, and Rodenticide Act is passed, prohibiting the sale of any pesticide shown to cause unreasonable, adverse effects to the environment during normal application. It is primarily a registration and labeling law.[25]

1948   Swiss chemist Carl Mueller is awarded the Nobel Prize for discovery of the insecticidal properties of DDT.[26]

1954   Residents living around Minamata Bay, Japan, develop neurological disorders, and children suffer from cerebral palsy, mental retardation, and other afflictions. For years, a nearby vinyl chloride factory has been regularly discharging mercury into the bay.

1962   Scientist and biologist Rachel Carson publishes *Silent Spring*.

1962–1969   U.S. military forces spray the defoliant Agent Orange over 3.6 million acres in Vietnam. They suspend spraying only when scientific reports conclude that Agent Orange could cause birth defects.[27]

1964   Despite evidence that the soil fumigant dibromochloropropane (DBCP) is toxic to testes and sperm, and pressured by DBCP manufacturers Shell and Dow Chemical Companies, the U.S. Department of Agriculture agrees to register it with the mild warning "not to breathe the vapors." Thousands of male agricultural and manufacturing workers subsequently become sterile from exposure to it. It is banned in the United States in 1985.

1966   On July 4, the federal Freedom of Information Act is signed after twelve years of contentious congressional battle to prevent its passage.[28] FOIA allows citizen access to government records on, among other things, chemicals and toxicity.

1969   Ohio's polluted Cuyahoga River, in 1881 called by the mayor of Cleveland an "open sewer through the center of the city," catches fire, igniting a public outcry about the state of the nation's environment.[29]

An oil rig blowout blackens the beaches of Santa Barbara, California, with oil.[30]

Humans set foot on the moon. The *Apollo 11* mission sends from space the first-ever photographs of the entire earth, showing it to be a "living planet of fantastic beauty and health . . . a wondrous ball of blue and green, rich in life."[31] The picture of the earth as a fragile ball in space was a significant catalyst regarding the need to protect the environmental resources.

1970   Some 20 million Americans turn out for events for the first Earth Day on April 21. Angry students wear gas masks down Fifth Avenue in New York City, pour oil in the reflecting pool at the Standard Oil building in San Francisco to protest pollution, and attend teach-ins at fifteen hundred colleges.[32]

A plan proposing the establishment of a federal environmental protection agency is submitted to Congress.[33]

The federal Environmental Protection Agency is created, with a budget of $900 million, fifty-four hundred employees, and a mission to improve and preserve the quality of the environment and protect human health.[10]

The Occupational Safety and Health Act is passed, with a stated goal "to assure so far as possible every working man and woman in the nation safe and healthful working conditions and to preserve our human resources."[34]

The Clean Air Act is passed to provide for establishment of clean air standards and regulations.

1972   The Federal Water Pollution and Control Act (the Clean Water Act) is passed to set a national clean water standard.

The Consumer Product Safety Act is passed.

DDT production is banned in the United States.

1974   The Safe Drinking Water Act is passed to ensure the safety of America's drinking water.

1976   The Toxic Substances Control Act is passed to identify and control chemical hazards to human health and the environment.

The Resource Conservation and Recovery Act is passed to deal with transport and disposal of hazardous wastes.

1978   Love Canal in Niagara Falls, New York, is the nation's first declared federal emergency for a nonnatural environmental disaster as a result of more than twenty thousand tons of toxic chemicals being dumped onto the site from 1942 to 1953.[35]

1980   The Comprehensive Environmental Response, Compensation and Liability Act (also known as Superfund) is passed to deal with hazardous waste sites.

1984   Worldwide headlines report a major chemical disaster in Bhopal, India. An explosion of the gas methyl isocynanate at a Union Carbide pesticide plant spreads over three hundred square miles, killing thousands and injuring hundreds of thousands.[36]

1986   The Superfund Amendments and Reauthorization Act is passed. Landmark right-to-know provisions are contained in Title III, which includes the Emergency Planning and Community Right-to-Know Act. It establishes and requires reporting of over three hundred chemicals to the Toxics Release Inventory.

California passes the Safe Drinking Water and Toxic Enforcement Act (Proposition 65), which includes a requirement for warning labels on products containing known cancer-causing or reproductive toxicants.

1989   The Massachusetts Toxics Use Reduction Act is passed and becomes the nation's first state law to require reporting of chemical use by facility.

1990   California's pesticide use reporting legislation is passed, the first state law of its kind to require reporting of agricultural pesticide applications.

The Pollution Prevention Act makes it federal policy to prevent toxic waste at the source.

1994   The EPA adds 286 new chemicals to the TRI. Ninety-four percent of the new chemicals have demonstrated chronic health hazards and/or environmental effects, including cancer or reproductive disorders.[37]

1996   The federal Food Quality Protection Act is passed, creating more rigorous health-based requirements for pesticide introduction and monitoring.

Taking Action: How to Assess Reproductive Threats at Home, in the Community, and in the Workplace

9

We cannot say with scientific certainty that exposure to a chemical at a certain time and in a certain dose will or will not cause a child to be born with a birth defect, a young woman to miscarry, or a husband to be sterile. Yet given the growing body of evidence pointing toward disturbing reproductive health outcomes and trends, the prudent path is to turn to the principles of precaution and prevention—principles not currently guiding all of our public policies but that regularly guide personal actions. Each day we instinctively decide how to act based on an understanding of perceived dangers and threats, such as deciding not to cross the street when a car is coming. Without thinking, we rely on the principles of probability and precaution. The same prudent course should guide individual activities regarding potentially harmful environmental exposures. As citizens, our activities need to go beyond personal behavior to support societal activities that prioritize pollution prevention.

No one should wait for a health crisis to occur, such as a miscarriage, or the birth of a child with a defect or developmental disturbance, to assess and deal with toxic threats. Undertaking a survey of one's own household, neighborhood, or workplace is one way to identify potential problem areas. In many cases, there are some very straightforward things to do to decrease immediate risks from chemical exposures. Other solutions are more complex and require that citizens try to effect changes in corporate and government policies.

This section provides some useful tools to help in protecting personal and community health. It is not meant to review all of the actions or choices available. Many excellent resources are available in this effort, some listed in appendix A.

Common Routes of Exposure

Most of us living in industrialized societies are exposed daily to untold numbers of chemicals. With some important exceptions, most of the doses outside of occupational settings are often low. However, the health risk may be cumulative due to repetitive and multiple exposures. For example, it is nearly impossible to avoid exposure to pesticides almost anywhere in the world because of their pervasive use. In an average day, a person might be exposed to pesticides at home, work, school, in the community, or even during a round of golf.

Systems thinking, which considers the interdependence of the individual parts of a system, requires a consideration of the many pathways of exposure that can result from production, use, and disposal of products. If you think of the life span of a product as the flow of a river, toxic exposures can occur upstream, during extraction of resources, manufacturing, or transportation, or downstream, during disposal, in addition to the actual time of use. For example, the use of a battery that contains mercury may not pose an immediate health risk, but its disposal in an incinerator contributes to a major source of airborne mercury.[1]

Proximity to a chemical does not necessarily qualify as an immediate health risk. There must also be a pathway of exposure—breathing it in the air, drinking it in water, absorbing it through the skin—and then it must make its way to a target tissue. Still, proximity to a chemical often increases the opportunity for an exposure pathway to develop. For example, an improperly contained hazardous waste site could contaminate the town's drinking water if the water supply is derived from local groundwater.

Sources of Information

Since the 1986 passage of the federal Emergency Planning and Community Right-to-Know Act (EPCRA), as well as other laws regulating health and safety in the workplace, Americans have had access to some information about emissions and potential health effects of toxic chemicals. This information, however, deals primarily with certain industrial emissions to the community and chemicals in the workplace; it does not address the connection between emissions and exposure or between exposure and potential health effects. In addition, federal law does not require labels on most consumer products to delineate toxic constituents or their possible health effects.

Agencies empowered to monitor, regulate, or investigate pollution or health threats in the workplace include the U.S. Department of Labor's Occupational Safety and Health Administration (OSHA), the Center for Disease Control's National Institute for Occupational Safety and Health (NIOSH), and state labor and health agencies. Community information and assistance might fall under a combination of agencies and laws, including local and state public health officials, state environmental protection agencies, Superfund, EPCRA, the Safe Drinking Water Act, the Consumer Product Safety Commission, and others. Laws governing access to information can also vary by state. For example, information is available to Massachusetts and New Jersey residents on industrial use of toxic chemicals because of state-implemented laws. The federal government has launched certain initiatives to help integrate information from a range of sources so that the public can more easily identify environmental hazards and assess health risks. It is important to be aware of what information is publicly available, where to find it, and how to interpret and use it.

Conducting Research Effectively

Advocacy organizations and resources can help expedite research on or investigation into toxic hazards at home, at work, or in the community. They can also often assist in clarifying technical information. Here are some guidelines for conducting research:

1. Begin at the local level. Seek information from the closest agencies and resources—for example:

   • In the workplace: Coworkers, union representatives, company health and safety officer, company physician

- In the community: Local board of health, fire department, water department; local or regional workers' rights advocacy organization; community advocacy organizations; health care institutions; universities
- State department of public health and/or department of labor and industries, department of environmental protection, elected officials
- Regional offices of federal environment and health and safety agencies (EPA, OSHA)
- Federal offices of those agencies, federal health agencies (NIOSH, ATSDR)

2. Be very specific with your request. If you want an understandable but scientifically referenced summary of the health effects of a certain chemical, ask for it in that way.
3. Try to be courteous during your inquiries, but if you are met with hostility, do not be deterred. You are entitled to public information (with some exceptions, such as that deemed confidential for national security), and if need be, you can file a Freedom of Information Act request to obtain it.
4. Document in one place the information you gather, chronologically if possible. Record dates, times, participants, and contents of conversations. Do not be discouraged if you fail to reach the proper department or information on your first try; you probably will not. Your next contact may provide the help you need or information you seek.
5. Workers' rights advocates advise acting with someone else in the workplace if you are seeking information. This protects you legally and provides you with a witness to your actions. If you believe your company does not adequately respond to your concern, try to consult with a worker advocate (union representative, advocacy group representative) who can help guide you through a formal complaint process. If you believe your job may be in jeopardy if you report the issue to a government agency, you can request anonymity, which may help prevent employer retaliation.

The Internet is by far the easiest way to access information on almost any subject. The EPA and other government agencies have organized large amounts of information for citizens and researchers in a user-friendly format on the World Wide Web. A certain quality control can be attributed to data on government sites in that they are publicly documented. Nevertheless, there are no standards or guidelines regarding the authenticity of information on the rest of the Internet; anyone can, and many do, post opinions on subjects. Moreover, numerous organizations and groups have been created that are funded and supported by industries and trade associations. They often adopt names that suggest they are independent advocates for the environment and health. Use common sense and request further information if you detect unreasonable bias.

How to File a Freedom of Information Act Request

The Freedom of Information Act (FOIA) was passed on July 4, 1966. Under this law, government records (with the exception of information deemed confidential) are required to be accessible to the public. Filing a FOIA request is relatively simple, but there may be a charge for copying costs. You may request an exemption and be granted a waiver of payment based on the fact that the information will be used for public benefit. In a brief letter you should include the following information:

• Your name, address, and telephone number
• The specific information you want
• How much you are willing to pay for the processing of the information, or if requesting a partial or total fee waiver, the purpose for which you will use the information (e.g., to conduct research on behalf of the public interest). The following sample letter is a guide:

Date
FOIA Officer
U.S. Environmental Protection Agency
401 M Street, SW
Washington, DC 20460

RE: Reproductive or Developmental Toxicity of _____

    Dear FOIA Officer _____ :

    I am interested in information regarding the potential effects of _____ on human reproduction and infant development.

    Pursuant to the Federal Freedom of Information Act (FOIA), I request the following documents:

---

(1) A risk assessment, if one has been performed, on _____ .

(2) Results of any animal toxicity testing on _____ in which possible reproductive or developmental effects were evaluated, including, but not limited to: reproductive or developmental toxicity studies, multigenerational studies, and continuous breeding studies. Please send the full report of the testing as well as any summaries of the studies done by the Agency, industry, or outside consultants.

(3) Any human studies or case reports, whether published or unpublished, concerning the possible effects of _____ on human reproduction or development.

(4) The text of any Agency summaries, internal or public documents, memoranda (including electronic mail), or other information relating to the reproductive or developmental toxicity of _____ .

I believe I am entitled to all the documents requested. However, should you withhold any document or portion of a document, please provide the remainder together with an index of the documents withheld and give a detailed explanation of EPA's grounds for nondisclosure.

In addition, since this request is made by an individual for no commercial purposes, and in the public interest, I request a waiver of fees. Should fees not be waived, I am willing to pay a fee of up to $____ . Please process this request within 10 working days.

Thank you for your cooperation.

Sincerely,

The Freedom of Information Act's Long Road to Daylight

On July 4, 1966, President Lyndon Baines Johnson finally released a landmark bill from more than a decade of bondage. There was no bill-signing ceremony, no commemorative pens were distributed, and no one from the small group of legislators, lawyers, and journalists who had championed its passage for twelve years were present. On his Texas ranch, Johnson quietly signed S. 1160, the Freedom of Information Act (FOIA), requiring government agencies to provide information about government activities to citizens on request. In a statement released by the White House, Johnson said: "This legislation springs from one of our most essential principles: a democracy works best when the people have all the information that the security of the nation permits."

Johnson's statement gave no inkling of the contentious debate that had ensued to try to block the bill's passage for more than a decade. The changes that would be imposed by FOIA were fought primarily on the grounds of power shifting from the executive to the legislative branch. Several memos from the Bureau of the Budget were sent in 1964 and 1965, citing strong opposition to the bill. Eventually, the most vehement opposition tempered as it became apparent that a veto would not be probable.

The FOIA had been initiated in the 1950s by the American Society of Newspaper Editors and the Society of Professional Journalists, then eventually shepherded through Congress. It would become the catalyst for disclosure laws in all fifty U.S. states, as well as an international model for citizen access to information. Although thirty years later it is still disliked by federal officials who must comply with its requests, FOIA remains a critical right-to-know tool used to maintain the vigilance Jefferson warned would be required to protect liberty in a new nation.[2,3]

Home Survey to Assess Threats

Routes of exposure to potential reproductive toxicants in the average household can stem not only from what is directly in the home but also from surrounding homes and the community's air, water, and land. Solvents, metals, pesticides, and endocrine-disrupting substances can be found not only in consumer and hobby products commonly known to contain harsh chemicals, such as pesticides, paints, varnishes, cleaners, and detergents, but also in more seemingly benign products such as nail polish remover, perfumes, markers, glue, shoe polish, and spot remover. Others are found in and on food and related products, such as packaging and plastic wraps. Still others are contained in building materials and finishes, drinking and bathing water, hobby materials, home office products, and auto maintenance products. Because people spend so much time at home, and with the increase in telecommuting and home-based businesses, household exposures can be a very important part of overall daily exposures.[4-6]

It is important for parents to remember that children are more vulnerable to potential health effects from toxic substances than average adults. They have higher levels of exposure from activities such as rolling on the floor and chewing on things; they have a higher respiratory rate, which increases their intake of air pollutants on a per pound basis; they drink more water and eat more per unit of body weight than do adults; and they eat substantially more of certain fruits and vegetables that may contain pesticides.[7,8]

Many common household exposures can be addressed reasonably easily by removal, substitution, or changes in behavior. An effort to identify potential trouble spots might start with a consideration of products and materials that seem to trigger an adverse physical reaction—a headache, dizziness, nausea—or that have a warning label on the side. Then think about all the human routes of exposure—breathing, eating, drinking, touching. Following is an exploration of the major substances of concern that people may come in contact with on a daily basis.

Lead Paint

Any home built and painted prior to 1978 probably has interior lead paint.[9] Lead paint may also contaminate the home's surrounding soil. The Residential Lead-Based Paint Hazard Reduction Act of 1992 requires those who are selling their homes to disclose any known information about lead paint in the dwelling but does not require that the paint be tested. Therefore, potential buyers should inspect surfaces for the presence of peeling paint, with special attention given to window frames and radiators. The best analytic technique is X-ray fluorescence with field instruments that can be brought into the home by a technician. Paint chips can also be sent to a certified lab for analysis. Home test kits for lead are available, but the EPA recommends that consumers not rely on these kits before doing renovations or to ensure safety.[10] Information about testing can be provided by a lead poisoning prevention unit in the local or state health department or the National Lead Information Center for a referral to a certified lab.[11]

Avoidance

Lead paint removal can be expensive. If it is not performed carefully by trained personnel, it can create even more problems, such as acute lead poisoning of the workers, dispersal of lead dust or fumes into the home, and contamination of the yard or garden. Some communities and states maintain excellent lead abatement programs and may even provide financial support for deleading. The U.S. Department of Housing and Urban Development (HUD) operates the Lead-Based Paint Hazard Control Grant Program, which provides financial support for lead control in private, low-income housing.[12]

If it is not possible to remove lead paint and the paint is peeling or flaking, then it is important to take some routine measures, such as regularly wiping surfaces and floors with a damp cloth and frequently washing children's hands and toys. It may also be possible to cover lead-painted surfaces with wallpaper, tiles, sheetrock, or paneling, but it is not effective simply to paint over lead paint.

In the garden, contaminated soil can be covered with a layer of loose dirt. Ideally, nonedible plants should be planted. Lead collects in the roots but does not travel into the shoots above the ground, so roots crops such as potatoes or carrots should not be planted in lead-contaminated soil.

Lead in Consumer Products

As Americans spend millions of dollars to delead their homes, manufacturers continue to sell consumer products—such as some lipstick, calcium supplement tablets, antacids, hair dyes, and miniblinds—that contain lead. A number of advocates have challenged the use of lead in these products. For example, the Natural Resources Defense Council has joined other groups in petitioning the FDA to initiate rule making concerning the presence of lead in dietary calcium supplements and antacids.[13] And a study reported in the *Journal of the American Pharmaceutical Association* urged pharmacists to advise that customers buy lead-free hair colorings and to stop selling dyes that contain lead. Unfortunately, the FDA responded with a statement that "the data in hand indicate that lead acetate containing hair products can be used safely." Lead is not included on the FDA list of "prohibited ingredients and other hazardous substances" for cosmetics.[14,15]

Avoidance
Check product labels for the presence of lead, and avoid the use of products such as hair dyes and lipsticks that may contain lead.

Drinking Water
The quality of the water supply can vary dramatically depending on environmental factors and on whether the supply is public or private. A public or community water supply is considered safe to drink if it meets the EPA's and states' maximum contaminant levels (MCLs) requirements,

---

which are levels for a wide range of substances, including volatile organic chemicals, pesticides, metals, radionuclides, and microbiological contaminants. Private wells and noncommunity water systems that serve fewer than twenty-five people or provide water to fewer than fifteen service connections are not subject to federal drinking water regulations. (These smaller "noncommunity" systems nevertheless serve over 22 million people in the United States).[16] A 1997 General Accounting Office report on the quality of drinking water in six states revealed that the most commonly exceeded standards in community water systems were those for total coliform bacteria, radiological elements, nitrate, and the herbicide atrazine. It also reported that little information is available on quality of water from private wells.[17]

Water can be contaminated in the ground or from household plumbing. Common contaminants that may affect reproduction include lead, organic solvents, pesticide residues, and trace amounts of chloroform and other chlorinated and brominated organic compounds. These chemicals are required to be monitored under the Safe Drinking Water Act (SDWA).

Anyone concerned about the community water supply can do the following:

• Request testing results from the local water department, which is also required to notify consumers every year about any violations of the SDWA in that water system.

• Ask the water department what public notices it has issued regarding violations over the past few years and what it has done to remedy the problem. Ask for copies of the monitoring results for water contaminants.

• Test for lead. Lead leaching from indoor pipes, water meters, and faucets within the house requires testing at the tap, which generally costs twenty to forty dollars. The U.S. EPA Drinking Water Hotline can identify a convenient certified testing laboratory.

Those who receive water from a small system or have a private well can contact the local or state health department about the quality of local groundwater. The EPA recommends regular testing of these systems and wells. Some areas may have testing programs. Special labs are now available that have considerably cut the costs for private testing for most of the possible pollutants from over $9,500 to around $300. Consumer kits are available to test water for the presence of certain chemicals such as metals but not for organic chemicals.[18] The EPA does not recommend home testing of water for toxic contaminants.[19]

Avoidance

For lead, run the tap water for a minute or two after any periods of disuse (such as overnight) to flush lead from pipes out of the system. Do not use hot water for consumption; use cold water and heat it. Some types of filters on drinking water taps can be an effective measure against lead in the water.[20] Lead is not a major concern in the shower because it is not absorbed through the skin and is not volatile.

Water filters can be useful in removing some solvents and other chemicals from drinking water. Solvents can also enter the body through skin absorption and inhalation in the shower. Shower heads are available that will filter out chlorine. But high-solvent-content wells should be closed.

It is important to purchase a water filter technically suited to achieve a specific goal. For example, carbon filters remove tastes, odors, and colors and also some organics such as chloroform, but are not usually considered an effective technology for removing metals. To remove dissolved metals requires other types of filters, including reverse osmosis.

Proper maintenance is essential to ensuring that a unit continues to remove contaminants. Activated carbon filters need to be changed periodically because bacteria grow in them and because in saturated units, a sudden release of concentrated chemicals can occur.

In buying water filters, *caveat emptor*. Unscrupulous manufacturers have sold units that can actually worsen water quality. In one instance, the Federal Trade Commission successfully prosecuted a large manufacturer of one treatment system that was leaching toxic methlyene chloride into the water in amounts exceeding safe EPA levels. Although notified of the problem, the company continued to sell more than 354,000 contaminated filters for nearly four years to the unwary public.[20]

Food

Food products can be contaminated with pesticides and other substances such as mercury, polychlorinated biphenyls, and dioxins that persist in the environment or that bioaccumulate and concentrate in foods higher up in the food chain (such as meat and fish).

Avoidance

Peeling and/or washing fruits and vegetables can remove surface residues of contaminants (special biodegradable fruit and vegetable washes are available in stores, or use a mild solution of soap and water). Here are some more guidelines.

- Purchase and eat organic foods that have not been sprayed with chemicals.
- Consider growing your own fruits and vegetables (after testing the soil for lead) to ensure greater quality control over your food source.
- Eat lower on the food chain to avoid toxicants that bioaccumulate in foods. A low-fat diet that consists mainly of fruits and vegetables is likely to be lower in most environmental toxicants.
- Try to eat fish that live in deep waters, which are likely to be less polluted than coastal waters. Do not eat freshwater fish from waters that have posted advisories for mercury or other contaminants. If you are unaware of whether an advisory has been posted, contact your state department of public health. Although the FDA recommends that pregnant women limit eating certain saltwater species such as swordfish, shark, and tuna to no more than two meals per month, we recommend that women who are pregnant or planning a pregnancy avoid these fish altogether due to the very high levels of mercury often found in these fish.[1]

Pesticides and Other Household Products

Pesticides comprise an important fraction of household products that may be hazardous to reproduction. Although the pesticide container is required to list the active ingredient (the one that actually poisons the pest), the so-called inert ingredients, which constitute up to 98 percent of the total ingredients, are not listed. These inert ingredients are often organic solvents, which may also be reproductive hazards.

Weed killers, fungicides, and other pesticides for use in the yard and garden can be tracked indoors on shoes and may persist for long periods of time in carpets.[21] Even one household's use of pesticides can provide a route of exposure to others in the neighborhood. Some states are considering legislation to require notification to neighbors when pesticide applications are imminent. Flea dips and other treatments for animals persist on the animal's fur for some time and may end up on the skin of anyone who pets the animal. Some pesticide products used on children to eliminate head lice contain the potent toxicant lindane. With more children in day care, head lice transmittal has become more common, and parents need to be aware that this cure could provoke side effects. There are substitutes for these toxic products and preventative measures.

Other household products also bear consideration. Household and bathroom deodorant products (e.g., toilet bowl deodorizers, air fresheners, surface disinfectants) and some cleaning products may contain solvents, suspected endocrine-disrupting chemicals such as alkylphenols, and sometimes pesticides. Some cosmetics contain formaldehyde or glycol ethers. Some mouthwashes and medicated skin products can contain phenols, and food-related packaging and plastic wrap can contain phthalates and bisphenol-A.

If a product does not list ingredients but there is a telephone number to call for the manufacturer, here is what to do:

1. Call the manufacturer and request the Material Safety Data Sheet (MSDS) for the product, which should contain the list of ingredients and some of the major potential health effects. MSDS information is also available from government agencies, on the Internet, from the Chemical Transportation Emergency Center, and from local poison control centers.

2. If you feel you are suffering a health effect from a product and cannot obtain an MSDS or equivalent, then have your physician contact the manufacturer. Most trade secret laws make an exemption for information obtained by a doctor for a patient who may be suffering from a health effect from a product.

3. Once you have the information, you may need assistance in understanding it. If you have not yet involved your health care provider, you can search for health effects in a variety of ways, including through the National Library of Medicine's on-line services, which include access to Medline and other health databases. Other government agencies, such as the Agency for Toxic Substances and Disease

Registry (ATSDR) through its ToxFAQ program, have published fact sheets on the health effects of more than two hundred chemicals. Hazardous substance fact sheets are available from the California Hazard Evaluation System and Information Service (HESIS) and the New Jersey Department of Health and Senior Services.

4. Discuss the information you have gathered with your physician or other health care provider. The provider may belong to a health care network that can provide even more information through internal mechanisms that provide risk research and analysis.

5. If you determine that a product could pose a risk you do not want to take, stop using it yourself. Then try to influence others to stop using it, ask the manufacturer to make the formula less toxic or stop making it, or convince the government to regulate it as a toxic substance.

Avoidance

For household products, choose from the wide variety of less or nontoxic alternatives that are available—for example baking soda, soap, and vinegar mixed with water for cleaning. Unfortunately, there are no federal standards for labeling "nontoxic" products, although a variety of private programs have tried to pattern themselves after labeling programs in other countries, such as Germany's Blue Angel and Canada's Environmental Choice. In addition, the Consumer Labeling Initiative is a voluntary, cooperative effort organized by the EPA to encourage companies to improve labels regarding environmental and health information.[22]

One well-known author of books on nontoxic products describes obtaining an MSDS sheet on a cleaning product advertised as "nontoxic, socially responsible, and otherwise committed to the environment." She found ethylene glycol monobutyl ether as the main ingredient, "a chemical so toxic that a major chemical company decided to no longer manufacture it." This chemical had been diluted down to 2 percent, and the toxicity studies then came out relatively harmless. But she questioned whether a product made from a diluted toxic chemical is really harmless.[5] This example points out the need for standards and disclosure in labeling and cautions that a product labeled "less toxic" or "nontoxic" may not actually be safe.

Toxic chemicals in products can affect humans through their life cycle. A standard industry argument for the continued use of toxic chemicals in products is that the public wants them. Consumers who choose less toxic products send a message to manufacturers that they do not want the toxic chemical cycle to continue.

Try not to use pesticides either indoors or on the lawn and garden. There are many alternatives to combat pests, from using simple substances like cayenne pepper to keep ants out of the house, to employing companion planting to discourage garden pests. Monocultures like grass encourage infestations of insects. More natural landscapes with many varieties of plants are much more pest resistant. (See appendix A for information on alternatives to pesticides.)

Finally, use the least amount possible of those toxic products you feel you must use for special problems, such as a major insect infestation. Choose products carefully, use them in a well-ventilated area (or open all windows if used indoors), avoid any skin contact, read the label instructions, and allow time for the area to air out before reentering. Children and pregnant women should never be in the vicinity when toxic chemicals are used.

Building and Finishing Products

Certain building and finishing products such as bonded wood products (plywood, particle board, chipboard), drywall compound, insulation, carpeting, draperies, vinyl molding, cabinets, and even furniture may contain and off-gas formaldehyde and other volatile organics.[23] Wallpaper and paints can contain fungicides.[24] Home renovation and new house construction may lead to exposures to higher concentrations of these chemicals. In one study, carpet lead levels increased thirty-fold during remodeling.[21] Although urea formaldehyde foam insulation is seldom used today, it is still available. No one should use it.

Avoidance

Pregnant women and young children should avoid home renovation and construction areas. Toxic fumes and dust containing substances such as lead and volatile organic compounds are commonly present during demolition and new construction. If it is impossible to avoid exposure during these times, then the use of protective masks and clothing should be employed.

- Although less toxic paints, finishes, and building materials may be more difficult to find, they do exist and are sold in stores and through

---

catalogs. Ask your builder for product specification sheets, and consider using alternatives such as solid woods instead of woods bonded with resins.

- Increase and improve ventilation during and after construction. Well-insulated, tightly built homes trap more air pollutants inside. Fans and circulating devices can draw in outside air.
- Remove shoes when entering the house. One study revealed that dust vacuumed from a living room carpet contained sixteen pesticides, some banned for years. Another showed that removing shoes can reduce lead levels in carpets by 90 percent.[25]
- Ask the carpet installer to air out carpets for several days prior to installation. Or choose natural fiber carpets constructed without chemicals. Area rugs can be an alternative to installed carpeting.
- If you have carpets, buy a better vacuum cleaner. A vacuum with an agitator picks up two to six times as much dust from a rug as a canister type.[26]

Hobbies

Those whose hobbies include potential exposures to lead or other metals, such as painters, potters, and stained glass window makers, and those who handle, shoot, and clean guns or make bullets should understand the health effects of lead. Hobbies that involve likely exposure to organic solvents include furniture refinishing, automobile repair, painting, and model building—in fact, any hobby involving the use of strippers, degreasers, non-water-based glues, or paints.

The nonprofit groups Arts, Crafts and Theater Safety (ACTS) and the Art and Creative Materials Institute (see appendix A) offer information about chemical hazards from arts and crafts work to artists and consumers.

Avoidance

It is virtually impossible to handle lead safely in the home. Lead dust remains on the skin, hands, and clothing and can contaminate the house or car, resulting in significant exposures to family members. Hobbies involving lead should not be practiced in living areas of the house.

Hobbies involving solvents should be practiced only in very well-ventilated areas. Avoid skin contact with solvents because they easily penetrate the skin and enter the body. Wear chemically resistant gloves, and never use solvents to clean paint or glue off the skin. Pregnant women should try to avoid all unnecessary exposures to solvents.

Some respirators filter out only particles; special masks designed to filter out particular chemicals are available and are rated by NIOSH for various uses. Check product labels for ratings.

Community Assessment

Various factors and pieces of information go into creating a community environmental health profile. Initial concerns may stem from awareness of local pollution, or from knowledge of chemical emissions into air and water from neighboring towns or regions. Or residents may be troubled by specific health conditions in their family or community, and may wonder whether they may stem from environmental contaminants.

Health Concerns

Perhaps several families in a neighborhood or town have had similar health problems, and they want to know if this is an unusual cluster. Citizen activism organized in response to community toxic hazards has become known as popular epidemiology. Stages of action after a health or pollution problem is noticed typically include information gathering and sharing of information with other residents and organizing discussions with government officials and scientific experts.[27]

They may consider conducting a preliminary neighborhood or community health survey and/or researching health statistics collected by cancer and birth defect registries or local hospitals or clinics. Undertaking even a preliminary survey can be a big job; a more comprehensive health survey can be complex and expensive because the survey must be designed properly, and information collected carefully, so that the results are as complete and accurate as possible and can be compared to those of other communities. Those who have serious concerns and feel the need to proceed in this manner should ask for assistance from a health researcher early in the design. Local or state public health officials should be able to provide assistance. Some of the questions in a survey might ask for basic demographic information, data on specific reproductive health outcomes, and risk factors including exposures to contaminated food or water in the community or toxic hazards on the job.

Although it may be extremely difficult to prove a direct cause and effect linking certain pollution to a health outcome (for all the reasons outlined in chapter 2), it is possible to achieve greater understanding of health problems and possible risk factors in the community. A good example of a successful use of popular epidemiology by citizens that eventually led to other studies by government agencies is the town of Woburn, Massachusetts (see chapter 4). Another is the case of Love Canal.

Love Canal and the Rise of Citizen Activism

When residents of Niagara Falls, New York, began to complain that unidentified substances were oozing into their basements in the late 1970s, they had no idea that their experience would become one of the most significant environmental flashpoints of the decade.

In the 1940s, the Hooker Chemical Corporation had used the abandoned Love Canal as a dump site for toxic chemicals.[28] Over a decade, it had dumped twenty thousand tons of toxic chemicals into the canal, including lindane, chlorobenzenes, and dioxin-contaminated trichlorophenol. In 1953, the city bought the land after the canal was sealed with a protective clay cap. Over time, homes and a school were built on and near the old canal site, and construction disrupted the clay cap. In the 1960s, a highway blocked the underground drainage from the canal area, and by the 1970s, contaminated water began seeping into homes.

Residents recognized that the liquid oozing into their homes was somehow contaminated and, concerned about potential health effects, called on the government. They received little assistance. Soon the Love Canal Homeowners' Association, led by Lois Gibbs, a local resident, began organizing neighborhood protests. Their actions brought national attention to the situation at Love Canal. In 1978, President Carter declared a federal emergency in the area, the first time ever for a nonnatural disaster. The governor arranged for the evacuation, purchase, and destruction of over two hundred homes closest to the canal. Many residents were not satisfied, and in 1980 a group of home owners held two EPA officials hostage for several hours to pressure the government to more action. Two days later, President Carter issued a second emergency declaration and purchased over five hundred more homes.

As a result of the experience at Love Canal, in 1980 the government passed the Comprehensive Environmental Response, Compensation, and Liability Act, also known as Superfund, a law designed to address the nation's hazardous waste problem. The Love Canal experience also stimulated the growth of citizen activism around issues of environmental contamination. Lois Gibbs went on to found the Citizens' Clearinghouse for Hazardous Waste, now known as the Center for Health, Environment and Justice, a major activist organization working to protect the public.

Pollution Concerns

Residents who are concerned about environmental contamination may want to identify sources of pollution by collecting specific information about the location, quantity, and nature of toxic chemical use and release, as well as the hazards and properties associated with each chemical. There are regulated and monitored sources of pollution such as hazardous emissions from industrial facilities (defined by the EPA as point sources), and nonpoint sources, such as chemical runoff from farms and golf courses. Hazardous waste from homes is also considered a nonpoint source. The average family uses approximately fifty-five gallons of hazardous household products each year.[29]

Industrial facilities such as incinerators and manufacturing plants may emit toxic chemicals into the surrounding soil, air, sewers, or local waterways. Raw materials, products, and waste may be trucked in and out, and accidents that result in chemical spills can occur on the roads or inside the factory. Facilities such as hospitals and airports can also be sources of pollution. Hospitals, for example, use products and practices that are polluting or hazardous. Medical waste incinerators are major sources of dioxin, mercury, and other dangerous toxic emissions.[30] Private and military airports can contribute to air pollution and hazardous waste from chemical spills or other activities such as deicing. Businesses such as dry cleaners and gas stations also need to be considered.

Many hazardous waste sites are no longer associated with a current industry but are contaminated from prior site activities. They might be vacant lots or located beneath new facilities. Over half a million sites with potential contamination have been reported to the EPA over the past fifteen years. More than 200,000 of them have yet to be cleaned up. (There is considerable debate on how "clean" is clean.) They include sites categorized under seven different programs, acts, or agencies, including Superfund, the Resource Conservation and Recovery Act, the leaking underground storage tank program, the Departments of Defense and Energy, and state sites. Site remediation in 1996 dollars is expected to cost $187 billion.[31]

Researching Pollution

Identifying the Suspected Hazard

An assessment of a community's environmental health should begin with local authorities and references, including public works, water, planning, health and fire officials, and the local library and newspaper. Local emergency planning committees (LEPCs), required by law to maintain records about toxic chemicals stored and released locally, are another good source of information. Over four thousand LEPC districts have been designated throughout the United States.[32] (Unfortunately, a survey of LEPCs found that only about 25 percent of the LEPCs strictly comply with their legal mandates, and about 20 percent are inactive or defunct.)[33] The State Emergency Response Commission provides information on which LEPCs cover which areas.

Potential sources of pollution in the neighborhood or community can be mapped by marking a map with sites of large and small industries, gas and bus stations, hospitals, airports—or any other potential source of land, water or air pollution. Information on emissions and waste sites can come from a number of databases, including the Toxics Release Inventory, the Comprehensive Environmental Response, Compensation and Liability Information System (for information on hazardous waste sites), and the Permit Compliance System (for data on more than seventy-five thousand water discharge permits). The EPA also maintains the Envirofacts relational database, which allows users to cross-reference these and other databases and to access the Facility Index System, a central inventory of over 675,000 facilities regulated or monitored by the EPA.[34] Some states, such as Massachusetts and New Jersey, have materials accounting laws requiring companies to release information about chemicals that are used, stored on site, and shipped in product. Information on small businesses not required to report to state or federal agencies will need to be collected directly from the companies or local authorities.

Some excellent computer software tools helpful for overlaying and synthesizing information. LandView from the EPA allows citizens and communities to combine information from several databases on detailed maps that include roads, rivers, railroads and landmarks, geographic boundaries, and census tracts.[35] The Environmental Defense Fund's on-line Chemical Scorecard adds another dimension to emissions and geographic information by providing information on health effects of chemicals. (See appendix A.)

Information unavailable through public records may need to come directly from representatives of industries about which residents have concerns. Most businesses have community liaisons, or at the least a public affairs department. For more serious problems, a community advisory council may be useful for continued dialogue and problem solving.

It can be very difficult to determine if emissions of a substance exceed a regulated limit or even under what body a substance may be regulated. There is no comprehensive such list for any given chemical. Water discharges and air emissions fall under different permits and laws. Individual facility permits need to be reviewed to determine violations. A more efficient approach may be to research the potential health effects of a substance that is being emitted or is contained in a contaminated site (see appendix A).

If residents believe there is a health hazard in the community from an as-yet-unidentified toxic waste site, or from emissions of hazardous substances, they can petition the state or federal government to conduct a site assessment. Local and state authorities are good for helping to build a case. State departments of health may be the first agencies to become involved and will likely work with federal agencies should that situation evolve. Site discoveries and investigations can be generated by communities or states or by a citizen report to the National Response Center, which is primarily designed to respond to emergency chemical spills but will also accept calls about suspected toxic sites.

A further step is to petition the federal government for a health study regarding a waste site. The Agency for Toxic Substances and Disease Registry (ATSDR) is the lead agency within the Public Health Service responsible for implementing the health-related provisions of Superfund. It is mandated to conduct public health assessments at every site on the National Priorities List of toxic waste sites and at other locations where petitions request an assessment. Anyone can petition ATSDR to conduct a health assessment, but the likelihood that this will happen is increased if residents can amass evidence of health risks, act in concert with others, or officially request an assessment through the community.

Taking Action

If the research points to a potential relationship between disease or health conditions, and chemical emissions into the air, water or land of the community, residents should bring their concerns to the board of health or other such agency and request assistance to press for further investigation by private, state, or federal public health organizations or agencies.

Community organizers should not wait for lengthy investigations before taking action. They can educate and organize their neighbors to address risk factors in homes, workplaces, and the community.

Community Sites of Concern

Some businesses in communities are of special concern because they are frequented by residents and typically use chemicals associated with reproductive toxicity. Among them are dry cleaners, gas stations, agricultural operations, and parks and golf courses.

Dry Cleaners

The way in which most professional cleaners go about cleaning clothes is not particularly safe. The dry cleaning machines are typically filled with perchloroethylene (PCE), a solvent that dissolves grease and oil from clothes without damaging the fibers. PCE, which can leave a pungent smell on clothes, has been linked to spontaneous abortion and infertility and is a suspected carcinogen.

The largest consumer exposures to PCE occur soon after the clothes are picked up because of chemical residue evaporation. Fumes from freshly dry cleaned clothes build up quickly in an enclosed car, potentially causing a narcotic effect on the central nervous system in as little as half an hour.[36] Newly cleaned clothes brought home and placed in a closet may result in PCE levels within the closet more than one hundred times the level the federal government considers safe for workers. The bedroom may have levels eight times as great, and even adjoining rooms may be contaminated at levels five times the worker safety standard.[37]

Workers and those who live near dry cleaning establishments are at even greater risk because of concentrated and frequent exposures. A 1995 study by Consumers Union of apartments in New York City located above dry cleaners found that most tested apartments contained average levels of PCE more than four times the health-based guideline set by the New York State Department of Health. Almost one-third of the apartments in the study registered levels of the toxic chemical more than ten times the standard, and at least one apartment was found to have an average PCE level 250 times greater than the Health Department's guideline.[38]

Alternatives to traditional solvent-based cleaning are available. Wet cleaning processes using soaps and a controlled application of water have been found to be extremely effective and in blind tests have enjoyed a customer satisfaction rate equal to or better than dry cleaning. There are as well other ways to reduce the risk of exposure:

- Avoid chemical dry cleaning whenever possible.
- Air out freshly dry cleaned clothes (preferably outdoors) before wearing or storing them indoors.
- If you live above or near a dry cleaner, have your home tested for PCE contamination.
- Women who are pregnant, trying to become pregnant, or nursing should avoid any exposure to PCE.
- Urge your local dry cleaner to switch to safer cleaning methods.

Gas Stations and Auto Repair Shops
Gasoline contains a complex mixture of organic solvents and in some areas may also contain manganese. The solvents in gasoline are very volatile and are an inhalation hazard when pumping the gas. Stations without vapor-lock systems allow considerable quantities of the vapors to escape into the air of the local neighborhood. For most people, a significant portion of their regular exposure to organic solvents comes from pumping gas into their car. Spills during fuel transfers also vaporize and may drift through the neighborhood. Finally, underground fuel storage tanks may leak and contaminate water supplies.

Auto repair and auto body shops may also emit significant quantities of volatile organic chemicals into the neighborhood, and chemicals or wastes stored on site may leak into the ground and eventually into the groundwater.

Avoidance

It is preferable to go to a gas station with a vapor-lock system in place (a rubber gasket, now required by many states, fits around the gas tank opening to prevent fumes from escaping) and also a lock that keeps the pump in the "on" position, allowing the person who is dispensing the gas to walk away during fueling. Full-serve stations reduce your exposure, but not that of the attendant.

Agricultural Operations

Nearby agricultural facilities present an exposure risk from pesticide use. Pesticides sprayed on a field may drift for some distance, depending on weather conditions. Those applied to a field may wash off after rain or irrigation and may end up in streams, lakes, and groundwater. Some crops are more pesticide intensive than others. Some are sprayed during a relatively brief time during the growing season; others undergo applications of pesticides year round. Greenhouses may be another area where pesticides are used.

California and, on a more limited scale, New York require more detailed reporting of pesticide use. The reporting system for commercial pesticide applications in California is quite complete.

Avoidance

Unless you move, you can probably do little about removing yourself from exposure to pesticides if they are used regularly on a nearby farm. However, you can talk to the farm operators about Integrated Pest Management (integrating natural pest control methods to reduce the use of pesticides) or organic farming techniques, or ask to be notified of spraying schedules so you can leave during those times.

Parks, Playgrounds, and Golf Courses

Golf courses, parks, and playgrounds may be heavily sprayed with pesticides, which may also contribute to pesticides in the groundwater. In a study of the San Francisco Recreation and Parks Department done between 1994 and 1995, no fewer than sixty different pesticides, including twenty suspected of causing reproductive harm, were used.[39] Park and playground equipment may also contain lead paint and/or wood preservatives that contain fungicides, which can be absorbed through the skin. Perhaps in part due to awareness about pesticides raised by the study, San Francisco has implemented an innovative program to target elimination of pesticide use in parks, playgrounds, and public buildings by the year 2000. It began with a ban on the use of the worst pesticides and includes provisions for four-day notification of spraying in all public buildings and an adoption of Integrated Pest Management as the official policy of the city.[40]

Avoidance

Inquire about your town's use of pesticides on parks, playgrounds, and public buildings, and work toward adoption of an ordinance similar to the one passed in San Francisco. At the least, you should request notification prior to spraying so you can avoid the areas during those times.

For recreational activities such as golf, try to find out from the golf course facilities management when the pesticides are applied. Many courses have almost continuous spraying programs. If at all possible, avoid playing directly after applications. You should encourage your golf course management to adopt IPM programs, if not pesticide-free maintenance.

Workplace Assessment

The work environment is a very important potential exposure source. Many jobs involve chemical exposures that often exceed levels commonly encountered in the home or community. Most of the well-documented reproductive health effects from chemical exposure derive from studies of workers.

Right to Know in the Workplace

Workers who handle chemicals or work near where they are used in the workplace should be familiar with the names of the chemicals and their potential health effects, as well as procedures for controlling exposures. The Occupational Safety and Health Administration (OSHA) requires employers to provide safe workplaces. All workers (except some state and city government workers) are covered by the Hazard Communication Standard (HazCom) of OSHA, which requires employers to label containers with chemical name or hazard warning, provide an MSDS for every chemical handled at the facility, and provide training on the health and safety hazards of toxic substances to which workers are exposed. An employer that does not provide MSDSs when asked or retaliates against an employee for requesting MSDSs is breaking the law. All employees should have copies of the MSDSs of all chemicals with which they are directly working.[41]

MSDSs are highly variable in quality; the information contained in them on health effects can be sketchy, incomplete, or even wrong. If an MSDS states that a chemical may be toxic to reproduction, there is likely quite good evidence that it is. If the MSDS says nothing about effects on reproduction or says that there are no effects, these statements may need to be further investigated.[42]

Employees who are concerned about their reproductive health may wish to do the following:

- Survey the workplace to find out where chemicals come into the plant and where they go, the adequacy of the ventilation system, how often that system is checked by the employer, and when (or if) air sampling is done. Also try to find out if any coworkers have health problems they believe to be work related.
- Use protective clothing and respirators if they are supplied. Work practices are an important piece of the exposure picture over which employees have some control. Washing up before eating, smoking, and going home are important to reduce personal and family exposures. If the exposures are high, however, even the most careful work practices may not provide sufficient protection.
- Find out whether the employer maintains a reproductive health policy that applies to pregnant workers, and may even extend to workers who are planning a pregnancy (in some cases, both sexes). By law, women cannot be denied jobs or be fired for being pregnant. It is illegal for an employer to ban women from certain jobs because the job could be unsafe for someone pregnant or contemplating pregnancy.[41] Rather, hazards should be removed or remediated to make jobs safe for all workers. Sometimes a pregnant worker provides a good incentive for the employer to reassess the exposures and exposure control practices in a certain work area and take steps to decrease or eliminate exposures for all workers.
- Discuss any concerns with a supervisor, company nurse, or physician.

If it is not possible or comfortable to approach your employer with questions, then take your concerns to your physician.

- Pregnancy and environmental hot lines are maintained in a number of states by the Organization of Teratology Information Services (OTIS). Call with questions.

If there are several employees with concerns about their reproductive health, and these concerns are not adequately addressed by the employer, several steps might be taken:

- You and your coworkers can file a formal complaint with the state health or labor agency or directly with the regional office of OSHA if you believe there is an OSHA standard violation. There are state-to-state variations regarding OSHA. Some states have approved OSHA state plans and have been granted authority to conduct enforcement actions. Others states do not have enforcement power but can provide consulting services through state occupational health and labor agencies, which can encourage OSHA to take action. The regional OSHA office or the state occupational health department (usually under the State Department of Health or Department of Labor) can determine which path to follow. Worker rights advocates suggest always acting in concert with another person. It would be wise to contact your union representative, or a workers' right advocate, to help guide you through the system. You can request anonymity if you fear retaliation from your employer.

- Following a complaint, state health agency inspectors may inspect the workplace and make recommendations for changes to increase safety. If OSHA becomes involved, it will look for violations of workplace standards and can order employers to remove hazards and impose fines if hazardous conditions are not corrected, but it is ill equipped to address reproductive toxicant hazards.

- Under the Health Hazard Evaluation Program (HHE) of the National Institute of Occupational Safety and Health (NIOSH), you can request a workplace health study, particularly if your concern is related to an uncommon or new problem. You must be an employee with two corroborating colleagues, a union representative, or an officer of the company to make a request. NIOSH will determine whether it will do a site visit to investigate your complaint.[43]

Office Workers

Office exposures can be similar to those at home. However, depending on ventilation, there may be solvent exposures from cleaning products, deodorizers, or office products such as markers, glues, and adhesives. Toners and other products from nearby copy centers or machines can contribute to higher levels of exposures. Particle board and carpets in the office may emit formaldehyde into the air. Pesticide residues from nighttime spraying may linger in the air and on surfaces. Most of these exposures are probably not as high as in a production workplace; however, it is often hard to control what is used in an office and it can even be hard to learn what has been used. The janitorial staff or building manager can be good resources for information concerning chemicals and pesticides.

Schools, Day Care Centers, and Other Public Buildings

Schools bridge the gap between home, community, and workplace; many of the same issues are relevant in these settings. Concerns about exposures at school need to consider the increased vulnerability of children to many toxic chemicals. Investigating potential risks can involve community and workplace authorities, as well as teachers, administrators, and the school board. School administrators and building managers are the first people to approach. Either the teachers' union (teacher) or the school board (parents) are logical next steps. Again, it is always useful to contact the local Coalition for Occupational Safety and Health (COSH) organization or other union or worker advocates for assistance.

Parents can get involved in their community's decision making process regarding policies on such issues as the spraying of pesticides inside and outside of schools and what types of cleaning chemicals are used. They can also request prior notification regarding pesticide use. There may be public health concerns regarding use of specific products; however, alternatives should always be considered. Sometimes antiquated policies remain in place simply because they have not been reviewed or challenged.

An example of an initiative that focuses on health and safety in schools is the Massachusetts Healthy Schools Network, which works to solve and prevent school indoor air quality problems. An important resource used in the program is the EPA's Indoor Air Quality Tools for Schools kit, designed to help interested people carry out indoor action plans. Included are checklists for school employees, fact sheets on indoor air pollution, and other materials. The EPA has other references available on such issues as lead in school drinking water and pest control in schools.

Libraries, town halls, and other public facilities may also be potential sources of indoor and outdoor chemical exposures to both workers and the public. Community health officials should be contacted first with any concerns about these facilities.

Consulting a Physician

Most health care practitioners have had little or no training in environmental health and may be poorly prepared to answer questions about environmental exposures. One way to deal with this situation is to seek out a specialist in occupational and environmental medicine (OEM) or ask for a referral to such a specialist. The Association of Occupational and Environmental Clinics (AOEC) can provide a referral. The other way is to be very well prepared with specifics regarding chemical concerns and exposures when consulting a health care provider. The key is to bring as much information as you can on chemicals about which you are concerned.

If the chemical is a consumer product or something used for a hobby, bring the product and, if possible, the MSDS. If the manufacturer will not provide the MSDS, then bring the manufacturer's telephone number and have the provider call for it because even trade secrets must be revealed to a physician under certain circumstances. If the exposure is in the workplace, bring the MSDSs of all chemicals of concern. If it is a local pollution problem, such as concerns about a nearby facility that may be releasing chemicals, then bring the most specific information possible about the chemicals and approximate quantities involved.

Ask your provider to seek out information about the possible effects of these chemicals. Sometimes this goes beyond their training and the time they are allotted to spend with each patient. Thus, it is helpful to bring information about the potential health effects of the chemicals, obtained from the MSDS or from other resources listed in this book.

The discussion should include a range of issues. If you are pregnant, your provider should ask about health problems, smoking, alcohol, medications, drugs, other environmental exposures, and family history. The goal is to identify any overall risk to the pregnancy, and then to try to address these risks.

An intake form at an AOEC clinic includes the following typical questions:

- Exposure history. Questions about exposures to chemicals in the home, workplace, or community and questions on all the types of jobs held (even summer jobs).
- Types of exposures. Questions about particularly heavy exposures to known toxicants.
- Community and home. Questions about pollution concerns in the community or products used in the home.

If a provider is interested but feels ill equipped to answer your questions, or seems inclined to brush off concerns without further investigation, show him or her chapter 10 in this book or ask for a referral to an OEM specialist.

Not all exposures are hazardous, and not all chemicals present in an area will result in significant exposures to workers or neighbors. Nevertheless, it is important to be aware of which chemicals show signs of being possible problems and to investigate any possible exposures before concluding that there is no significant risk.

Conclusion

There are multiple daily sources of exposures in all of our lives in this industrialized society. Many of these exposures are at low levels, and the health risk from any one exposure is probably small; yet some of these exposures are to substances that most people can do something about. Eliminating even one exposure in a day, or decreasing several exposures, will make a difference in overall exposures and may thereby decrease potential health risks.

Primer for the Clinician

10

A twenty-seven-year-old woman sees you when she is twelve weeks pregnant with her first child. She is concerned that her job as a laboratory technician may be hazardous to her fetus.

A young couple comes to your office with concerns about fertility. They were married almost two years ago and have not yet succeeded in conceiving a child.

One of your patients is a three-year-old child who is beginning to worry you. He seemed healthy as an infant but now is behind the normal developmental milestones for his age. You are considering having him evaluated by a specialist.

All health care workers in direct patient care should be alert to potential occupational or environmental exposures. Reproductive and developmental toxicants threaten pregnant women, women and men of reproductive age, and infants and children, whose bodies are still developing. A brief environmental and occupational history is an important part of the routine medical history. Appropriate follow-up of pertinent positive responses is equally important.

The Occupational and Environmental History

As part of every full history and physical examination, the patient should answer the following questions:

- What work do you do now, and what work have you done in the past?

- What are your hobbies?
- Are you exposed to any fumes, dusts, or chemicals at work or in your hobbies?
- Do you have any concerns about exposures in your home or community?

If the answers to these questions are negative, it may be appropriate to stop unless the index of suspicion of an exposure problem is high. A positive response to either of the last two questions, or a job or hobby that is unfamiliar to the clinician or may entail exposures to chemicals or physical agents (radiation, excessive heat), does require follow-up questioning tailored to the patient's response.

Workplace Exposures

If the patient's exposure concerns center around the workplace, appropriate follow-up questions may include:

- Describe exactly what you do at work and what chemicals you use.
- How long have you been doing this job?
- Are coworkers or their families experiencing similar problems?
- Do you get symptoms in the workplace? Do they persist on weekends or vacations?
- Is your work area well ventilated?
- Do you use personal protective equipment (gloves, respirator, coveralls, etc.)?
- Do you eat, drink, or smoke in the workplace?
- Are there adequate hand-washing and/or shower facilities at work?
- Do you wear your work clothes home? Who washes them?
- Has your employer, OSHA, or the state taken air exposure measurements?
- Have there been any accidents or exposure incidents on the job?

The laboratory technician who is twelve weeks pregnant works alone in a tiny quality control room, where she dissolves small amounts of various chemicals in a solvent. She washes glassware with the solvent and has cleaned up numerous small spills on the countertop and floor. She reports poor ventilation and says she develops a headache and burning eyes by the end of the day; these symptoms do not occur on weekends. She wears latex gloves and an apron but no respirator.

Hobbies

Some people have hobbies that may pose an exposure risk. Common hobbies with potential exposures to reproductive toxicants include auto repair (lead, solvents); ceramics, painting, and handling of firearms (lead); furniture refinishing (solvents); and gardening (pesticides). In exploring these exposures, the follow-up questions should be similar to those about occupational exposures.

The father of the three-year-old boy reports that his hobby is fishing. He goes to local lakes every weekend and brings his catch home to the family dinner table. The family eats local fish at least twice a week. He is not aware of any state fish advisories, but you suspect that he may not read English.

Household Exposures

In addition to hobbies, other household exposures may occur from chemical use in or around the home, take-home exposures from work, or cultural, social, or religious practices in the home. Some questions that may help elicit relevant information include:

- What household chemicals, cleaning agents, or pesticides do you use?
- What is your spouse's job? Does he or she wear work clothes home?
- When was your house built? Have there been recent renovations?
- Is your water supplied from a well or from a municipal supply?
- Do you use any folk remedies or nonprescribed medications?
- What types of cosmetics do you use?

Some folk remedies contain lead or mercury. Mercury may be in skin-lightening creams sold in other countries or imported illegally. Metallic mercury may also be used in religious practices. Lead is found in cosmetics, including lipstick and hair dye sold in the United States and eye makeup from other countries, such as kohl, which is commonly used in the Middle East.

The young infertile couple lives on a farm. The husband sprays pesticides as often as once a week during certain seasons. He does not wear coveralls when he sprays, and his wife washes his work clothes with the rest of their clothes. They have a private well and have never had the water tested.

Community Exposures

If the patient's concerns center around exposure in the community, some appropriate follow-up questions may include:

- Are other family members or neighbors experiencing similar problems?
- Do you smell chemical odors in your house or neighborhood?
- Is there an industrial facility, toxic waste site, or incinerator near your house?
- From which direction do the winds blow? Is the site uphill or downhill?

The mother of the three-year-old boy is anxious about pollution in their neighborhood. The industrial facility one block away from their house frequently emits irritating smoke and odors. In the mornings, they sometimes find a layer of greasy soot on the car. She knows that the name of the facility is Bayview Smelting.

Pitfalls to Avoid

There are some mistakes that clinicians may make when evaluating an occupational or environmental concern. One is to cease follow-up questioning after encountering an unfamiliar job title, whereas the appropriate next step is to ask for more information about what that job entails. Other pitfalls include dismissal of a patient's concerns without thorough evaluation, or overreaction by telling a patient to leave the job or move out of the neighborhood without first fully evaluating whether such extreme advice is really called for.

An important role for the clinician is to provide reassurance and to remove unwarranted anxiety. This reassurance must be reserved for situations where it is appropriate, such as after a careful evaluation of the nature of potential exposures, the extent of the exposures, and the evidence for potential health effects. Both glib reassurance and uninformed overreaction are a disservice to the patient.

When seeing a patient with a reproductive health concern, it is important not to neglect the rest of a good medical history, including questions about personal habits that can affect reproductive function or interact with other exposures. The occupational and environmental history is not a replacement for a thorough general medical history tailored to the patient's chief complaint.

How to Obtain More Information

In the event of a concern about a potential occupational or environmental exposure or illness, the next step is to collect further evidence regarding the identity of the possible exposure. If the identity of the chemical(s) to which the patient may be exposed is not known, then identification may require some of the following steps:

- Have the patient request Material Safety Data Sheets (MSDSs) from the employer. These contain the identity of the chemical and some information about health effects (see figure 10.1).
- If the source is a consumer product, phone the manufacturer for an MSDS. Alternatively, poison control centers can frequently provide this information.
- If the chemical is not identified on an MSDS because the ingredient is a trade secret, a physician can access this information but may first have to sign a confidentiality agreement.
- If the concern centers on a nearby industrial facility, the Toxics Release Inventory may be useful. (See chapter 9.)

As you continue the infertility workup of the young couple, you investigate the possibility of pesticide overexposures. The husband comes for his next visit with the cans and bottles of the pesticides he uses most often. The ingredients include 2,4-D, diazinon, carbaryl, vinclozolin, and chlorpyrifos. You note that the labels report that over 97 percent of the ingredients are inert ingredients, without further specification. You consider calling the manufacturers for more information about these "inerts."

The laboratory technician returns for her second visit with some Material Safety Data Sheets. Although numerous chemicals are used in the laboratory, she points to the one that she uses every day in large quantities: N-methyl-2-pyrrolidone. The MSDS does not mention any effects on reproduction.

You spend a few minutes on the Internet, pulling up the Toxics Release Inventory emissions data for Bayview Smelting, the company that worries the mother of the three-year-old boy. The facility turns out to be the second largest emitter of lead in the state.

Once the identity of the exposure is known, some qualitative information about the level (dose) of exposure can come from the patient

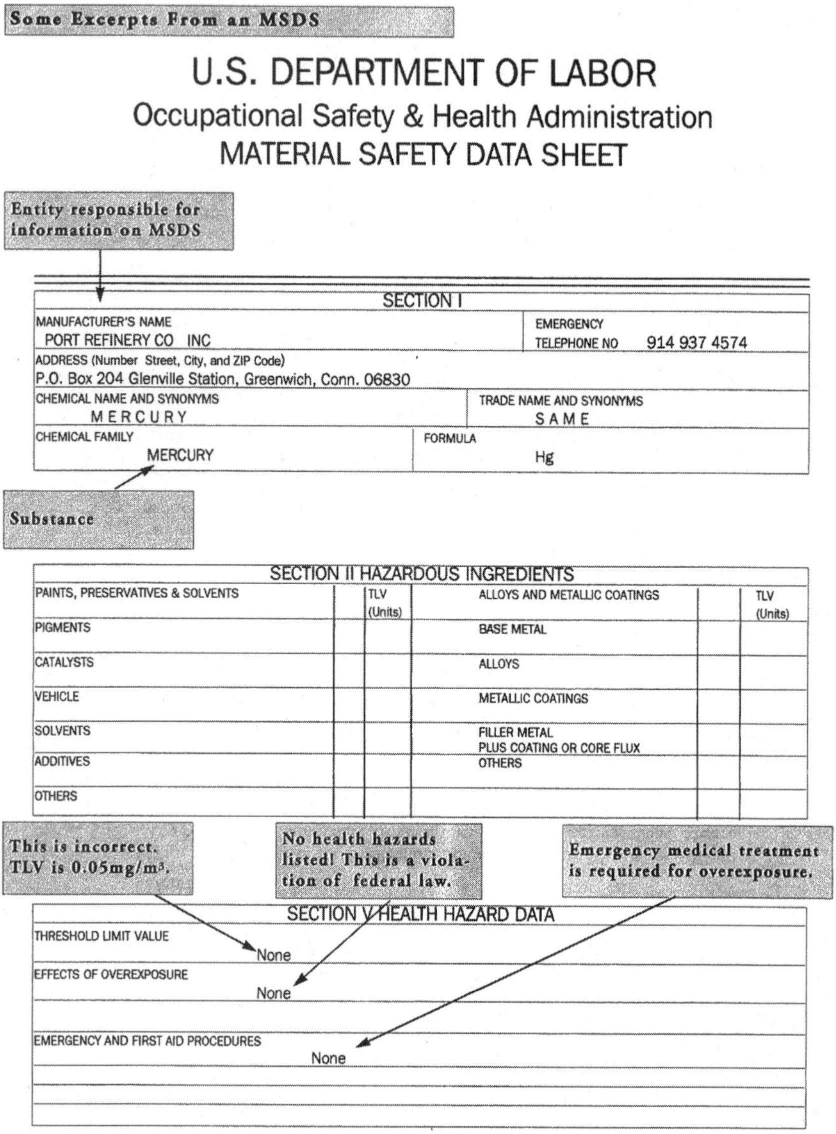

U.S. DEPARTMENT OF LABOR
Occupational Safety & Health Administration
MATERIAL SAFETY DATA SHEET

Entity responsible for
information on MSDS

SECTION I		
MANUFACTURER'S NAME	EMERGENCY	
PORT REFINERY CO INC	TELEPHONE NO	914 937 4574
ADDRESS (Number Street, City, and ZIP Code)		
P.O. Box 204 Glenville Station, Greenwich, Conn. 06830		
CHEMICAL NAME AND SYNONYMS	TRADE NAME AND SYNONYMS	
MERCURY	SAME	
CHEMICAL FAMILY	FORMULA	
MERCURY	Hg	

Substance

SECTION II HAZARDOUS INGREDIENTS				
PAINTS, PRESERVATIVES & SOLVENTS	TLV (Units)	ALLOYS AND METALLIC COATINGS	TLV (Units)	
PIGMENTS		BASE METAL		
CATALYSTS		ALLOYS		
VEHICLE		METALLIC COATINGS		
SOLVENTS		FILLER METAL PLUS COATING OR CORE FLUX		
ADDITIVES		OTHERS		
OTHERS				

This is incorrect.
TLV is 0.05mg/m³.

No health hazards
listed! This is a viola-
tion of federal law.

Emergency medical treatment
is required for overexposure.

SECTION V HEALTH HAZARD DATA	
THRESHOLD LIMIT VALUE	None
EFFECTS OF OVEREXPOSURE	None
EMERGENCY AND FIRST AID PROCEDURES	None

Figure 10.1   Example of a poor quality Material Safety Data Sheet

history. The type of environment, ventilation, and the patient's symptoms may all be helpful in this regard. Biological monitoring for residues of the chemical may be useful to confirm or quantify estimates of exposure. (See table 10.1.) Unfortunately, biological monitoring for many chemicals has not been developed, is poorly validated, or may not be available in most laboratories. Thus, it is not always possible to confirm or quantify exposure in this way.

If there is a need to determine the level of exposure more exactly, environmental monitoring may be necessary. Monitoring may be done by state health departments, the federal Occupational Safety and Health Administration (OSHA), or the National Institute for Occupational Safety and Health (NIOSH) for workplace exposures. Alternatively, employers may have monitoring capabilities and may be willing to share this information. Occupational physicians can sometimes arrange a workplace visit with an industrial hygienist to observe the working conditions and potentially measure exposures. Community exposures may be evaluated by state or local health departments, departments of environmental protection, or the federal Agency for Toxic Substances and Disease Registry (ATSDR) (see appendix A).

Because the farm couple receives all their water from a well, you recommend that they call the state health department to check into the possibility of getting the drinking water tested.

The employer of the laboratory technician denies having done any air monitoring in the quality control laboratory. The patient asks you not to call OSHA or the state department of labor because she is afraid of retaliation from the employer. On further history, you learn that when the patient cleans up solvent spills, her latex gloves partially dissolve, resulting in direct skin contact. In addition, the room she works in is small and without good ventilation. You assume she may be significantly exposed.

After determining that the three-year-old child eats fish from local lakes, you seek more information. The local health department is happy to provide you with a list of lakes in the region with fish advisories. Fish from numerous local lakes have been found to be significantly contaminated with mercury.

Once the identity of the chemical is known, and there is evidence, even qualitative, that the patient may be exposed, the next step is to

Table 10.1   Biological monitoring for exposure to potential reproductive toxicants

Substance	Biological monitoring test
Metals	
Arsenic	Urine (S), hair, nails (L)
Lead	Blood (S), urine, bone (L)
Mercury	Urine (S), blood, hair (L)
Cadmium	Blood (S), urine (L)
Manganese	(Urine, blood)
Solvents	
Benzene	Benzene in exhaled air, blood S-phenylmercapturic acid, muconic acid in urine, (phenol in urine)
Xylene	Methyl hippuric acid in urine, xylene in blood
Toluene	Toluene in blood, exhaled air, (hippuric acid in urine)
Styrene	Phenylglyoxylic acid in urine, styrene in exhaled air, blood, (mandelic acid in urine)
Phenol	(Urine)
Perchloroethylene (PCE)	PCE in exhaled breath, blood, trichloroacetic acid in urine
Trichloroethylene (TCE)	Trichloroacetic acid, trichloroethanol in urine, TCE in exhaled breath, free trichloroethanol in blood
Chloroform	Exhaled breath, blood
Ethylene glycol monoethyl ether	Ethoxyacetic acid in urine
Methylene chloride	Carboxyhemoglobin, methylene chloride in blood, exhaled breath
Pesticides	
Organophosphates	Red blood cell cholinesterase, plasma cholinesterase; DMP, DMTP, DEP, DETP in urine
Parathion	p-nitrophenol, DEP, DETP in urine
Malathion	Alpha-monocarboxylic acid, dicarboxylic acid, DMP, DMTP in urine
Chlorpyrifos	3,5,6-trichloro-2-pyridinol, DEP, DETP in urine
Carbaryl	1-naphthol in urine or blood
Propoxur	Isopropoxyphenol in urine
EBDCs	Ethylene thiourea (ETU) in urine
Lindane	Lindane, tri- and tetra-chlorophenol isomers in serum
DDT	Serum DDE
Chlordane/heptachlor	Trans-nonachlor, heptachlor epoxide, oxychlordane, and heptachlor in serum
2,4-D	2,4-dichlorophenoxyacetic acid in urine
Methyl bromide	Bromide ion in blood
Pentachlorophenol (PCP)	Pentachlorophenol in urine
PCBs, dioxins	Blood, fat

Note: Tests in parentheses are poorer choices in terms of accuracy, sensitivity, wide interindividual variability, or poor correlation with toxicity.
(S) short-term (recent) exposures, (L) long-term exposures.
EBDCs = ethylene bisdithiocarbamates (maneb, zineb, nabam, metiram, mancozeb); DEP = diethylphosphate; DETP = diethylthiophosphate; DMP = dimethylphosphate; DMTP = dimethylthiophosphate.

---

gather information regarding potential health effects. Unfortunately, of the over seventy-five thousand chemicals that may be in use today, only a small fraction has been tested for effects on reproduction and development. Thus, the clinician may face a challenging task in attempting to answer a question about the risks to a patient from a particular exposure. Useful resources for reproductive health effects information include pregnancy risk hot lines or poison control centers, textbooks of occupational and environmental medicine, computer databases available through the National Library of Medicine (Medline, Toxline, ReproTox, RTECS, or others), and specialists in occupational and environmental medicine. Additional information is available in chapters 3 through 6 of this book and the resource guide in appendix A.

A call to poison control and review of several textbooks fails to reveal any information about infertility and pesticide exposure. A Medline search, however, locates several articles reporting on a possible association between unspecified mixed pesticide exposure and infertility in males and females. You are unable to determine which, if any, of the pesticides may be a particular problem for the farm couples.

A Medline search on solvents reveals numerous studies linking a wide range of solvents with spontaneous abortions. A couple of studies mention N-methyl-2-pyrrolidone, which the laboratory technician uses, specifically in connection with an increased risk of miscarriage.

A general textbook of occupational medicine has a chapter on methyl mercury and its effects on the fetus and children. You realize that it is possible to test for mercury levels in the three-year-old's blood, urine, and hair and make arrangements through a good laboratory to do biological monitoring of the entire family. You also order a blood lead test on the child.

Follow-up

If a potential health threat is identified, the clinician must follow up to ensure that the situation is adequately addressed. In some states there is a legal requirement to report occupational illness to the appropriate state agency. The clinician must present the information and alternatives to the patient in as clear and thorough a manner as possible. Further action may involve these avenues:

- Close medical monitoring
- Risk reduction through behavioral modification (smoking cessation, changes in hobbies or work practices, dietary changes)
- Contacting the employer to advocate for changes in the worksite, job modification, worker removal, or to put the patient on disability leave
- Contacting state or federal agencies to urge investigation of a worksite or a community exposure situation
- Medical referral to an occupational and environmental medicine specialist

In a workplace, there are often several options for exposure reduction. The best option is generally process change. This requires evaluation of the job to eliminate the use of the particular chemical or to substitute a safer alternative. If a process change is not possible or practical, then enclosing the process and improving ventilation (perhaps through a fume hood) is the next best alternative. Personal protective equipment is generally least helpful because it may fit poorly, be misused, or be insufficient to protect against the exposure. Transfer of the worker to a different job is a possibility during pregnancy, but does not address the health hazard.

Individual patients may be in a difficult situation where they are facing a decision between a potential reproductive health hazard and economic insecurity. Risk of losing a job leads many people into taking health risks; others are unable to afford to renovate housing to remove peeling lead paint or to move out of a community near a source of environmental pollution. In these situations, the clinician may provide information, advocacy, and assistance but may not always be able to provide the optimal solution.

The preliminary workup of the farm couple shows that the husband has a borderline low sperm count, but there are no other clear abnormalities. You cannot say that their problem is necessarily pesticide related, but you plan to treat the infertility while also recommending exposure reduction. You discuss minimization of pesticide use, use of proper protective clothing, and avoidance of washing contaminated clothing with clean clothing. You refer the couple to a nonprofit organization that provides information about pesticide use reduction.

You write a letter to the employer of the laboratory technician asking for an exposure assessment of the quality control lab and recommending that the patient be

moved to a job without chemical exposures until her workstation has been evalu-
ated by an industrial hygienist. A follow-up telephone call to the employer ascer-
tains that the employer is planning to comply. You follow the patient closely
throughout the remainder of her pregnancy. The employer eventually installs a
fume hood above the lab bench and eliminates a solvent filtration step that was
responsible for most of the exposure.

The three-year-old child has a mildly elevated blood organic mercury level, but
it is unclear if this level is high enough to result in clinically delayed development.
The child's lead level is 19 μg/dl, which is above the Centers for Disease Control
level of concern of 10 μg/dl. You ask the parents to stop eating fish from the
local lakes, arrange for lead testing of the home, and plan to refer the child to a
consultant for further guidance.

Further Reading

Numerous excellent articles and books can help guide the clinician in taking a
more thorough, yet concise, occupational and environmental history. Some of the
most helpful resources follow.

Occupational and Environmental History

Goldman RH. Suspecting occupational disease. In: McCunney RJ (ed), *A Practical
Approach to Occupational and Environmental Medicine*. Boston: Little, Brown,1994.

Goldman RH, Peters JM. The occupational and environmental health history.
JAMA 246(24):2831–2836, 1981.

Occupational and Environmental Health Committee of the American Lung Association. Taking the occupational history. *Ann Intern Med* 99:641–651, 1983.

Paul M. Clinical evaluation and management. In: Paul M (ed), *Occupational and
Environmental Reproductive Hazards: A Guide for Clinicians*. Philadelphia: Williams
and Wilkins, 1993.

Paul M, Himmelstein J. Reproductive hazards in the workplace: What the prac-
titioner needs to know about chemical exposures. *Obstet Gynecol* 71:921–938,
1988.

Material Safety Data Sheets

Several previous authors have appropriately critiqued the clinical utility of Material
Safety Data Sheets (MSDSs). The following resources demonstrate that although
MSDSs may be useful for determining ingredients information, their utility as a

source of toxicity information is extremely limited, particularly with regard to effects on reproduction and development.

Kolp P, Sattler B, Blayney M, Sherwood T. Comprehensibility of material safety data sheets. *Am J Ind Med* 23:135–141, 1993.

Lerman SE, Kipen HM. Material safety data sheets: Caveat emptor. *Arch Intern Med* 150:981–984, 1990.

Paul M, Kurtz S. Analysis of reproductive hazard information on Material Safety Data Sheets for lead and the ethylene glycol ethers. *Am J Ind Med* 25:403–415, 1994.

Biological Monitoring and General Information

Biological monitoring is an exciting and rapidly evolving area with the potential for great future utility for clinical medicine. More detailed discussions of biological monitoring strategies for specific chemicals may be found in some of the following resources.

Coye MJ, Lowe JA, Maddy KJ. Biological monitoring of agricultural workers exposed to pesticides: II. Monitoring of intact pesticides and their metabolites. *J Occup Med* 28(8):628–636, 1986.

Lauwerys R, Hoet P. *Industrial Chemical Exposure: Guidelines for Biological Monitoring*, 2d ed. Boca Raton, FL: CRC Press, 1993.

Rosenberg J, Harrison RJ. Biologic monitoring. In: LaDou J (ed), *Occupational and Environmental Medicine*. Stamford, CT: Appleton & Lange, 1997.

Reflections and Recommendations

11

And, oftentimes, to win us to our harm,
The instruments of darkness tell us truths,
Win us with honest trifles, to betray us
In deepest consequence.
—*Macbeth* I,3

Exposures to toxic chemicals without the informed consent of those exposed emerges as a dominant theme throughout this book. Sometimes evidence of exposures or toxicity is disregarded, concealed, or not communicated to those who bear the risks. In other cases, the risks are unknown and unstudied, but exposures continue. Many factors conspire to maintain a system that fails to incorporate basic principles of public health practice or to acknowledge fundamental human rights: a fragmented and reactive approach to problem solving, the perception of science as completely objective, an uninformed public, an economic system that encourages rapid development and marketing of new chemical products, and the curious notion that corporations have the right to expose people to untested chemicals.

The complexities of modern, industrial society encourage specialization, fragmented thinking, and a reductionist worldview. We divide complex issues into pieces to manage them more efficiently, neglecting to reassemble them for more integrated analysis. This tendency is pervasive in science and is reflected in civil and political institutions. Physicians specialize and subspecialize until few are able to address or even see the spectrum of health concerns affecting the whole person and his or her living environment. Scientists focus on narrow questions, losing sight of the broader implications of their work. Local, state, and federal governments divide the world of their responsibility into pieces, distributing them to agencies and departments for oversight and regulation. Institutional subcultures compete for turf, perpetuating artificial barriers.

A reductionist worldview encourages us to dissect complex ecosystems into component parts, examining air, soil, water, individuals, cells, enzymes, and DNA. Relationships and reverberations, positive and negative feedback loops, and unexpected cascades of events, apparent only through the lens of integrated analysis, are lost from view. Regulatory policies reflect this reductionist thinking. Separately established programs monitor individual chemical contaminants in air, water, or food. We regulate single chemicals rather than the complex mixtures that characterize actual exposures. The health effects of pesticides are predicted from tests of individual chemicals in animals of certain species, ignoring epidemiological evidence of harm at current exposure levels. There is no comprehensive examination of pesticide effects on populations, species diversity, and ecosystems; of the viability of nonchemical alternatives; and of the sustainability of current agricultural practices. Every regulatory proposal is studied and contested for years until it theoretically provides acceptable public health protection from one chemical and one route of exposure. Meanwhile, monitoring studies disclose a dizzying array of chemical toxicants in breast milk, blood, urine, and breath.

When regulators in 1923 allowed tetraethyl lead to be added to gasoline, they took comfort in the small amount per gallon and predicted safe air levels of lead. But public health professionals, including some of the most eminent of the day, openly worried that blanketing the nation with a fine dusting of lead was likely to be a serious health hazard. The Ethyl Corporation argued that there was no conclusive proof of risk. In fact, leaded gasoline did poison workers in industries from petroleum processing to auto repair. Airborne lead was deposited on land, in water, and on food. It lodged in human bone and was remobilized with increased bone turnover during pregnancy, causing prenatal damage to children's brains. A cautionary public health position was overwhelmed by a naive understanding of how lead moves through ecosystems, poor understanding of subtle developmental effects of low-dose lead exposure, and corporate profit motives, setting the stage for failure to protect the health of the public.

Science: Fragmentation and Political Influence

Scientific inquiry provides the basis for much of environmental health policy. Scientific practice is often held as being so objective that many policymakers promote it as a tool to mediate among diverse worldviews and priorities. Science, they say, rather than emotion or special interests, will guide public policy. This is particularly true in matters of public health. But if science is to be given that task, it is important to examine whether science and policy are as cleanly separated as our reductionist tendencies lead us to believe. A spectrum of opinions has emerged.

At one extreme are those who view the scientific enterprise as totally objective and independent of social forces. At the other are those who believe that scientific information is always socially constructed. For them, understanding the structure of social, political, and cultural institutions is essential for understanding the design, execution, and interpretation of scientific studies.

Every culture has its own language, practices, and standards. Science is no different. The scientific method, with established conventions, assumptions, logic, and standards of proof, is well known. A specialized language has obvious value within the culture, but it also tends to exclude participation by outsiders, contributing to an apparent isolation of the scientific endeavor from the politics of society and fostering a closed membership. What may not be immediately obvious are the biases that influence how science is practiced and interpreted for the purpose of setting public policy.

We have seen how particular agendas are served by publicly promoting the objectivity of science while economic and political forces subtly influence scientific investigation. Funding decisions determine which questions get asked and who gets to answer them. Editorial decisions determine what and where studies are published. Scientists who stray too far into policy debates with opinions unfavorable to their institutional supporters may find their research funding drying up. Even the seemingly objective choice of a null hypothesis or the decision to use a *p*-value of 0.05 to determine statistical significance is a political decision that may affect public health. Commonly, social and political debate is restricted to alternative ideas about what to do with information that science has provided. Less often, the public insists on scrutinizing the details of how that scientific information is collected.

Certain science-based techniques easily lend themselves to biased manipulation. When the overlap between science and policy is not explicitly identified, it can be abused by those with a specific political or economic agenda. For example, risk assessment and cost-benefit analysis are tools that are supposed to be useful for quantifying risks, costs, and benefits. But obscured in the complex mathematical formulas and apparent precision of the results are simplifications, assumptions, and value judgments that usually go unnoticed and unchallenged. A focus on cancer risk assessment diverts our attention from the fact that we do not have the slightest idea how to estimate the risks of many other health outcomes, including most discussed in this book. Estimated costs and benefits of cost-benefit analyses are primarily supplied by chemical manufacturers and are easily biased toward supporting continued chemical production and use.

The Soundness of Science

The voices of industry have been among the loudest promoting the idea that science can and must be practiced totally independent of political and economic interests. They have argued that policy decisions must be based on sound science rather than emotion or ideology. What they have in mind is a particular interpretation of science focusing on understanding molecular mechanisms, specific definitions of proof and causation, and narrow definitions of adverse effects. They imply that cautionary health-protective actions, taken without full proof of causation and harm, are scientifically unsound and unjustified.

In one of the more egregious examples of co-opting the cultural language of science for self-serving purposes, the tobacco industry argued for years that there was no proof that cigarette smoking caused cancer, based on lack of full understanding of the mechanism by which the disease developed. Yet the epidemiological evidence that smokers are at markedly increased risk of lung cancer was well established and overwhelming. In this instance, the strict definition of causality, which requires knowledge of a plausible biological mechanism, was used to argue against health-protective policies and provided cover to politicians who profited from tobacco industry support. Even today, the lack of proof of harm, where proof is defined with full scientific rigor, continues to be used as an excuse for manufacturing and marketing harmful products. Used in this way, science, promoted as the impartial mediator, contributes to a failure to protect public and environmental health.

The Right to Know

The reviews of the medical and scientific literature in this book show that people are being exposed to harmful chemicals with resulting adverse reproductive and developmental health effects. In many cases they are not aware of their exposure or the associated risks. Many workers are exposed to harmful chemicals on the job where they have keen incentives not to investigate too thoroughly. Others are exposed during the use of inadequately labeled consumer products. Political institutions have failed to inform or protect them. In the field of medicine, it is unethical to treat people with drugs or subject them to experimentation without their full and informed consent. This ethical principle is generally recognized throughout the world as a matter of international law. Yet as we have documented, people are regularly exposed to hazardous and inadequately tested substances without being informed and without their consent. Their right to refuse has been denied. Environmental justice is not even close to being recognized as a fundamental human right.

Rather ominously, another rights-based argument has received much more vocal support. It is based on property rights and reflects the underlying philosophy that most chemical manufacturers and processors have the right to freely produce and market chemicals without defining public health and environmental risks. For example, under the Toxic Substances Control Act, before undertaking any action to control production and use of potentially harmful industrial chemicals, the EPA must demonstrate that the risks outweigh both the costs to industry and the lost benefits of unrestricted use of the chemical. Should the risks be unknown, as is the case with the vast majority of chemicals in commercial use, the agency has extraordinarily limited authority even to require meaningful testing. This statute clearly puts the rights of the manufacturer above the rights of workers, consumers, or exposed community members.

Right-to-know laws accomplish two purposes. First, and foremost, they acknowledge the public's right to information necessary to make informed choices that affect their personal autonomy. Secondly, these laws provide information that may shed light on the interconnections between exposures and diseases. Unfortunately, disease registries, chemical use and release data, environmental and biomonitoring data, and regulatory information are not well integrated. There is no systematic effort to put the disparate pieces together for individuals, communities, populations, and ecosystems. They are also not nearly complete in their accounting of releases and locations of toxic chemicals. Right-to-know laws need to be expanded to provide more information. They also need to be more fully integrated so that individuals with reproductive health concerns may easily find information about exposures in the workplace, in the community, and the home from water, air, and food, and match this information to medical records. With these improvements, individuals will be better able to generate a systematic analysis for their families or communities and to begin to identify ways to better protect themselves and others.

An Integrated Public Health Approach

Acknowledging the complexities of the ecosystem into which synthetic chemicals are released is an important first step toward an integrated public health approach. Systematic analyses have taught us that system inputs often have effects that are uncertain and unpredictable. Industrial discharge of mercury into Minamata Bay was not expected to cause any health effects; a narrow analysis assumed that the mercury would be diluted in the ocean and would not present a public health threat. In fact, bacteria in the sediments of the bay biotransformed the mercury, and it accumulated in the food chain, resulting in a public health disaster. In northern Quebec, hydroelectric dam construction created the conditions for sediment buildup, providing a habitat for bacteria that biotransformed mercury, which had been atmospherically transported from fossil fuel burning and trash incinerators thousands of miles away. Organic mercury entered the food chain of native people who depended on fish from these waters as a dietary staple. As a result, the brains of developing fetuses in this remote area are regularly exposed to this central nervous system toxicant.

It is arrogant to argue that production and use of a chemical are safe when it has not been fully tested and the results revealed to the public. And it is arrogant to suppose that we can confidently predict the effects of releasing a persistent chemical into the environment. If we hope to protect public health, we need to develop a cautionary, systematic approach to science and policy. Exclusively narrow, fragmented methods for evaluating public health issues are guaranteed to fail to be protective. Certainly specialization has a role, but, it must be part of a broad assessment of biological, social, and ecological interrelationships. A systematic public health approach will involve redefining goals and priorities. To do that we must use scientific tools better suited to the complexity of the tasks and ask more elegant and relevant questions.

Redefining Goals

The Precautionary Approach

A precautionary approach is essential to systematic protection of public health and the environment. Guided by an awareness of uncertainties, a precautionary approach serves as an incentive to generate information. It demands proof of safety rather than proof of harm, requiring that chemicals be well studied for health and environmental effects before being released into the environment.

In order to set appropriate goals, reevaluate priorities, and measure progress, biological and environmental monitoring should be more systematic and thorough. When chemical residues are detected in human tissues or in environmental media, the safety of those exposures should either be known or thoroughly evaluated. However, exposure-reduction efforts should not be contingent on proof of harm. Information about potential environmental and occupational hazards must be readily and publicly available, so that people may make more informed decisions.

Selecting Scientific Tools

A precautionary approach requires selection of new scientific tools. For example, reversing the null hypothesis in scientific investigations when there is preliminary evidence of toxicity would create a requirement to assume, as the hypothesis that must be rejected, that a chemical is harmful. The burden of proof is then on the manufacturer to show that it is safe, or the conditions under which it is safe, as necessary prior to commercial use. This is the approach required in the pharmaceutical industry, which must conduct human clinical trials before a new drug can be marketed. Yet with a few exceptions, for the large majority of industrial chemicals to which people, wildlife, and ecosystems are regularly exposed, no such demonstration of safety is required. A rigorous definition of statistical significance ($p < 0.05$) can be extraordinarily useful for keeping us from wrongly asserting something to be true. However, it contributes to failure to protect public health if it causes us to ignore important trends in community-based studies where it is difficult to control for multiple variables. A particular definition of statistical significance is no more than a mathematical tool whose use is established by convention and may not be appropriate in all circumstances.

In the field of epidemiology, exposure assessment is a critical part of any analysis, yet it is the piece that is often missing. It is essential to gather good data on the extent and distribution of human exposures to a wide range of chemicals in the environment. Without exposure information, we are poorly equipped to detect causal exposure-disease relationships, monitor trends, recognize disproportionately affected communities, or determine if interventions are effective. Similarly, improved surveillance is a fundamental requirement for recognizing disease trends and generating truly relevant and informed questions. Stakeholder involvement in research and research agendas should also be a priority. Communities must be involved as equal partners and help define the questions being asked, develop protocols, participate in analysis and interpretation of results, and disseminate research findings.

Medical education should teach physicians how to consider their patients in the context of their physical and social environment. Scientists should be rewarded for interdisciplinary work and work that places individual scientific questions in a broader ecological or social context.

Asking Better Questions

Life cycle analyses and the long-term sustainability of practices with public health and environmental impacts are essential considerations in all policy decisions. For example, although styrofoam may not be a public health hazard when consumers use it, analyses of benzene and styrene exposures to workers during manufacturing, the combustion by-products from waste incineration, and their ecological and public health impacts are all important for understanding true risks.

At some point, it is essential to ask whether a chemical proposed for manufacture and use is necessary or if there are safer alternatives. This question must be asked and answered by a broader constituency than those who have vested economic interests in a particular class of chemicals. For example, many people question whether current agricultural practices, which rely on a constant flow of pesticides and fertilizers to the economic benefit of a few corporations, are really necessary, or if they should be replaced by sustainable agricultural practices that preserve the quality of our water, air, and land.

In the context of toxicity testing, relevant questions include assessment of the toxicity of chemical mixtures, a search for subtle effects of widespread low-dose exposures, and testing for a wider range of end points. Traditional toxicity testing may overlook or mask numerous important but subtle health effects. For example, at high doses originally used to test for developmental effects, PCBs were fetotoxic. This was assumed to be the relevant end point, and extrapolations were made from there to estimate a safe dose for humans. Careful testing at lower doses, however, ultimately revealed subtle effects on neurobehavioral development of offspring at doses more than 100-fold lower than those originally believed safe.

The generations born into the chemical age have been born into an age of specialization, fragmentation, and uncertainty. Periods of seeming quietude are regularly punctuated by health crises among workers and other highly exposed groups. Epidemics of reproductive harm, such as those due to mercury, lead, DBCP, and glycol ethers, briefly capture the attention of scientists, regulators, and the public. Yet each incident is perceived narrowly, viewed as isolated, considered irrelevant to most people, apparently solved, and then forgotten.

To some physicians who see ongoing individual tragedies, there is no illusion of quietude. To some scientists, public health professionals, environmental advocates, and alert members of the public, there are worrisome trends that are becoming clearer with time. With our limited understanding and in the interest of public and environmental health, we must respect the complexity of the ecosystems of which we are part, insist on the public right to information, and adopt a precautionary approach to protect the generations to come.

Appendix A

Resource Guide

There is no one comprehensive source for all information on a particular toxic hazard. It is important to remember that public agencies and private organizations may have very different goals and agendas, and that the way information is interpreted and presented must be analyzed and scrutinized for subjectivity and vested interest based on the stated goals of the agency or organization. The Internet is now by far the easiest way to access information on most subjects. (Note that all World Wide Web addresses must be preceded by http://.) The resources that follow were selected based on currency and usefulness of information, as well as reliability to the best of our knowledge. We do not in any way provide a complete list or imply an endorsement of any organization or product, but merely offer pathways to research.

Federal Government Agencies: Environment, Health, and Workplace Protection

Environmental Protection Agency

United States Environmental Protection Agency (EPA)
401 M Street, SW
Washington, DC 20460

(202) 260-2090 (directory assistance)

www.epa.gov

There are ten regional EPA offices throughout the country. Call for the contact numbers. *Access EPA* (publication number EPA 220-B-93-008), a guide to EPA's environmental services and databases, is available from the Government Printing Office at (202) 512-1800.

Selected EPA Internet Sites

www.epa.gov/epahome/r2k.htm

Excellent community right-to-know page with links to food, air, water, and land issues and databases such as the Toxics Release Inventory. Includes a link entitled "Concerned Citizens at the Workplace."

www.epa.gov/opptintr/tri (Toxics Release Inventory home page)

Database that provides information to the public about releases of toxic chemicals to the air, water, and land from some manufacturing facilities. The Toxics Release Inventory User Support Service, at (202) 260-1531, helps citizens locate and access TRI data. It provides general information about the TRI and support for access to any of the data formats; comprehensive search assistance for the TRI on-line and CD-ROM applications; and referrals to EPA regional and state TRI contacts and libraries where TRI is available.

EPA and Other Federal or Government-Contracted Information Lines and Hot Lines

Consumer Product Safety Commission Hotline. (800) 638-2772.

Emergency Planning and Community Right-to-Know Hotline. (800) 535-0202. Provides fact sheets on Toxics Release Inventory (TRI) state releases and includes state TRI contacts.

Environmental Publications and Information, National Center. (800) 490-9198.

Food and Drug Administration's Office of Cosmetics and Colors Automated Information Line. (800) 270-8869.

Government Accounting Office (GAO). (202) 512-6000. For copies of GAO reports www.gao.gov.

Indoor Air Quality Information Clearinghouse. (800) 438-4318, www.epa.gov/iaq/index.html.

Publications available free through the IAQ Info Line include: *The Inside Story: A Guide to Indoor Air Quality* (April 1995, IAQ-0029); *Carpet and Indoor Air Quality Fact Sheet* (October 1992, IAQ-0040), and *Indoor Air Pollution: An Introduction for Health Professionals* (1994, IAQ-0052).

Lead Information Center, National Safety Council. (800) LEAD-FYI, to obtain an information package. (800) 424-5323, to speak to an information specialist.

National Pesticide Telecommunications Network, EPA and Oregon State University. (800) 858-7378. Provides information on a range of issues regarding pesticides and provides referrals for more information. Documents available include *Citizens Guide to Pest Control and Safety* and the *EPA Catalog on Pesticide Publications*. http://ace.orst.edu/info/nptn provides links to numerous databases, including EXTOXNET.

National Response Center Hotline. (800) 424-8802. To report a chemical spill or a new hazardous waste site. This hot line is operated by the National Response Team, fifteen federal agencies with the EPA as chair and the U.S. Coast Guard, which staffs the hot line, as vice chair.

RCRA/Superfund Hotline. (800) 424-9346. Information on solid and hazardous waste issues and Superfund sites.

U.S. EPA Safe Drinking Water Hotline. (800) 426-4791. Information on the Clean Water Act and also on filters, state drinking water offices, and testing.

TSCA Hotline. (202) 554-1404. Questions pertaining to the Toxic Substances Control Act. Or e-mail to tsca-hotline@epamail.epa.gov.

Other Federal Agencies

Agency for Toxic Substances and Disease Registry (ATSDR)
U.S. Department of Health and Human Services
1600 Clifton Rd., NE
Atlanta, GA 30333
(800) 447-1544, (404) 639-6315 (fax)
atsdr1.atsdr.cdc.gov:8080/atsdrhome.html

Conducts public health assessments of waste sites, maintains health surveillance and registries, educates and trains on hazardous substances. Provides fact sheets on more than one hundred toxic chemicals under the ToxFAQs program. Maintains HazDat, an on-line database with information on hazardous waste sites and community health impacts.

Centers for Disease Control and Prevention (CDC)
National Institute of Occupational Safety and Health (NIOSH)
4676 Columbia Parkway
Cincinnati, OH 45226
(800) 356-4674
www.cdc.gov/niosh/homepage.html

Responsible for investigating and assessing potential workplace hazards for the Occupational Safety and Health Administration. Call for information to obtain a health hazard evaluation of your workplace. For the *NIOSH Pocket Guide to Chemical Hazards,* with information on 677 chemicals, order by fax: (513) 533-8573; e-mail: pubstaft@ NIOSDT1.em.cdc.gov; or mail: NIOSH Publications, Mailstop C-13, 4676 Columbia Parkway, Cincinnati, OH 45225. One copy is free.

National Library of Medicine
8600 Rockville Pike
Bethesda, MD 20894
(800) 638-8480
www.nlm.nih.gov

A gold mine of free health, toxicological, chemical, and chemical release information accessible through MEDLARS (MEDical Literature Analysis and Retrieval System), comprising two computer subsystems (ELHILL and TOXNET) on which reside over forty online databases containing about 18 million references. Databases include: MEDLINE, NLM's premier bibliographic database with references from more than thirty-seven hundred international biomedical journals; ChemID (Chemical Identification) and CHEMLINE (Chemical Dictionary Online); HSDB (Hazardous Substances Databank); RTECS (Registry of Toxic Effects of Chemical Substances); DART (Developmental and Reproductive Toxicology); IRIS (Integrated Risk Information System); TRI (Toxics Release Inventory); and TRIFACTS (Toxics Release Inventory Fact Sheets).

Occupational Safety and Health Administration (OSHA)
U.S. Department of Labor
200 Constitution Avenue
Washington, DC 20210
(202) 219-8148
www.osha.gov/

Entrusted with overseeing worker protection and the enforcement agency for workplace environments. Works with state agencies. Web site offers information and links on programs and services, compliance assistance, standards, and technical information.

REPROTOX
Reproductive Toxicology Center
2440 M Street, NW, Suite 217
Washington, DC 20037-1404
(202) 293-5137; fax: (202) 778-6199

Provides summaries of reproductive and developmental data for chemical and physical agents. Available only to health care providers, with several levels of subscriptions available.

Nongovernmental Medical and Health Organizations

American Association of Occupational Health Nurses
2920 Brandywine Road, Suite 100
Atlanta, GA 30341
(800) 241-8014; fax: (770) 455-7271
www.aaohn.org

Professional organization of occupational health nurses. Provides educational activities and standards of care and practice.

American College of Obstetricians and Gynecologists
409 Twelfth Street, SW
Washington, DC 20024
(800) 673-8444; (202) 863-2518 to reach Resource Center
www.acog.org

Professional organization of obstetricians and gynecologists. Provides continuing medical education and referrals.

American College of Occupational and Environmental Medicine
55 West Seegers
Arlington Heights, IL 60005
(847) 228-6850

Professional organization of occupational medicine physicians.

American Lung Association
1740 Broadway
New York, NY 10019-4374
(212) 315-8700; fax (212) 315-8872
www.lungusa.org

Provides a variety of printed resources on air pollution and chemical hazards in the air.

Association of Occupational and Environmental Health Clinics
1010 Vermont Avenue, NW, Suite 513
Washington, DC 20005
(202) 347-4976; fax: (202) 347-4950
occ-env-med.mc.duke.edu/oem/aoec.htm

An association of 55 clinics and 255 individual members in the United States and Canada specializing in occupational and environmental health issues. Provides referrals to clinics for medical advice and care, conducts educational activities, and maintains a lending library including slide shows and videotapes.

Consortium for Environmental Education in Medicine
44 Bromfield Street, 5th Floor
Boston, MA 02108
(617) 292-0094
www.2nature.org/ceem/health.html

A nonprofit organization working to infuse environmental health perspectives into medical education and practice. Focuses on faculty development, curriculum development, and student development.

MedWeb
Emory University Health Sciences Center Library
www.gen.emory.edu/MEDWEB/medweb.html

Numerous on-line links to public and private environment and health resources.

National Women's Health Network
514 10th Street, NW, Suite 400
Washington, DC 20004
(202) 628-7814 (Information Clearinghouse); fax: (202) 347-1168
Women's health advocacy group with general women's health information and a resource center. Publication available: *Turning Things Around: A Woman's Occupational and Environmental Health Resource Guide* (1990), $9.95.

Organization of Teratology and Information Services (OTIS)
2128 Elmwood Avenue
Buffalo, NY 14207
(716) 874-4747, ext. 477
Maintains pregnancy and environmental hot lines throughout the country to answer questions regarding prenatal exposures.

Environmental Advocacy, Right to Know, Community Action, and Workers' Rights

Organizations
CCHW (Center for Health, Environment and Justice)
P.O. Box 6806
Falls Church, VA 22040
(703) 237-CCHW
www.essential.org
Assistance and organizing on toxic hazards and cleanups. Publishes *Everyone's Backyard* and *Environmental Health Monthly*.

Children's Environmental Health Network
5900 Hollis Street, Suite E
Emeryville, CA 94608
(510) 450-3818; fax: (510) 450-3773
www.cehn.org
A national project dedicated to pediatric environmental health, it is the only national multidisciplinary project whose sole purpose is to protect

the health of children as it relates to environmental hazards. Three areas of concentration are education, research, and policy.

Coalition for Occupational Safety and Health (COSH)
(Call information for the nearest COSH.)

Nonprofit committees with twenty-three offices throughout the country that provide assistance to workers on health and safety issues and offer a variety of resources, including education and technical assistance on job hazards, a hot line, library, and publications and videos including *Confronting Reproductive Health Hazards on the Job: A Guide for Workers* (1992) (90 pages).

Environmental Defense Fund (EDF)
257 Park Avenue South
New York, NY 10010
(800) 684-3322 (membership and public information)
www.edf.org
www.scorecard.org (to access the EDF Chemical Scorecard)

A nonprofit membership organization of more than 300,000 founded in 1967 and dedicated to protecting the environmental rights of all people, including future generations. Maintains offices in five cities. The on-line EDF Chemical Scorecard is an extremely valuable interactive right-to-know tool that allows citizens to identify sources of pollution at the local level via linkages to maps and chemical release information; research scientific, health effects, and regulatory information on chemicals; learn about pollution prevention; and even send faxes to top-ranked polluters in a geographic area.

Environmental Research Foundation
P.O. Box 5036
Annapolis, MD 21403
(410) 263-1584; fax: (410) 263-8944
www.monitor.net/rachel/

Publishes the respected environmental publication, *Rachel's Environment and Health Weekly;* on-line subscription available by sending e-mail to: rachel-weekly-request@world.std.com. Also publishes *How to Research Chemicals: A Resource Guide* (1995), an excellent guide on strategies for researching toxic chemicals and their effects.

Institute for Global Communications (IGC)
P.O. Box 29904
San Francisco, CA 94129-0904
(415) 561-6100
www.igc.apc.org

Progressive Internet network provider includes EcoNet, WomensNet, LaborNet, ConflictNet, PeaceNet. Includes links to hundreds of progressive member organizations.

INFORM, Inc.
120 Wall Street, 16th Floor
New York, NY 10005-4001
(212) 361-2400; fax: (212) 361-2412
www.informinc.org

A nonprofit environmental research organization that has published numerous excellent publications, including *Preventing Industrial Toxic Hazards: A Guide for Communities* and *Toxics Watch 1995*.

JSI Center for Environmental Health Studies
44 Farnsworth Street
Boston, MA 02110
(617) 482-9485
www.jsi.com

Provides technical assistance, training, and public education on environmental health to community, labor, and environmental organizations. Performs community health surveys. Recent publication and training course: *Every Community's Right to Know: A Guide for Community Outreach and Education on Environment and Health Information.* Tutorial on *Environment and Health.*

Natural Resources Defense Council
40 West 20th Street
New York, NY 10011
(212) 727-2700
www.nrdc.org

A nonprofit with more than 350,000 members nationwide with the mission of preserving the environment, protecting the public health, and ensuring the conservation of wilderness and natural resources. Has four offices nationwide that pursue NRDC goals through research, advocacy, litigation, and education.

Northeast Organic Farming Association (NOFA)
c/o Julie Rawson
411 Sheldon Road
Barre, MA 01005
(978) 355-2853

A coalition of consumers, gardeners, and farmers working together for cleaner food and a safer, healthier environment. Educational conferences, newsletter.

Physicians for Social Responsibility
1101 Fourteenth Street, NW, Suite 700
Washington, DC 20005
(202) 898-0150; fax: (202) 898-0172
www.psr.org

A national organization of over fifteen thousand physicians, health practitioners, and supporters dedicated to the elimination of nuclear and other weapons of mass destruction, protection of the environment and its attendant public health issues, and the prevention of violence. Regional and state chapter affiliates nationwide conduct public education, research, and policy work related to these issue areas. PSR is the national affiliate of the Nobel Prize–winning International Physicians for the Prevention of Nuclear War.

Right-to-Know Network
1742 Connecticut Avenue, NW
Washington, DC 20009
(202) 234-8494; fax: (202) 234-8584
www.rtk.net

Established to empower citizen involvement in community and government decision making. Provides free access to numerous databases including the TRI and IRIS, information on EPA enforcement actions and fines, chemical production, company pollution discharge permits, chemical effects, corporation environmental impacts, population statistics, and chemical accidents. Contains graphics files containing area maps, the Computer-Aided Management of Community Operations

(CAMEO) worst-case accident scenario modeling program, and discussion groups. RTK Net staff can assist. Excellent resource.

U.S. Public Interest Research Group
218 D Street, SE
Washington, DC 20003-1900
(202) 546-9707; fax: (202) 546-2461
www.pirg.org

A nonprofit organization dedicated to serving as a watchdog for the nation's citizens and environment, with more than 1 million members nationwide. Affiliated state PIRGs in thirty-five states investigate public health and environmental concerns, craft solutions, educate the public, and offer citizens meaningful opportunities for civic participation.

Washington Toxics Coalition
4649 Sunnyside Avenue N, Suite 540E
Seattle, WA 98103
(206) 632-1545; fax: (206) 632-8661
www.accessone.com/~watoxics

A nonprofit organization dedicated to protecting public health and the environment by identifying and promoting alternatives to toxic chemicals.

Working Group on Community Right-to-Know
Hosted by the U.S. PIRG Education Fund
218 D Street, SE
Washington, DC 20003
(202) 546-9586

Coordinates a community right-to-know program for public interest groups concerned with chemical hazards and toxic pollution.

Publications

Chemical Alert: A Community Action Handbook, edited by Marvin Legator and Sabrina Strawn. Austin: University of Texas Press, 1993. (512) 471-4032. Written for citizens, activists, and medical professionals, provides information on the health effects of chemicals and discusses strategies for communities to conduct their own health surveys. An update to the popular and very useful *Health Detective's Handbook.*

Get to Know Your Local Polluter. 1993. Citizens for a Better Environment, 3255 Hennepin, Minneapolis, MN 55408. (612) 824-8637, (612) 824-0506 (fax). Provides a great example of how to use information on toxic chemicals in a way that produces results.

No Safe Place: Toxic Waste, Leukemia and Community Action, by Phil Brown and Edwin J. Mikkelson. Berkeley: University of California Press, 1990. The story of how the citizens of Woburn, Massachusetts, used popular epidemiology to link a cluster of childhood leukemia to industrial chemicals in the public water supply, and other community responses to toxic hazards.

The Toxics Free Neighborhoods Guide. The Environmental Health Coalition, (619) 235-0281. Guide to toxics use reduction, prevention, and cleanup around communities. Call to order.

Toxic Nation: The Fight to Save Our Communities from Environmental Contamination, by Fred Setterberg and Lonny Shavelson. New York: Wiley, 1993. How grass-roots groups throughout the country are fighting back against toxic chemical contamination.

Work on Waste. Ellen and Paul Connett. 82 Judson Street, Canton, NY 13617. (315) 379-9200, (315) 379-0448 (fax). Weekly reporter on rational resource management; information on dangers of incineration.

MSDS Sheets and Hazardous Substance Sheets

California Occupational Health Program

Hazard Evaluation and Information Service

2151 Berkeley Way, Annex 11, 3d Floor

Berkeley, CA 94704

(510) 540-3138

Provides fact sheets and other information on the health effects of chemicals in the workplace.

New Jersey Department of Health and Senior Services

Right-to-Know Program

Cn 368

Trenton, NJ 08625-0368

(609) 984-2202

www.state.nj-us/health/eoh/rtkweb/rtkhome.htm

Provides hazardous substance fact sheets.

Chemicals and Pesticides

Arts, Crafts and Theater Safety
181 Thompson Street, #23
New York, NY 10012
(212) 777-0062
www.caseweb.com/acts/

Information on the safe use of toxic art materials, safe work environments, and chemical composition of art materials.

Art and Creative Materials Institute, Inc.
100 Boylston Street, Suite 1050
Boston, MA 02116
(617) 426-6400

Sponsors a certification program for art materials and can provide a list of certified nontoxic products.

Bio-Integral Resource Center (BIRC)
P.O. Box 7414
Berkeley, CA 94707
(510) 524-2567; fax: (510) 524-1758
www.igc.org/birc/

Provides information on all aspects of environmentally sound pest management practices.

National Coalition Against the Misuse of Pesticides
701 E Street, SE, Suite 200
Washington, DC 20003
(202) 543-5450; fax: (202) 543-4791
www.ncamp.org

Coalition of groups working for pesticide reform. Information on alternative pest management.

Pesticide Action Network of North America
49 Powell Street, 6th Floor
San Francisco, CA 94102

(415) 981-1771; fax: (415) 981-1991

www.panna.org/panna/

Provides information and activism for pesticide reform.

Pesticide Education Center

P.O. Box 420870

San Francisco, CA 94142

(415) 391-8511

www.pesticides.org/pesticides

Educates the public about the use and health effects of pesticides.

School Issues

Massachusetts Healthy Schools Network

MassCOSH

555 Amory St.

Boston, MA 02130

(617) 524-6686, ext. 19

A collaborative initiative with other nongovernmental organizations and government agencies that works on issues of healthy school environments.

U.S. EPA

Indoor air quality "Tools for Schools" kits available (document 055-000-00503-6) through:

Superintendent of Documents

P.O. Box 371954

Pittsburgh, PA 15250-7954

(202) 512-1800

General Resources for Consumers

Co-op America

1612 K Street NW, Suite 600

Washington, DC 20006

(202) 872-5307

www.coopamerica.org

A nonprofit member organization dedicated to linking consumers with socially and environmentally responsible businesses in a nationwide green marketplace through such vehicles as the National Green Pages.

Sustaining the Earth: Choosing Consumer Products That Are Safe for You, Your Family, and the Earth, by Debra Dadd-Redalia. New York: Morrow, 1994. A wealth of information from the author of another good resource: *The Nontoxic Home and Office.* Los Angeles: Jeremy P. Tarcher, 1992.

Appendix B

Reproductive Outcomes Associated with Chemical Exposures

Chemical	Health effects*	Where used/ Where found†
Solvents		
1,1,1-TCA	SA, SBD	CW, IP, WP/S
Benzene	C, LBW, MA, SBD, O (childhood leukemia)	CP, CW, IP, WP/S, O (gasoline)
Chloroform	LBW, SBD	AP, CW, CF, IP
Epichlorohydrin	C, MI	H/S, IP, WP/S, O (wines stored in contaminated vats)
Formaldehyde	MA, SA	CP, IP, O (bldg materials)
Glycol ethers	FI, LBW, MI, SA, SBD	CP, H/S, IP, WP/S
N-methyl-pyrrolidone (NMP)	SA, O (stillbirth)	CP, H/S, IP, WP/S
Perchloroethylene (PCE)	FI, MI, SA, O (infant jaundice)	CW, IP, WP/S
Phenol	C, LBW, O (infant jaundice)	AP, CP, CW, IP
Styrene	H, MA, MI	AP, CP, CW, IP, O (firefighting)
Toluene	FD, H, LBW, MI, SA, SBD	CP, CW, IP, O (cigarette smoke, gasoline)

Chemical	Health effects*	Where used/ Where found†
Solvents		
Trichloroethylene (TCE)	LBW, FD, H, SA, SBD, O (childhood leukemia)	CW, IP, WP/S
Xylene	LBW, SA, SBD	CW, H/S, IP, WP/S, O (gasoline)
Metals		
Arsenic	SA, SBD, FD, LBW, O (hearing loss)	CF, CW, IP, O (wood products)
Cadmium	FD, LBW, MI, SBD, O (lung damage, placental toxicity)	CF, CW, IP, O (cigarette smoke)
Lead	FD, FI, H, LBW, MI, SA, SBD	CF, CS, CW, O (paint)
Manganese	MI, FD, LBW	IP, AP, O (gasoline)
Mercury (inorganic)	MA, SA, O (acrodynia)	CP, IP, O (dental fillings)
Mercury (organic)	FD, SBD	CF
Pesticides		
2,4-D	C, MI, SBD	AG/P, AP, CW, HG/P
Atrazine	H, LBW, SA, SBD	AG/P, AP, CW, HG/P
Benomyl	MI, SA, SBD	AG/P, CF, HG/P
Bromoxynil	SBD	AG/P, CW
Chlorpyrifos	SBD	AG/P, AP, CF, CW, HG/P
Cyanazine	LBW, SA, SBD	AG/P, CW
Cypermethrin	FD, H, LBW	AG/P, CF
Diazinon	C, H, MI, SA, SBD	AG/P, AP, CF, HG/P, CW
Dicofol	FD, H, MI	AG/P, CF
Dimethoate	H, MI	AG/P, CF
Dithiocarbamate	FD, H, MI, SA, SBD	AG/P, CF, O (rubber, plastics)
Endosulfan	H, MI	AG/P, CF, HG/P
Ethylene dibromide	MI	CW, WP/S, IP
Ethylene oxide	C, MI, SA	IP, O (sterilizing medical equipment)
Lindane	H, MA, MI, SA	AG/P, AP, CF, HG/P O (body lice treatment)
Linuron	SA, LBW	AG/P, CW

Chemical	Health effects*	Where used/ Where found†
Pesticides		
Malathion	H	AG/P, AP, CF, CP, HG/P, O (flea/tick dip)
Metam sodium	SBD	AG/P, AP
Methoxychlor	FD, FI, H, SA	AG/P, CF, HG/P
Methyl bromide	MI	AG/P, IP
Parathion	C, FD, MI, SA	AG/P, CF
Propargite	O (bone developmental abnormalities)	AG/P, AP
Resmethrin	SA, LBW, H	AG/P, AP, HG/P
Vinclozolin	H, SBD	AG/P, CF
Other		
Alkylphenols	H	CP, CF, CW, IP
Bisphenol A	H, O (enlarged prostate)	CP, CF, IP, O (dental sealants)
Dioxin	H, FD, MI, SA, SBD, O (altered sex ratio, endometriosis in primates)	AP, CF, CS, IP
PCBs	H, FD, FI, LBW, MA, SA	CF, CS, IP (banned in United States)
Phthalates	H, MI, SA, SBD	CP, CF, CW, IP

Notes: The effects have been derived from at least two animal studies and/or human studies. Evidence is consistently found for some effects, inconsistently for others. An absence of an effect may represent negative studies, a single positive animal study, or absence of data.

The table does not contain information about the level of human risk attributable to each exposure, dose-response data, or human exposure levels.

Pesticides listed are in common use commercially or in the home (e.g., lawn and garden products, pet applications) and are found in a variety of products, many of them available over the counter.

* C = chromosome damage; FD = functional defects; FI = female infertility; H = hormonal; LBW = low birth weight; MA = menstrual abnormalities; MI = male infertility and/or spermatotoxicity; O = other (specified); SA = spontaneous abortion; SBD = structural birth defects.

† AG/P = agricultural pesticide; AP = air pollution (including incineration); CF = contaminated food; CP = consumer products; CS = contaminated soil; CW = contaminated water; H/S = household solvents; HG/P = home/garden pesticide; IP = industrial processes (including dry cleaning); O = other (specified); WP/S = workplace solvents.

Notes

Introduction

1. Mosher W, Pratt W. *Fecundity, infertility, and reproductive health in the United States, 1982*. DHHS (PHS) 87-1990. Hyattsville, MD: National Center for Health Statistics, 1987.

2. Wilcox AJ, Weinberg CR, O'Connor JF, et al. Incidence of early loss of pregnancy. *N Engl J Med* 319:189–194, 1988.

3. Shepard TH, Fantel AG, Mirkes PE. Developmental toxicology: Prenatal period. In Paul M (ed.), *Occupational and Environmental Reproductive Hazards: A Guide for Clinicians*. Baltimore: Williams & Wilkins, 1993.

4. Carlsen E, Giwercman A, Keiding N, Skakkebaek NE. Evidence for decreasing quality of semen during the past 50 years. *Br Med J* 305:609–613, 1992.

5. Bleyer WA. What can be learned about childhood cancer from "Cancer Statistics Review 1973–1988." *Cancer* 71 (Suppl):3229–3236, 1993.

Chapter 1

1. Carr BR, Blackwell RE (eds). *Textbook of Reproductive Medicine*. Norwalk, CT: Appleton and Lange.

2. Polin RA, Fox WW (eds). *Fetal and Neonatal Physiology*. 2d ed. Philadelphia: WB Saunders, 1998.

3. Wilson JD, Foster DW (eds). *Williams' Textbook of Endocrinology*. 8th ed. Philadelphia: WB Saunders, 1992.

4. Paul M (ed). *Occupational and Environmental Reproductive Hazards: A Guide for Clinicians*. Baltimore: Williams & Wilkins, 1993.

5. Swaab DF, Hofman MA. Sexual differentiation of the human hypothalamus: Ontogeny of the sexually dimorphic nucleus of the preoptic area. *Dev Brain Res* 44:314–18, 1988.

6. Kimmel CA, Buelke-Sam J (eds). *Developmental Toxicology*. 2nd ed. New York: Raven Press, 1994, p 69.

7. Sharpe RM, Fisher JS, Millar MM, et al. Gestational and lactational exposure of rats to xenoestrogens results in reduced testicular size and sperm production. *Environ Health Perspect* 103(12):1136–1143, 1995.

8. Witorsch RJ (ed). *Reproductive Toxicology*. 2d ed. New York: Raven Press, 1995, pp. 112–113.

9. Witorsch RJ (ed). *Reproductive Toxicology*. 2d ed. New York: Raven Press, 1995, p. 203.

10. Kimmel CA, Buekle-Sam J (eds). *Developmental Toxicology*. 2d ed. New York: Raven Press, pp. 337–339.

11. Nelson K, Holmes LB. Malformations due to presumed spontaneous mutations in newborn infants. *New Engl J Med* 320:19–23, 1989.

12. Mably TA, Moore RW, Peterson RE. In utero and lactational exposure of male rats to 2,3,7,8-tetrachlorodibenzo-p-dioxin: 1. Effects on androgenic status. *Toxicol Appl Pharmacol* 114:97–107, 1992.

13. Kimmel CA, Buelke-Sam J (eds). *Developmental Toxicology*. 2nd ed. New York: Raven Press, 1994, p. 54.

14. Shenefelt RE. Morphogenesis of malformations in hamsters caused by retinoic acid: Relation to dose and stage of treatment. *Teratology* 5:103–118, 1972.

15. Kimmel CA, Buelke-Sam J (eds). *Developmental Toxicology*. 2d ed. New York: Raven Press, 1994, p. 254.

16. Wilcox AJ, Weinberg CR, O'Connor JF, et al. Incidence of early loss of pregnancy. *N Engl J Med* 319(4):189–194, 1988.

17. Dobbing J, Sands J. Comparative aspects of the brain growth spurt. *Early Hum Dev* 3(1):79–83, 1979.

18. Marcus M, Silbergeld E, Mattison D, et al. A reproductive hazards research agenda for the 1990s. *Environ Health Perspect* 101 (suppl 12):175–180, 1993.

Chapter 2

1. Daston GP. Do thresholds exist for developmental toxicants? In: Kalter H (ed), *Issues and Reviews in Teratology,* Vol. 6, New York: Plenum Press, 1993.

2. Mcmaster SB. Developmental toxicity, reproductive toxicity, and neurotoxicity as regulatory endpoints. *Toxicol Lett* 68:225–230, 1993.

3. Cummings A, Harris S. Carbendazim and maternally mediated early pregnancy loss in the rat. *Biol Repro* 42:66, 1990.

4. Herbst AL, Scully RE. Adenocarcinoma of the vagina in adolescence: A report of 7 cases including 6 clear-cell carcinomas. *Cancer* 25:745–757, 1970.

5. Hill AB. The environment and disease. *Proc Roy Soc Med* 58:295–300, 1965.

6. Fox GA. Practical causal inference for ecoepidemiologists. *J Toxicol Environ Health* 33:359–373, 1991.

7. Connor T. *Burdens of Proof: Science and Public Accountability in the Field of Environmental Epidemiology, with a Focus on Low Dose Radiation and Community Health Studies*. Columbia, SC: Energy Research Foundation, 1997.

8. Rehder H, Sanchioni F, Cefes G, Gropp A. Pathological-embryological investigations in cases of abortion related to the Seveso accident. *J Swiss Med* 108:1817–1825, 1978.

9. Mastroiacovo P, Spagnolo A, Marni E, Meazza L, Bertollini R, Segni G. Birth defects in the Seveso area after TCDD contamination. *JAMA* 259:1668–1672, 1988.

10. Mocarelli P, Brambilla P, Gerthoux PM, Patterson DG, Needham LL. Change in sex ratio with exposure to dioxin. *Lancet* 348:409, 1996.

11. Edmonds LD, Layde PM, James LM, Flynt JW, Erickson JD, Oakley GP. Congenital malformations surveillance: Two American systems. *Int J Epidemiol* 10(3):247–252, 1981.

12. Holtzman NA, Khoury MJ. Monitoring for congenital malformations. *Annu Rev Public Health* 7:237–266, 1986.

13. Klingberg MA, Papier CM, Hart J. Birth defects monitoring. *Am J Ind Med* 4:309–328, 1983.

14. Texas Department of Health. *An Investigation of a Cluster of Neural Tube Defects in Cameron County, Texas*. Texas, 1992.

15. American companies across border blamed for birth defects in Texas. *Environ Health Lett,* August 1997, p. 110.

16. Cedillo Becerril LA, Harlow SD, Sanchez RA, Sanchez Monroy D. Establishing priorities for occupational health research among women working in the maquiladora industry. *Int J Occup Environ Health* 3:221–230, 1997.

17. Blatter BM, van der Star M, Roeleveld N. Review of neural tube defects: Risk factors in parental occupation and the environment. *Environ Health Perspect* 102(2):140–145, 1994.

18. Shaw GM, Jensvold NG, Wasserman CR, Lammer EJ. Epidemiologic characteristics of phenotypically distinct neural tube defects among 0.7 million California births, 1983–1987. *Teratology* 49:143–149, 1994.

Chapter 3

1. Clarkson TW, Nordberg GF, Sagar PR. Reproductive and developmental toxicity of metals. *Scand J Work Environ Health* 11:145–154, 1985.

2. Goyer RA. Transplacental transport of lead. *Environ Health Perspect* 89:101–105, 1990.

3. Silbergeld EK. Implications of new data on lead toxicity for managing and preventing exposure. *Environ Health Perspect* 89:49–54, 1990.

4. Brody DJ, Pirkle JL, Kramer RA, et al. Blood lead levels in the US population. *JAMA* 272:277–283, 1994.

5. Pirkle JL, Brody DJ, Gunter EW, et al. The decline in blood lead levels in the United States. *JAMA* 272:284–291, 1994.

6. Crocette AF, Mushak P, Schwartz J. Determination of numbers of lead-exposed women of childbearing age and pregnant women: An integrated summary of a report to the U.S. Congress on childhood lead poisoning. *Environ Health Perspect* 89:121–124, 1990.

7. Winder C. Reproductive and chromosomal effects of occupational exposure to lead in the male. *Reprod Toxicol* 3:221–233, 1989.

8. Thomas JA, Brogan WC. Some actions of lead on the sperm and on the male reproductive system. *Am J Ind Med* 4:127–134, 1983.

9. Rom W. Effects of lead on the female and reproduction: A review. *Mt Sinai J Med* 43:542–552, 1976.

10. Lancranjan I, Popescu HI, Gavanescu O, Klepsch I, Serbanescu, M. Reproductive ability of workmen occupationally exposed to lead. *Arch Environ Health* 30:127–132, 1975.

11. Hu WY, Wu SH, et al. A toxicological and epidemiological study on reproductive functions of male workers exposed to lead. *J Hyg Epi Micro* 36:25–30, 1992.

12. Alexander BH, Checkoway H, van Netten C, et al. Semen quality of men employed at a lead smelter. *Occup Environ Med* 53:411–416, 1996.

13. Cullen MR, Kayne RD, Robins JM. Endocrine and reproductive dysfunction in men associated with occupational inorganic lead intoxication. *Arch Environ Health* 39:431–440, 1984.

14. Braunstein GD, Dahlgren J, Loriaux DL. Hypogonadism in chronically lead-poisoned men. *Infertility* 1:33–51, 1978.

15. Uzych L. Teratogenesis and mutagenesis associated with the exposure of human males to lead: A review. *Yale J Biol Med* 58:9–17, 1985.

16. Agency for Toxic Substances and Disease Registry (ATSDR). *Toxicological Profile for Lead*. Atlanta, GA: U.S. Department of Health and Human Services, ATSDR, April 1993.

17. Lindbohm M-L, Taskinen H, Kyyronen P, Sallmen M, Anttila A, Hemminki K. Effects of parental occupational exposure to solvents and lead on spontaneous abortion. *Scand J Work Environ Health* 18 (Suppl 2):37–39, 1992.

18. Murphy M, Graziano J, Popovac D. Past pregnancy outcomes among women living in the vicinity of a lead smelter in Kosovo, Yugoslavia. *Am J Public Health* 80:33–35, 1990.

19. Hu H. Knowledge of diagnosis and reproductive history among survivors of childhood plumbism. *Am J Public Health* 81:1070–1072, 1991.

20. McMichael AI, Vimpani GV, Robertson EF, Baghurst PA, Clark PD. The Port Pirie Study: Maternal blood lead and pregnancy outcome. *J Epidemiol Community Health* 40:18–25, 1986.

21. Needleman HL, Rabinowitz M, Leviton A, Linn S, Schoenbaum S. The relationship between prenatal exposure to lead and congenital anomalies. *JAMA* 251: 2956–2959, 1984.

22. Bellinger D, Sloman J, Leviton A, Rabinowitz M, Needleman HL, Waternaux C. Low-level lead exposure and children's cognitive function in the preschool years. *Pediatrics* 87:219–227, 1991.

23. Dietrich KN, Succop PA, Berger OG, Hammond PB, Bornschein RL. Lead exposure and the cognitive development of urban preschool children: The Cincinnati Lead Study cohort at age 4 years. *Neurotoxicol Teratol* 13:203–211, 1991.

24. Needleman HL, Schell A, Bellinger D, Leviton A, Allred EN. The long-term effects of exposure to low doses of lead in childhood: An 11-year follow-up report. *N Engl J Med* 322:83–88, 1990.

25. Davis JM, Svendsgaard DJ. Lead and child development. *Nature* 329:297–300, 1987.

26. Needleman HL, Gatsonis G. Low-level lead exposure and the IQ of children: A meta-analysis of modern studies. *JAMA* 263:673–678, 1990.

27. Needleman HL, Reiss JA, Tobin MJ, Biesecker GE, Greenhouse JB. Bone lead levels and delinquent behavior. *JAMA* 275:363–369, 1996.

28. Wasserman GA, Staghezza-Jaramillo B, Shrout P, et al. The effect of lead exposure on behavior problems in preschool children. *Am J Public Health* 88:481–486, 1998.

29. Rice DC. Lead-induced changes in learning. *Neurotoxicology* 14:167–178, 1993.

30. Goyer RA. Toxic effects of metals. In: Amdur MO, Doull J, Klaassen CD (eds), *Casarett and Doull's Toxicology: The Basic Science of Poisons,* 4th ed. New York: McGraw-Hill, 1993.

31. Agency for Toxic Substances and Disease Registry (ATSDR). *Draft Toxicological Profile for Mercury.* Atlanta, GA: U.S. Department of Health and Human Services, ATSDR, October 1992.

32. West CR, Smith CM (eds). *Mercury in Massachusetts: An Evaluation of Sources, Emissions, Impacts, and Controls*. Draft. Boston: Massachusetts Department of Environmental Protection, November 1995.

33. Bakir F, Damluji SF, Amin-Zaki L, et al. Methylmercury poisoning in Iraq. *Science* 181:230–241, 1973.

34. Marsh DO, Myers GJ, Clarkson TW, Amin-Zaki L, Tikriti S, Majeed MA. Fetal methylmercury poisoning: Clinical and toxicological data on 29 cases. *Ann Neurol* 7:348–355, 1980.

35. Harada H. Congenital Minamata disease: Intrauterine methylmercury poisoning. *Teratology* 18:285–288, 1978.

36. Cox C, Clarkson TW, Marsh DO, Amin-Zaki L, Tikriti S, Meyers GG. Dose-response analysis of infants prenatally exposed to methyl mercury: An application of a single compartment model to single-strand hair analysis. *Environ Res* 49:318–332, 1989.

37. Burbacher T, Rodier R, Weiss B. Methylmercury developmental neurotoxicity: A comparison of effects in humans and animals. *Neurotoxicol Teratol* 12:191–202, 1990.

38. Rodier PM, Aschner M, Sager PR. Mitotic arrest in the developing CNS after prenatal exposure to methylmercury. *Neurobehav Toxicol Teratol* 6:379–385, 1984.

39. Myers GJ, Davidson PW, Cox C, et al. Summary of the Seychelles child development study on the relationship of fetal methylmercury exposure to neurodevelopment. *Neurotoxicology* 16:711–716, 1995.

40. Myers GJ, Marsh DO, Davidson PW, et al. Main neurodevelopmental study of Seychellois children following in utero exposure to methylmercury from a maternal fish diet: Outcome at six months. *Neurotoxicology* 16:653–664, 1995.

41. Grandjean P, Weihe P, White RF, et al. Cognitive deficit in 7-year-old children with prenatal exposure to methylmercury. *Neurotoxicol Teratol* 19(6):417–428, 1997.

42. Miller RK, Bellinger D. Metals. In: Paul M (ed), *Occupational and Environmental Reproductive Hazards: A Guide for Clinicians*. Baltimore: Williams & Wilkins, 1993.

43. Lauwerys R, Roels H, Genet P, Toussaint G, Bouckaert A, De Cooman S. Fertility of male workers exposed to mercury vapor or to manganese dust: A questionnaire study. *Am J Ind Med* 7:171–176, 1985.

44. Alcser KH, Birx KA, Fine LJ. Occupational mercury exposure and male reproductive health. *Am J Ind Med* 15:517–529, 1989.

45. Cordier S, Deplan F, Mandereau L, Hemon D. Paternal exposure to mercury and spontaneous abortions. *Br J Ind Med* 48:375–381, 1991.

46. Ericson A, Kallen B. Pregnancy outcome in women working as dentists, dental assistants, or dental technicians. *Int Arch Occup Environ Health* 61:329–333, 1989.

47. Sikorsky R, Juszkiewicz T, Paszkowski T, Szprengier-Juszkiewicz T. Women in dental surgeries: Reproductive hazards in occupational exposure to metallic mercury. *Int Arch Occup Environ Health* 59:551–557, 1987.

48. Agency for Toxic Substances and Disease Registry (ATSDR). *Case Studies in Environmental Medicine: Cadmium Toxicity.* Atlanta, GA: U.S. Department of Health and Human Services, ATSDR, June 1990.

49. Rom WN. *Environmental and Occupational Medicine.* 2d ed. Boston: Little, Brown, 1988.

50. Parizek J, Zahor Z. Effects of cadmium salts on testicular tissue. *Nature* 177: 1036–1038, 1956.

51. Gunn SA, Gould TC, Anderson WAD. Zinc protection against cadmium injury to rat testis. *Arch Pathol* 71:274–281, 1961.

52. Saksena SK, Dahlgren L, Lau IF, Chang MC. Reproductive and endocrinological features of male rats after treatment with cadmium chloride. *Biol Reprod* 16: 609–613, 1977.

53. Mason HJ. Occupational cadmium exposure and testicular endocrine function. *Hum Exp Toxicol* 9:91–94, 1990.

54. Benoff S, Hurley IR, Barcia M, Mandel FS, Cooper GW, Hershlag A. A potential role for cadmium in the etiology of varicocele-associated infertility. *Fertil Steril* 67(2):336–347, 1997.

55. Levin A, Miller RK. Fetal toxicity of cadmium in the rat: Decreased uteroplacental blood flow. *Toxicol Appl Pharmacol* 58:297–306, 1981.

56. Levin AA, Plautz JR, Di Sant'Agnese PA, Miller RK. Cadmium: Placental mechanisms of fetal toxicity. *Placenta* 3:303–318, 1981.

57. Weir PJ, Miller RK, Maulik D, Di Sant'Agnese PA. Toxicity of cadmium in the perfused human placenta. *Toxicol Appl Pharmacol* 105:156–171, 1990.

58. Bryce-Smith D, Despande R, Hughes J, Waldron HA. Lead and cadmium levels in stillbirths. *Lancet* 1:1159, 1977.

59. Daston G. Toxic effects of cadmium on the developing lung. *J Toxicol Environ Health* 9:51–61, 1982.

60. Agency for Toxic Substances and Disease Registry. *Draft Toxicological Profile for Cadmium.* Atlanta, GA: U.S. Department of Health and Human Services, ATSDR, October 1991.

61. Ragan HA, Mast TJ. Cadmium inhalation and male reproductive toxicity. *Rev Environ Contam Toxicol* 114:1–22, 1990.

62. Ali MM, Murthy RC, Chandra SV. Developmental and longterm neurobehavioral toxicity of low level in-utero cadmium exposure in rats. *Neurobehav Toxicol Teratol* 8(5):463–468, 1986.

63. Le XC, Cullen WR, Reimer KJ. Human urinary arsenic excretion after one-time ingestion of seaweed, crab, and shrimp. *Clin Chem* 40:617–624, 1994.

64. Goldsmith JR, Deane M, Thom J, Gentry G. Evaluation of health implications of elevated arsenic in well water. *Water Res* 6:1133–1136, 1972.

65. Valentine JL, Hang HK, Spivey G. Arsenic levels in human blood, urine, and hair in response to exposure via drinking water. *Environ Res* 20:24–31, 1979.

66. Hopenhayn-Rich C, Smith AH, Goeden HM. Human studies do not support the methylation threshold hypothesis for the toxicity of inorganic arsenic. *Environ Res* 60:161–177, 1993.

67. Kerr HD, Saryan LA. Arsenic content of homeopathic medicines. *Clin Toxicol* 24:451–459, 1986.

68. Holland RH, McCall MS, Lanz HC. A study of inhaled ^{74}As in man. *Cancer Res* 19:1154–1156, 1959.

69. Wester RC, Maibach HI, Sedik L, Melendres J, Wade M. In vivo and in vitro percutaneous absorption and skin decontamination of arsenic from water and soil. *Fundam Appl Toxicol* 20:336–340, 1993.

70. Lindgren A, Danielsson BRG, Dencker L, Vahter M. Embryotoxicity of arsenite and arsenate: Distribution in pregnant mice and monkeys and effects on embryonic cells *in situ*. *Acta Pharmacol Toxicol* 54:311–320, 1984.

71. Hanlon DP, Ferm VH. Concentration and chemical status of arsenic in the blood of pregnant hamsters during critical embryogenesis. 1. Subchronic exposure to arsenate utilizing constant rate administration. *Environ Res* 40:372–379, 1986.

72. Tabacova S, Baird DD, Balabaeva I, Lolova D, Petrov I. Placental arsenic and cadmium in relation to lipid peroxides and glutathione levels in maternal-infant pairs from a copper smelter area. *Placenta* 15:873–881, 1994.

73. Reproductive and Cancer Hazard Assessment Section. *Evidence on Developmental and Reproductive Toxicity of Inorganic Arsenic*. Sacramento: Office of Environmental Health Hazard Assessment, California Environmental Protection Agency, October 1996.

74. Vahter M. Metabolism of arsenic. In: Fowler BA (ed), *Biological and Environmental Effects of Arsenic,* vol. 6, pp. 171–198. Amsterdam: Elsevier, 1983.

75. Peters HA, Croft WA, Woolson EA, Darcey BA, Olson MA. Seasonal arsenic exposure from burning chromium-copper-arsenate-treated wood. *JAMA* 251(18): 2393–2396, 1984.

76. Beaudoin AR. Teratogenicity of sodium arsenate in rats. *Teratology* 10:153–158, 1974.

77. Hood RD. Effects of sodium arsenite on fetal development. *Bull Environ Contam Toxicol* 7:216–222, 1972.

78. Ferm VH, Hanlon DP. Arsenate-induced neural tube defects not influenced by constant rate administration of folic acid. *Pediatr Res* 20:761–762, 1986.

79. Schroeder HA, Mitchener M. Toxic effects of trace elements on the reproduction of mice and rats. *Arch Environ Health* 23:102–106, 1971.

80. Willhite CC. Arsenic-induced axial skeletal (dysraphic) disorders. *Exp Mol Pathol* 34:145–158, 1981.

81. Osswald H, Goerttler K. Arsenic-induced leucoses in mice after diaplacental and postnatal application. *Verh Dtsch Ges Pathol* 26:289–293, 1971.

82. Borzsonyi M, Bereczky A, Rudnai P, Csanady M, Horvath A. Epidemiological studies on human subjects exposed to arsenic in drinking water in southeast Hungary. *Arch Toxicol* 66:77–78, 1992.

83. Zierler S, Theodore M, Cohen A, Rothman KJ. Chemical quality of maternal drinking water and congenital heart disease. *Int J Epidemiol* 17:589–594, 1988.

84. Aschengrau A, Zierler S, Cohen A. Quality of community drinking water and the occurrence of spontaneous abortion. *Arch Environ Health* 44:283–290, 1989.

85. Nordstrom S, Beckman L, Nordenson I. Occupational and environmental risks in and around a smelter in northern Sweden. I. Variations in birth weight. *Hereditas* 88:43–46, 1978.

86. Nordstrom S, Beckman L, Nordenson I. Occupational and environmental risks in and around a smelter in northern Sweden. III. Frequencies of spontaneous abortion. *Hereditas* 88:51–54, 1978.

87. Nordstrom S, Beckman L, Nordenson I. Occupational and environmental risks in and around a smelter in northern Sweden. V. Spontaneous abortion among female employees and decreased birth weight in their offspring. *Hereditas* 90:291–296, 1979.

88. Nordstrom S, Beckman L, Nordenson I. Occupational and environmental risks in and around a smelter in northern Sweden. VI. Congenital malformations. *Hereditas* 90:297–302, 1979.

89. Earnest NM, Hood RD. Effects of chronic prenatal exposure to sodium arsenite on mouse development and behavior. *Teratology* 24:53A, 1981.

90. Nagaraja TN, Desiraju T. Regional alterations in the levels of brain biogenic amines, glutamate, GABA, and GAD activity due to chronic consumption of inorganic arsenic in developing and adult rats. *Bull Environ Contam Toxicol* 50:100–107, 1993.

91. Nagaraja TN, Desiraju T. Effects on operant learning and brain acetylcholine esterase activity in rats following chronic arsenic intake. *Hum Exp Toxicol* 13:353–356, 1994.

92. Tabacova S. Maternal exposure to environmental chemicals. *Neurotoxicology* 7:421–440, 1986.

93. Gray LE, Laskey JW. Multivariate analysis of the effects of manganese on the reproductive physiology and behavior of the male house mouse. *J Toxicol Environ Health* 6:861–867, 1980.

94. Laskey JW, Rehnberg GL, Hein JF, Carter SD. Effects of chronic manganese (Mn₃O₄) exposure on selected reproductive parameters in rats. *J Toxicol Environ Health* 8:677–687, 1982.

95. Gennart J-P, Buchet J-P, Roels H, Ghyselen P, Ceulemans E, Lauwerys R. Fertility of male workers exposed to cadmium, lead, or manganese. *Am J Epidemiol* 135:1208–1219, 1992.

96. Allessio L, Apostoli P, Ferioli A, Lombardi S. Interference of manganese on neuroendocrinal system in exposed workers: Preliminary report. *Biol Trace Elem Res* 21:249–253, 1989.

97. Webster WS, Valois AA. Reproductive toxicology of manganese in rodents. *Neurotoxicology* 8:437–444, 1987.

98. Cotzias GC, Miller ST, Papavasiliou PS, Tang LC. Interactions between manganese and brain dopamine. *Med Clin North Am* 60:729–738, 1976.

99. Kostial K, Kello D, Jugo S, Rabar I, Maljkovic T. Influence of age on metal metabolism and toxicity. *Env Health Perspect* 25:81–86, 1978.

100. Tabacova S. Maternal exposure to environmental chemicals. *Neurotoxicology* 7:421–440, 1986.

101. Webster WS, Valois AA. Reproductive toxicology of manganese in rodents. *Neurotoxicology* 8:437–444, 1987.

102. Sanchez DJ, Domingo JL, Llobet M, Keen CL. Maternal and developmental toxicity of manganese in the mouse. *Toxicol Lett* 69:45–52, 1993.

103. Tsuchiya H, Shima S, Kurita H, et al. Effects of maternal exposure to six heavy metals on fetal development. *Bull Environ Contam Toxicol* 38:580–587, 1987.

104. Kilburn CJ. Manganese, malformations and motor disorders: Findings in a manganese-exposed population. *Neurotoxicology* 8:421–430, 1987.

105. Lown BA, Morganti JB, D'Agostino R, Stineman CH, Massaro EJ. Effects on the postnatal development of the mouse of preconception, postconception, and/or suckling exposure to manganese via maternal inhalation exposure to MnO$_2$ dust. *Neurotoxicology* 5:119–131, 1984.

Chapter 4

1. MacFarland HN. Toxicology of solvents. *Am Ind Hyg Assoc J* 47:704–707, 1986.

2. Dowty BJ, Laseter JL, Storer J. The transplacental migration and accumulation in blood of volatile organic constituents. *Pediatr Res* 10:696–701, 1976.

3. Fisher J, Mahle D, Bankston L, Greene R, Gearhart J. Lactational transfer of volatile chemicals in breast milk. *Am Indust Hyg Assoc J* 58:425–431, 1997.

4. Weisel CP, Jo W-K. Ingestion, inhalation, and dermal exposures to chloroform and trichloroethene from tap water. *Environ Health Perspect* 104:48–51, 1996.

5. Lindbohm ML, Taskinen H, Sallmen M, Hemminki K. Spontaneous abortions among women exposed to organic solvents. *Am J Ind Med* 17:449–463, 1990.

6. Lindbohm ML, Taskinen H, Kyyronen P, Sallmen M, Anttila A, Hemminki K. Effects of parental occupational exposure to solvents and lead on spontaneous abortion. *Scand J Work Environ Health* 18 (Suppl 2):37–39, 1992.

7. Taskinen H, Kyyronen P, Hemminki K, et al. Laboratory work and pregnancy outcome. *J Occup Med* 36:311–319, 1994.

8. Agnesi R, Valentini F, Mastrangelo G. Risk of spontaneous abortion and maternal exposure to organic solvents in the shoe industry. *Int Arch Occup Environ Health* 69:311–316, 1997.

9. Lipscomb JA, Fenster L, Wrensch M, Shusterman D, Swan S. Pregnancy outcomes in women potentially exposed to occupational solvents and women working in the electronics industry. *J Occup Med* 33:597–604, 1991.

10. Pastides H, Calabrese EJ, Hosmer DW Jr, Harris DR Jr. Spontaneous abortion and general illness symptoms among semiconductor manufacturers. *J Occup Med* 30:543–551, 1988.

11. Schenker MB, Gold EB, Beaumont JJ, et al. Association of spontaneous abortion and other reproductive effects with work in the semiconductor industry. *Am J Ind Med* 28(6):639–659, 1995.

12. Correa A, Gray RH, Cohen R, et al. Ethylene glycol ethers and risks of spontaneous abortion and subfertility. *Am J Epidemiol* 143(7):707–717, 1996.

13. Windham GC, Shusterman D, Swan SH, et al. Exposure to organic solvents and adverse pregnancy outcome. *Am J Ind Med* 20:241–259, 1991.

14. Kyyronen P, Taskinen H, Lindbohm ML, Hemminki K, Heinonen OP. Spontaneous abortions and congenital malformations among women exposed to tetrachloroethylene in dry cleaning. *J Epidemiol Community Health* 43:346–351, 1989.

15. Ng TP, Foo SC, Yoong T. Risk of spontaneous abortion in workers exposed to toluene. *Br J Ind Med* 49:804–808, 1992.

16. Deane M, Swan SH, Harris JA, Epstein DM, Neutra RR. Adverse pregnancy outcomes in relation to water contamination, Santa Clara County, California, 1980–1981. *Am J Epidemiol* 129:894–904, 1989.

17. Wrensch M, Swan S, Lipscomb J, et al. Pregnancy outcomes in women potentially exposed to solvent-contaminated drinking water in San Jose, California. *Am J Epidemiol* 131:283–300, 1990.

18. Bosco MG, Figa-Talamanca I, Salerno S. Health and reproductive status of female workers in dry cleaning shops. *Int Arch Occup Environ Health* 59:295–301, 1986.

19. Waller K, Swan SH, DeLorenze, Hopkins B. Trihalomethanes in drinking water and spontaneous abortion. *Epidemiology* 9:134–140, 1998.

20. Goldberg SJ, Lebowitz MD, Graver EJ, Hicks S. An association of human congenital cardiac malformations and drinking water contaminants. *J Am Coll Cardiol* 16:155–164, 1990.

21. Sever LE. Congenital malformations related to occupational reproductive hazards. *Occ Med Rev* 9:471–494, 1994.

22. Swan SH, Shaw G, Harris JA, Neutra RR. Congenital cardiac anomalies in relation to water contamination, Santa Clara County, California, 1981–1983. *Am J Epidemiol* 129:885–893, 1989.

23. Bove FJ, Fulcomer MC, Klotz JB, Esmart J, Dufficy EM, Savrin JE. Public drinking water contamination and birth outcomes. *Am J Epidemiol* 41(9):850–862, 1995.

24. Holmberg PC, Nurminen M. Congenital defects of the central nervous system and occupational factors during pregnancy. *Am J Ind Med* 1:167–176, 1980.

25. Holmberg PC. Central nervous system defects in children born to mothers exposed to organic solvents during pregnancy. *Lancet* 2:177–179, 1979.

26. Holmberg PC, Hernberg S, Kurppa K, Rantala K, Riala R. Oral clefts and organic solvent exposure during pregnancy. *Int Arch Occup Environ Health* 50:371–376, 1982.

27. Kurppa K, Holmberg PC, Hernberg S, Rantala K, Riala R, Nurminen T. Screening for occupational exposures and congenital malformations. *Scand J Work Environ Health* 9:89–93, 1983.

28. Cordier S, Ha MC, Ayme S, Goujard J. Maternal occupational exposure and congenital malformations. *Scand J Work Environ Health* 18:11–17, 1992.

29. Cordier S, Bergeret A, Goujard J, et al. Congenital malformations and maternal occupational exposure to glycol ethers. Occupational Exposure and Congenital Malformations Working Group. *Epidemiology* 8:355–363, 1997.

30. McDonald JC, LaVoie J, Cote R, McDonald D. Chemical exposures at work in early pregnancy and congenital defect: A case-referent study. *Br J Ind Med* 44:527–533, 1987.

31. Tikkanen J, Heinonen OP. Cardiovascular malformations and organic solvent exposure during pregnancy in Finland. *Am J Ind Med* 14:1–8, 1988.

32. Tikkanen J, Heinonen OP. Occupational risk factors for congenital heart disease. *Int Arch Occup Environ Health* 64:59–64, 1992.

33. Lagakos S, Wessen BJ, Zelen M. An analysis of contaminated well water and health effects in Woburn, Massachusetts. *J Am Statistical Assoc* 81:583–596, 1984.

34. Magee CA, Loffredo CA, Correa-Villaseñor A, Wilson PD. Environmental factors in occupations, home, and hobbies. In: Ferencz C, Rubin JD, Magee CA, Loffredo CA (eds), *Epidemiology of Congenital Heart Disease: The Baltimore-Washington Infant Study, 1981–1989.* Armonk, NY: Futura Publishing, 1993.

35. Sallmen M, Lindbohm ML, Kyyronen P. Reduced fertility among women exposed to organic solvents. *Am J Ind Med* 27:699–713, 1995.

36. Smith EM, Hammonds-Ehlers M, Clark MK, Kirchner HL, Fuortes L. Occupational exposures and risk of female infertility. *J Occup Environ Med* 39(2):138–147, 1997.

---

37. Kramer MD, Lynch CF, Isacson P, et al. The association of waterborne chloroform with intrauterine growth retardation. *Epidemiology* 3(5):407–413, 1992.

38. Berry M, Bove F. Birth weight reduction associated with residence near a hazardous waste landfill. *Environ Health Perspect* 105:856–861, 1997.

39. Savitz DA, Andrews KW, Pastore LM. Drinking water and pregnancy outcome in central North Carolina: Source, amount, and trihalomethane levels. *Environ Health Perspect* 103(6):592–596, 1995.

40. Eskenazi, B, Bracken M, Holford TR, Grady J. Exposure to organic solvents and hypertensive disorders of pregnancy. *Am J Ind Med* 14:177–188, 1988.

41. Hollenberg NK. Vascular injury to the kidney. In: Braunwald E, Isselbacher KJ, Petersdorf RG, et al. (eds), *Harrison's Principles of Internal Medicine,* 11th ed. New York: McGraw-Hill, 1987, 1204.

42. Wess JA. Reproductive toxicity of ethylene glycol monomethyl ether, ethylene glycol monoethyl ether and their acetates. *Scand J Work Environ Health* 18 (Suppl 2):43–45, 1992.

43. Daniell WE, Vaughan TL. Paternal employment in solvent related occupations and adverse pregnancy outcomes. *Br J Ind Med* 45:193–197, 1988.

44. Hoglund CV, Iselius EL, Knave BG. Children of male spray painters: Weight and length at birth. *Br J Ind Med* 49:249–253, 1992.

45. Brender JD, Suarez L. Paternal occupation and anencephaly. *Am J Epidemiol* 131:517–521, 1990.

46. Olsen J. Risk of exposure to teratogens amongst laboratory staff and painters. *Dan Med Bull* 30:24–28, 1983.

47. Strakowski SM, Butler MG. Paternal hydrocarbon exposure and Prader-Willi Syndrome. *Lancet* 2:1458, 1987.

48. Eskenazi B, Wyrobek AJ. A study of the effect of perchlorethylene exposure on semen quality in dry cleaning workers. *Am J Ind Med* 20:575–591, 1991.

49. Kelsey KT, Wiencke JK, Little FF, Baker EL Jr, Little JB. Sister chromatid exchange in painters recently exposed to solvents. *Environ Res* 50:248–255, 1989.

50. Taskinen H, Anttila A, Lindbohm ML, Sallmen M, Hemminki K. Spontaneous abortions and congenital malformations among the wives of men occupationally exposed to organic solvents. *Scand J Work Environ Health* 15:345–352, 1989.

51. Eskenazi B, Fenster L, Hudes M, et al. A study of the effect of perchlorethylene exposure on the reproductive outcomes of wives of dry-cleaning workers. *Am J Ind Med* 20:593–600, 1991.

52. Welch LS, Schrader SM, Turner TW, Cullen MB. Effects of exposure to ethylene glycol ethers on shipyard painters: II. Male reproduction. *Am J Ind Med* 14: 509–526, 1988.

53. National Institute for Occupational Safety and Health. *Methylene Chloride.* Current Intelligence Bulletin 46. Atlanta, GA: U.S. Department of Health and Human Services, NIOSH, 1986.

54. Cook RR, Bodner KM, Kolesar RC, et al. A cross-sectional study of ethylene glycol monomethyl ether process employees. *Arch Environ Health* 37:346–351, 1982.

55. Gold EB, Sever LE. Childhood cancers associated with parental occupational exposures. *Occup Med Rev* 9:495–539, 1994.

56. Peters J, Preston-Martin S, Yu MC. Brain tumors in children and occupational exposure of parents. *Science* 213:235–236, 1981.

57. Olsen JH, de Nully Brown P, Schulgen G, Jensen OM. Parental employment at time of conception and risk of cancer in offspring. *Eur J Cancer* 27:958–965, 1991.

58. Peters JM, Garabrant DH, Wright WE, Bernstein L, Mack TM. Uses of a cancer registry in the assessment of occupational cancer risks. *Natl Cancer Inst Monogr* 69:157–161, 1985.

59. Fabia J, Thuy TD. Occupation of father at time of birth of children dying of malignant diseases. *Br J Prev Soc Med* 28:98–100, 1974.

60. Kantor AF, Curnen MGM, Meigs JW, Flannery JT. Occupation of fathers of patients with Wilms' tumor. *J Epidemiol Community Health* 33:253–256, 1979.

61. Kwa SL, Fine LJ. The association between parental occupation and childhood malignancy. *J Occup Med* 22:792–794, 1980.

62. Vianna NJ, Kovasznay B, Polan A, Ju C. Infant leukemia and paternal exposure to motor vehicle exhaust fumes. *J Occup Med* 26:679–682, 1984.

63. Shu XO, Gao YT, Brinton LA, et al. A population-based case-control study of childhood leukemia in Shanghai. *Cancer* 62:635–644, 1988.

64. Lowengart RA, Peters JM, Cicioni C, et al. Childhood leukemia and parents' occupational and home exposures. *J Natl Cancer Inst* 79:39–46, 1987.

65. O'Leary LM, Hicks AM, Peters JM, London S. Parental occupational exposures and risk of childhood cancer: A review. *Am J Ind Med* 20:17–35, 1991.

66. Cutler JJ, Parker GS, Rosen S, Prenney B, Healey R, Caldwell GG. Childhood leukemia in Woburn, Massachusetts. *Public Health Rep* 101:201–205, 1986.

67. Durant JL, Chen J, Hemond HF, Thilly WG. Elevated incidence of childhood leukemia in Woburn, Massachusetts: NIEHS Superfund Basic Research Program Searches for Causes. *Environ Health Perspect* 103(Suppl 6):93–98, 1995.

68. Beavers JD, Himmelstein JS, Hammond SK, Smith TJ, Kenyon EM, Sweet CP. Exposure in a household using gasoline-contaminated water. *J Occup Environ Med* 38:35–38, 1996.

69. Ahlborg G, Hogstedt C, Bodin L, Barany S. Pregnancy outcome among working women. *Scand J Work Environ Health* 15:227–233, 1989.

70. Lindbohm ML, Hemminki K, Kyyronen P. Parental occupational exposure and spontaneous abortions in Finland. *Am J Epidemiol* 120:370–378, 1984.

71. Taskinen H. Effects of parental occupational exposure on spontaneous abortion and congenital malformations (Review). *Scand J Work Environ Health* 16:297–314, 1990.

72. Baker EL. A review of recent research on health effects of human occupational exposure to organic solvents. *J Occup Med* 36:1079–1092, 1994.

73. *Draft Hazard Identification of the Developmental and Reproductive Toxic Effects of Benzene.* Sacramento: Office of Environmental Health Hazard Assessment, California Environmental Protection Agency, September 1997.

74. Michon S. Disturbances of menstruation in women working in an atmosphere polluted with aromatic hydrocarbons [Abstract]. *Pol Tyg Lek* 20:1648–1649, 1965.

75. Mikhailova LM, Kobyets GP, Lyubomudrov VE, Braga GF. The influence of occupational factors on diseases of the female reproductive organs. *Pediatriya Akusherstvo Ginekologiya* 33:56–58, 1971.

76. Witkowski KM, Johnson NE. Organic solvent water pollution and low birth weight in Michigan. *Soc Biol* 39(1–2):45–54, 1992.

77. Louik C, Mitchell AA. *Occupational Exposures and Birth Defects: Final Performance Report.* Cincinnati: National Institute for Occupational Safety and Health, May 28, 1992.

78. Feingold L, Savitz DA, John EM. Use of a job-exposure matrix to evaluate parental occupation and childhood cancer. *Cancer Causes Control* 3:161–169, 1992.

79. Van Steensel-Moll HA, Valkenburg HA, Van Zanen GE. Childhood leukemia and parental occupation: A register-based case-control study. *Am J Epidemiol* 121: 216–224, 1985.

80. Magnani C, Pastore G, Luzzatto L, Terracini B. Parental occupation and other environmental factors in the etiology of leukemias and non-Hodgkin's lymphomas in childhood: A case control study. *Tumor* 76:413–419, 1990.

81. Johnson CC, Annegers JF, Frankowski RF, et al. Childhood nervous system tumors—an evaluation of the association with paternal occupational exposure to hydrocarbons. *Am J Epidemiol* 126:605–613, 1987.

82. Wilkins JR, Sinks TH. Occupational exposures among fathers of children with Wilms' tumor. *J Occup Med* 26:427–435, 1984.

83. Hakulinen T, Salonen T, Teppo L. Cancer in the offspring of fathers in hydrocarbon-related occupations. *Br J Prev Soc Med* 30:138–140, 1976.

84. Agency for Toxic Substances and Disease Registry (ATSDR). *Draft Toxicological Profile for Chloroform.* Atlanta, GA: U.S. Department of Health and Human Services, ATSDR, February 1996.

85. IARC. Chloroform. *Internation Agency for Research on Cancer Monographs on the Evaluation of the Carcinogenic Risk of Chemicals to Humans,* 20:401–427, 1979.

86. Schwetz BA, Leong BKJ, Gehring PJ. Embryo- and fetotoxicity of inhaled chloroform in rats. *Toxicol Appl Pharmacol* 28:442–451, 1974.

87. Murray FJ, Schwetz BA, McBride JG, Staples RE. Toxicity of inhaled chloroform in pregnant mice and their offspring. *Toxicol Appl Pharmacol* 50:515–522, 1979.

88. Baeder C, Hofmann T. *Inhalation Embryotoxicity Study of Chloroform in Wistar Rats*. Frankfurt: Pharma Research Toxicology and Pathology, Hoechst Aktiengesellschaft, 1988. As discussed in: *Agency for Toxic Substances and Disease Registry. Draft Toxicological Profile for Chloroform*. Atlanta, GA: U.S. Dept of Health and Human Services, ATSDR, February 1996.

89. Ruddick JA, Villeneuve DC, Chu I. A teratological assessment of four trihalomethanes in the rat. *J Environ Sci Health* B18(3):333–349, 1983.

90. Thompson DJ, Warner SD, Robinson VB. Teratology studies on orally administered chloroform in the rat and rabbit. *Toxicol Appl Pharmacol* 29:348–357, 1974.

91. Palmer AK, Street AE, Roe FJC, Worden AN, Van Abbe NJ. Safety evaluation of toothpaste containing chloroform. II. Long-term studies in rats. *J Environ Pathol Toxicol* 2:821–833, 1979.

92. Land PC, Owen EL, Linde HW. Mouse sperm morphology following exposure to anesthetics during early spermatogenesis. *Anesthesiology* 51:259, 1979.

93. Land PC, Owen EL, Linde HW. Morphologic changes in mouse spermatozoa after exposure to inhalation anesthetics during early spermatogenesis. *Anesthesiology* 54:53–56, 1981.

94. Gulati DK, Hope E, Mounce RC, et al. *Chloroform: Reproduction and Fertility Assessment in CD-1 Mice When Administered by Gavage*. Report by Environmental Health Research and Testing, Inc., Lexington, KY, to National Toxicology Program, National Institute of Environmental Health Sciences, Research Triangle Park, NC, 1988.

95. Jorgenson TA, Rushbrook CJ. *Effects of Chloroform in the Drinking Water of Rats and Mice: Ninety-day Subacute Toxicity Study*. Report by SRI International, Menlo Park, CA, to Health Effects Research Laboratory, Office of Research and Development, USEPA, Cincinnati, OH, 1980.

96. National Cancer Institute. *Report on Carcinogenesis Bioassay of Chloroform*. Bethesda, MD: Carcinogenesis Program, NCI, 1976.

97. Heywood R, Sortwell RJ, Noel PRB, et al. Safety evaluation of toothpaste containing chloroform. III. Long-term study in beagle dogs. *J Environ Pathol Toxicol* 2:835–851, 1979.

98. Tylleskar-Jensen J. Chloroform—a cause of pregnancy toxaemia? *Nordisk Medicin* 77:841–842, 1967. As discussed in: Welch LS. Organic solvents. In: Paul M (ed), *Occupational and Environmental Reproductive Hazards: A Guide for Clinicians*. Baltimore: Williams & Wilkins, 1993.

99. Aschengrau A, Zierler S, Cohen A. Quality of community drinking water and the occurrence of late adverse pregnancy outcomes. *Arch Environ Health* 48:105–113, 1993.

100. Kanitz S, Franco Y, Patrone V, et al. Association between drinking water disinfection and somatic parameters at birth. *Environ Health Perspect* 104:516–520, 1996.

101. Wallace LA, Pellizzari ED, Hartwell TD, et al. The influence of personal activities on exposure to volatile organic compounds. *Environ Res* 50:37–55, 1989.

102. Morris RD, Audet A-M, Angelillo IF, Chalmers TC, Mosteller F. Chlorination, chlorination by-products, and cancer: A meta-analysis. *Am J Public Health* 82: 955–963, 1992.

103. John JA, Quast JF, Murray FJ, Calhoun LG, Staples RE. Inhalation toxicity of epichlorohydrin: Effects on fertility in rats and rabbits. *Toxicol Appl Pharmacol* 68:415–423, 1983.

104. Slott VL, Suarez JD, Simmons JE, Perreault SD. Acute inhalation exposure to epichlorohydrin transiently decreases rat sperm velocity. *Fundam Appl Toxicol* 15:597–606, 1990.

105. Dabney BJ. Cytogenetic findings in employees with potential exposure to epichlorohydrin. *Prog Clin Biol Res* 207:59–73, 1986.

106. Milby TH, Whorton MD, Stubbs HA, Ross CE, Joyner RE, Lipshultz LI. Testicular function among epichlorohydrin workers. *Br J Ind Med* 38:372–377, 1981.

107. Venable JR, McClimans CD, Flake RE, Dimick DB. A fertility study of male employees engaged in the manufacture of glycerine. *J Occup Med* 22:87–91, 1980.

108. Landrigan PJ. Formaldehyde. In: Rom WN (ed), *Environmental and Occupational Medicine*. Boston: Little, Brown, 1992.

109. Majumder PK, Kumar VL. Inhibitory effects of formaldehyde on the reproductive system of male rats. *Indian J Physiol Pharmacol* 39(1):80–82, 1995.

110. Shah BM, Vachharajani KD, Chinoy MJ, Chowdhury AR. Formaldehyde-induced changes in testicular tissue of rats. *J Reprod Biol Comp Endocrinol* 7:42–52, 1987.

111. Chowdhury AR, Gautam AK, Patel KG, Trivedi HS. Steroidogenic inhibition in testicular tissue of formaldehyde exposed rats. *Indian J Physiol Pharmacol* 36: 162–168, 1992.

112. Maronpot RR, Miller RA, Clarke WJ, Westerberg RB, Decker JR, Moss OR. Toxicity of formaldehyde vapor in B6C3F1 mice exposed for thirteen weeks. *Toxicology* 41:253–266, 1986.

113. Discussed in: Formaldehyde. Dabney BJ (ed), *Reprotext Database*. Tomes+ Vol. 32, CD-Rom ed. Englewood, CO: Micromedex, 1996.

114. Marks TA, Worthy WC, Staples RE. Influence of formaldehyde and Sonacide on embryo and fetal development in mice. *Teratology* 22:51–58, 1980.

115. Gofmekler VA. Effect on embryonic development of benzene and formaldehyde in inhalation experiments. *Hyg Sanit* 33:327–332, 1968.

116. Hurni H, Ohder H. Reproduction study with formaldehyde and hexamethylenetetramine in beagle dogs. *Food Cosmet Toxicol* 11:459–462, 1973.

117. Griesemer RA, Ulsamer AG, Arcos JC, et al. Report of the federal panel on formaldehyde. *Environ Health Perspect* 43:139–168, 1982.

118. Shumilina AV. Menstrual and reproductive functions of workers with occupational exposure to formaldehyde. *Gig Tr Prof Zabol* 12:18–21, 1975.

119. John EM, Savitz DA, Shy CM. Spontaneous abortions among cosmetologists. *Epidemiology* 5:147–155, 1994.

120. Petrelli G, Traina ME. Glycol ethers in pesticide products: A possible reproductive risk? *Reprod Toxicol* 9:401–402, 1995.

121. *Glycol Ethers.* HESIS Fact Sheet No. 8. Berkeley, CA: Hazard Evaluation System and Information Service, January 1989.

122. Foster PMD, Creasy DM, Foster JR, Thomas LV, Cook MW, Gangolli SD. Testicular toxicity of ethylene glycol monomethyl and monoethyl ether in the rat. *Toxicol Appl Pharmacol* 69:385–399, 1983.

123. Chapin RE, Dutton SL, Ross MD, Lamb JC. Effects of ethylene glycol monomethyl ether (EGME) on mating performance and epididymal sperm parameters in F344 rats. *Fundam Appl Toxicol* 5:182–189, 1985.

124. Lamb JC, Gulati DK, Russell VS, Hommel L, Sabharwal PS. Reproductive toxicity of ethylene glycol monoethyl ether tested by continuous breeding of CD-1 mice. *Environ Health Perspect* 57:85–90, 1984.

125. Linder RE, Strader LF, Slott VL, Suarez JD. Endpoints of spermatotoxicity in the rat after short duration exposures to fourteen reproductive toxicants. *Reprod Toxicol* 6(6):491–505, 1992.

126. Hardin BD, Bond GP, Sikov MR, Andrew FD, Beliles RP, Niemeier RU. Testing of selected workplace chemicals for teratogenic potential. *Scand J Work Environ Health* 7 (Suppl 4):66–75, 1981.

127. Doe JE. Ethylene glycol monoethyl ether and ethylene glycol monoethyl ether acetate teratology studies. *Environ Health Perspect* 57:33–41, 1984.

128. Tyl RW, Pritts IM, France KA, Fisher LC, Tyler TR. Developmental toxicity evaluation of inhaled 2-ethoxy-ethanol acetate in Fischer 344 rats and New Zealand white rabbits. *Fundam Appl Toxicol* 10:20–39, 1988.

129. Nelson BK, Brightwell WS, Burg JR, Massari VJ. Behavioral and neurochemical alterations in the offspring of rats after maternal or paternal inhalation exposure to the industrial solvent 2-methoxyethanol. *Pharmacol Biochem Behav* 20:269–279, 1984.

130. National Institute for Occupational Safety and Health. *Health Hazard Evaluation Report: Precision Castparts Corporation.* HETA 85-415-1688. 1986.

131. Bolt HM, Golka K. Maternal exposure to ethylene glycol monomethyl ether acetate and hypospadias in offspring: A case report. *Br J Ind Med* 47:352–353, 1990.

132. Swan SH, Beaumont JJ, Hammond SK, et al. Historical cohort study of spontaneous abortion among fabrication workers in the Semiconductor Health Study: Agent-level analysis. *Am J Ind Med* 28(6):751–769, 1995.

133. Eskenazi B, Gold EB, Lasley BL, et al. Prospective monitoring of early fetal loss and clinical spontaneous abortion among female semiconductor workers. *Am J Ind Med* 26:833–846, 1995.

134. Eskenazi B, Gold EB, Samuels SJ, et al. Prospective assessment of fecundability of female semiconductor workers. *Am J Ind Med* 28:817–831, 1995.

135. Welch LS. Organic solvents. In: Paul M (ed), *Occupational and Environmental Reproductive Hazards: A Guide for Clinicians.* Baltimore: Williams & Wilkins, 1993.

136. Hooper K, LaDou J, Rosenbaum JS, Book SA. Regulation of priority carcinogens and reproductive or developmental toxicants. *Am J Ind Med* 22:793–808, 1992.

137. Rioux JP, Myers RAM. Methylene chloride poisoning: A paradigmatic review. *Emerg Med* 6:227–238, 1988.

138. Stewart RD, Hake CL. Paint-remover hazard. *JAMA* 235(4):398–401, 1976.

139. Gabrielli A, Layon AJ. Carbon monoxide intoxication during pregnancy: A case presentation and pathophysiologic discussion, with emphasis on molecular mechanisms. *J Clin Anesth* 7:82–87, 1995.

140. Sorokin Y. Asphyxiants. In: Paul M (ed), *Occupational and Environmental Reproductive Hazards: A Guide for Clinicians.* Baltimore: Williams & Wilkins, 1993.

141. Fechter LD, Annau Z. Toxicity of mild prenatal carbon monoxide exposure. *Science* 197:680–682, 1977.

142. Mactutus CF, Fechter LD. Moderate prenatal carbon monoxide exposure produces persistent, and apparently permanent, memory deficits in rats. *Teratology* 31:1–12, 1985.

143. Singh J, Scott LH. Threshold for carbon monoxide induced fetotoxicity. *Teratology* 30:253–257, 1984.

144. Astrup P, Trolle D, Olsen HM, Kjeldsen K. Effect of moderate carbon monoxide exposure on fetal development. *Lancet* 2:1220–1222, 1972.

145. Bailey LJ, Johnston MC, Billet J. Effects of carbon monoxide and hypoxia on cleft lip in A/J mice. *Cleft Palate—Craniofacial J* 32(1):14–19.

146. Ginsburg MD, Myers RE. Fetal brain damage following maternal carbon monoxide intoxication: An experimental study. *Acta Obstet Gynecol Scand* 53:309–317, 1974.

147. Ginsburg MD, Myers RE. Fetal brain injury after maternal carbon monoxide intoxication. *Neurology* 26:15–23, 1976.

148. Singh J, Smith CB, Moore-Cheatum L. Additivity of protein deficiency and carbon monoxide on placental carboxyhemoglobin in mice. *Am J Obstet Gynecol* 167(3):843–846, 1992.

149. Schwetz BA, Leong BK, Gehring PJ. The effect of maternally inhaled tri-chloroethylene, perchloroethylene, methyl chloroform, and methylene chloride on embryonal and fetal development in mice and rats. *Toxicol Appl Pharmacol* 32:84–96, 1975.

150. Bornschein RL, Hastings L, Manson JM. Behavioral toxicity in the offspring of rats following maternal exposure to dichloromethane. *Toxicol Appl Pharmacol* 52:29–37, 1980

151. Hardin BD, Manson JM. Absence of dichloromethane teratogenicity with inhalation exposure to rats. *Toxicol Appl Pharmacol* 52:22–28, 1980.

152. Kelly M. Case reports of individuals with oligospermia and methylene chloride exposure. *Reprod Toxicol* 2:13–17, 1988.

153. Taskinen H, Lindbohm M-L, Hemminki K. Spontaneous abortions among women working in the pharmaceutical industry. *Br J Ind Med* 43:199–205, 1986.

154. Norman CA, Halton DM. Is carbon monoxide a workplace teratogen? A review and evaluation of the literature. *Br Occup Hyg Soc* 34:335–347, 1990.

155. Koren G, Sharav T, Patuszak A, et al. A multicenter prospective study of fetal outcome following accidental carbon monoxide poisoning in pregnancy. *Reprod Toxicol* 5:397–403, 1991.

156. Akhter SA, Barry BW. Absorption through human skin of ibuprofen and flurbiprofen: Effect of dose variation, deposited drug films, occlusion, and the penetration enhancer N-methyl-2-pyrrolidinone. *J Pharm Pharmacol* 37:27–37, 1985.

157. Schmidt R. Tierexperimentelle Untersuchungen zur embryotoxischen und teratogenen Wirkung von N-methyl-pyrrolidon (NMP). *Biol Rundsch* 14:35–41, 1976.

158. Zeller H, Peh J. *BASF Corporation Report on Testing of N-Methylpyrrolidone for Possible Mouse Teratogenicity.* EPA/OTS, Doc 88-920003050. Washington, DC: Environmental Protection Agency, 1970.

159. Becci PJ, Knickerbocker MJ, Reagan EL, Parent RA, Burnette LW. Teratogenicity study of N-methylpyrrolidone after dermal application to Sprague-Dawley rats. *Fund Appl Toxicol* 2:73–76, 1982.

160. Letter from BASF Corp submitting information on a report on the testing of N-methylpyrrolidone for its possible teratogenic effects on Sprague-Dawley rats. EPA/OTS, Doc 88-920003049.

161. Jakobsen BM, Hass U. Prenatal toxicity of N-methylpyrrolidone inhalation in rats: A teratogenicity study. Presentation at the 18th conference of the European Teratology Society. *Teratology* 42:18A–19A, 1990.

162. Lee KP, Chromey NC, Culik R, Barnes JR, Schneider PW. Toxicity of N-methyl-2-pyrrolidone (NMP): Teratogenic, subchronic, and two-year inhalation studies. *Fund Appl Toxicol* 9:222–235, 1987.

163. Exxon Biomedical Sciences Inc. *Multi-Generational Rat Reproduction Study on N-Methyl-2-Pyrrolidone.* EPA/OTS, Doc 89-900000099. Washington, DC: Environmental Protection Agency, 1991.

164. Letter from GAF Chem Corp to US EPA submitting preliminary results of N-methyl-2-pyrrolidone developmental toxicity study with attachments. EPA/OTS, Doc 89-910000217, Washington, DC: Environmental Protection Agency, 1991.

165. Letter from BASF Corp submitting results of studies of the prenatal toxicity of N-methylpyrrolidone. EPA/OTS, Doc 89-920000111. Washington, DC: Environmental Protection Agency, 1992.

166. Hass U, Lund S, Elsner J. Effects of prenatal exposure to N-methylpyrrolidone on postnatal development and behavior in rats. *Neurotoxicol Teratol* 16:241–249, 1994.

167. Solomon GM, Morse EP, Garbo MJ, Milton DK. Stillbirth after occupational exposure to N-methyl-2-pyrrolidone. *J Occup Environ Med* 38(7):705–716, 1996.

168. Wallace D, Groth E. *Upstairs, Downstairs: Perchlorethylene in the Air in Apartments above New York City Dry Cleaners.* Yonkers, NY: Consumers Union, October 1995.

169. Popp W, Muller G, Baltes-Schmitz B, et al. Concentrations of tetrachloroethene in blood and trichloroacetic acid in urine in workers and neighbors of dry-cleaning shops. *Int Arch Occup Environ Health* 63:393–395, 1992.

170. Aggazzotti G, Fantuzzi G, Predieri G, Righi E, Moscardelli S. Indoor exposure to perchloroethylene (PCE) in individuals living with dry cleaning workers. *Sci Total Environ* 156:133–137, 1994.

171. Aschengrau A, Ozonoff K, Paulu C, et al. Cancer risk and tetrachloroethylene-contaminated drinking water in Massachusetts. *Arch Environ Health* 48:284–292, 1993.

172. Rachootin P, Olsen J. The risk of infertility and delayed conception associated with exposures in the Danish workplace. *J Occup Med* 253:394–402, 1983.

173. Olsen J, Hemminki K, Ahlborg G, et al. Low birthweight, congenital malformations, and spontaneous abortions among dry-cleaning workers in Scandinavia. *Scand J Work Environ Health* 16:163–168, 1990.

174. Ahlborg G. Pregnancy outcome among women working in laundries and dry-cleaning shops using tetrachlorethylene. *Am J Ind Med* 17:567–575, 1990.

175. Bagnell PC, Ellenberger HA. Obstructive jaundice due to a chlorinated hydrocarbon in breast milk. *Can Med J* 5:1047–1048, 1977.

176. Schreiber JS. Predicted infant exposure to tetrachlorethylene in human breastmilk. *Risk Anal* 13:515–524, 1993.

177. van der Gulden JW, Zielhuis GA. Reproductive hazards related to perchlorethylene: A review. *Int Arch Occup Environ Health* 61:235–242, 1989.

178. Korshunov SF. Early and late embryotoxic effects of phenol (Experimental data). *Gig Tr Sostoyanie Spetsificheskikh funkts RAB Neftekhm Khim. Promsti.* 149–153, 1974. (as cited in *Chem Abstr* 87-16735).

179. Minor JL, Becker BA. A comparison of the teratogenic properties of sodium salicylate, sodium benzoate, and phenol. *Toxicol Appl Pharmacol* 19:373, 1971.

180. Jones-Price C, Ledoux TA, Reel FR, Fisher PW, Langhoff-Paschke L. *Teratologic Evaluation of Phenol* (CAS No. 108-95-2) *in CD Rats*. NTIS PB83-247726. Washington, DC: National Technical Information Service, 1983.

181. Jones-Price C, Ledoux TA, Reel FR, Langhoff-Paschke L, Maur MC, Kimmel CA. *Teratologic evaluation of phenol* (CAS No. 108-95-2) *in CD-1 Mice*. NTIS PB85-104451. Washington, DC: National Technical Information Service, 1983.

182. Bulsiewicz H. The influence of phenol on chromosomes of mice *Mus musculus* in the process of spermatogenesis. *Folia Morphol* (Warsz.) 36(1):13–22, 1977.

183. Kolesnikova TN. Effect of phenol on sexual cycle of animals in chronic inhalation poisoning. *Gig Sanit* 37(1):105–106, 1972.

184. Scow K, Goyer M, Payne E, et al. *An Exposure and Risk Assessment for Phenol*. Washington, DC: Office of Water Regulations and Standards, US EPA, 1981.

185. Wysowski DK, Flynt JW, Goldfield M, Altman R, Davis AT. Epidemic neonatal hyperbilirubinemia and use of a phenolic disinfectant detergent. *Pediatrics* 61: 165–170, 1978.

186. Doan MH, Keith L, Shennan AT. Phenol and neonatal jaundice. *Pediatrics* 64:324–325, 1979.

187. *Sci News*, Sept 17, 1994: 191.

188. Izumova AS. The action of small concentrations of styrol on the sexual function of rats. *Gig Sanit* 37:29–30, 1972, as discussed in: Brown NA (ed), Reproductive and developmental toxicity of styrene. *Reprod Toxicol* 5:3–29, 1991.

189. Brown NA. Reproductive and developmental toxicity of styrene. *Reprod Toxicol* 5:3–29, 1991.

190. Ragule N. The problem of the embryotropic action of styrol. *Gig Sanit* 85-6, 1974, as discussed in: Brown NA (ed), Reproductive and developmental toxicity of styrene. *Reprod Toxicol* 5:3–29, 1991.

191. Vergieva T, Zaikov KH, Palatov S. Study of the embryotoxic action of styrene. *Khig Zdraveopaz* 22:39–43, 1979, as discussed in: Brown NA (ed), Reproductive and developmental toxicity of styrene. *Reprod Toxicol* 5:3–29, 1991.

192. Vainio H, Hemminki K, Elovaara E. Toxicity of styrene and styrene oxide on chick embryos. *Toxicology* 8:319–325, 1977.

193. Shigeta S, Maiyake K, Aikawa H, Misawa T. Effects of postnatal low-levels of exposure to styrene on behavior and development in rats. *J Toxicol Sci* 14(4): 279–286, 1989.

194. Lindbohm M-L. Effects of styrene on the reproductive health of women: A review. In: Sorsa M, Vainio H, Hemminki K (eds), *Butadiene and Styrene: Assessment of Health Hazards*. IARC Scientific Publications no. 127. Lyon: International Agency for Research on Cancer, 1993.

195. Harkonen H, Holmberg PC. Obstetric histories of women occupationally exposed to styrene. *Scand J Work Environ Health* 8:74–77, 1982.

196. Harkonen H, Tola S, Korkala ML, Hernberg S. Congenital malformations, mortality, and styrene exposure. *Ann Acad Med Singapore* 13(2 Suppl):404–407, 1984.

197. Hemminki K, Lindbohm ML, Hemminki T, Vainio H. Reproductive hazards and plastics industry. *Prog Clin Biol Res* 141:79–87, 1984.

198. Lindbohm ML, Hemminki K, Kyyronen P. Spontaneous abortion among women employed in the plastics industry. *Am J Ind Med* 8:579–586, 1985.

199. All summarized in Brown NA. Reproductive and developmental toxicity of styrene. *Reprod Toxicol* 5:3–29, 1991.

200. Mutti A, De Carli S, Ferroni C, Franchini I. Adverse reproductive effects of styrene exposure. In: Hogstedt C, Reuterwall C (eds), *Progress in Occupational Epidemiology*. Amsterdam: Elsevier, 1988.

201. Lemasters GK, Hagen A, Samuels SJ. Reproductive outcomes in women exposed to solvents in 36 reinforced plastic companies, 1: Menstrual dysfunction. *J Occup Med* 27:490–494, 1985.

202. Mutti A, Vescovi PP, Falzoi M, Arfini G, Valenti G, Franchini I. Neuroendocrine effects of styrene on occupationally exposed workers. *Scand J Environ Health* 10:225–228, 1984.

203. Jelnes JE. Semen quality in workers producing reinforced plastics. *Reprod Toxicol* 2:209–212, 1988.

204. Sethi N, Srivastava RK, Singh RK. Safety evaluation of a male injectible antifertility agent, styrene maleic anhydride, in rats. *Contraception* 39:217–226, 1989.

205. Brown NA. Reproductive and developmental toxicity of styrene. *Reprod Toxicol* 5:3–29, 1991.

206. Bjÿrge C, Brunborg G, Wiger R, et al. A comparative study of chemically induced DNA damage in isolated human and rat testicular cells. *Reprod Toxicol* 10:509–519, 1996.

207. Wilkins-Haug L. Teratogen update: Toluene. *Teratology* 55:145–151, 1997.

208. Donald JM, Hooper K, Hopenhayn-Rich C. Reproductive and developmental toxicity of toluene: A review. *Env Health Perspect* 94:237–244, 1991.

209. Shigeta S, Aikawa H, Misawa T, Yoshida T, Momotani H, Suzuki K. Learning impairment in rats following low-level toluene exposure during brain development—a comparative study of high avoidance rats and Wistar rats. *Ind Health* 24:203–211, 1986.

210. Kostas J, Hotchin J. Behavioral effects of low-level perinatal exposure to toluene in mice. *Neurobehav Teratol Toxicol* 3:467–469, 1981.

211. Jones HE, Balster RL. Neurobehavioral consequences of intermittent prenatal exposure to high concentrations of toluene. *Neurotoxicol Teratol* 19:305–313, 1997.

212. Thiel H, Chahoud I. Postnatal development and behaviour of Wistar rats after prenatal toluene exposure. *Arch Toxicol* 71:258–265, 1997.

213. Hersh JH, Podruch PE, Rogers G, Weisskopf B. Toluene embryopathy. *J Pediatr* 106:922–927, 1985.

214. Goodwin TM. Toluene abuse and renal tubular acidosis. *Obstet Gynecol* 71: 715–718, 1988.

215. Svensson B-G, Nise G, Erfurth E-M, Nilsson A, Skerfving S. Hormone status in occupational toluene exposure. *Am J Ind Med* 22:99–107, 1992.

216. Suzuki T, Kashimura S, Umetsu K. Thinner abuse and aspermia. *Med Sci Law* 23:199–202, 1983.

217. Ono A, Sekita K, Ogawa Y, et al. Reproductive and developmental toxicity studies of toluene. II. Effects of inhalation exposure on fertility in rats. *J Environ Pathol Toxicol Oncol* 15:9–20, 1996.

218. Office of Environmental Health Hazard Assessment. *Safe Drinking Water and Toxic Enforcement Act of 1986 (Prop. 65): Status Report.* Sacramento: California Environmental Protection Agency, January 1994.

219. Wallace LA, Pellizzari ED, Leaderer B, et al. Emissions of volatile organic compounds from building materials and consumer products. *Atmos Environ* 21:385–395, 1987.

220. Wallace LA, Pellizzari ED, Hartwell TD, et al. The influence of personal activities on exposure to volatile organic compounds. *Environ Res* 50:37–55, 1989.

221. Andelman JB. Human exposures to volatile halogenated organic chemicals in indoor and outdoor air. *Environ Health Perspect* 62:313–318, 1985.

222. Andelman JB. Inhalation exposure in the home to volatile organic contaminants of drinking water. *Sci Total Environ* 47:443–460, 1985.

223. Manson JM, Murphy M, Richdale N, Smith MK. Effect of oral exposure to trichloroethylene on female reproductive function. *Toxicology* 32:229–242, 1984.

224. Land PC, Owen EL, Linde HW. Morphologic changes in mouse spermatozoa after exposure to inhalation anesthetics during early spermatogenesis. *Anesthesiology* 54:53–56, 1981.

225. Zenick H, Blackburn K, Hope E, et al. Effects of trichloroethylene exposure on male reproductive function in rats. *Toxicology* 31:237–250, 1984.

226. Healy TEJ, Poole TR, Hopper A. Rat fetal development and maternal exposure to trichloroethylene 100 p.p.m. *Br J Anaesth* 54:337–341, 1982.

227. Dorfmueller MA, Henne SP, York RG, Bornschein RL, Molina G, Manson JM. Evaluation of teratogenicity and behavioral toxicity with inhalation exposure of maternal rats to trichloroethylene. *Toxicology* 14:153–166, 1979.

228. Beliles RP, Brucik DJ, Mecler FJ. *Teratogenic-Mutagenic Risk of Workplace Contaminants: Trichloroethylene, Perchloroethylene, and Carbon Disulfide.* Contract no. 210-77-0047. Washington, DC: U.S. Department of Health, Education and Welfare, 1980.

229. Hardin BD, Bond GP, Sikov MR, et al. Testing of selected workplace chemicals for teratogenic potential. *Scand J Work Environ Health* 7(Suppl 4):66–75, 1981.

230. Schwetz BA, Leong KJ, Gehring PJ. The effect of maternally inhaled trichloroethylene, perchloroethylene, methyl chloroform, and methylene chloride on embryonal and fetal development in mice and rats. *Toxicol Appl Pharmacol* 32:84–96, 1975.

231. Cosby NC, Dukelow WR. Toxicology of maternally ingested trichloroethylene (TCE) on embryonal and fetal development in mice and of TCE metabolites on *in vitro* fertilization. *Fundam Appl Toxicol* 19:268–274, 1992.

232. Fort DJ, Stover EL, Rayburn JR, et al. Evaluation of the developmental toxicity of trichloroethylene and detoxification metabolites using Xenopus. *Teratog Carcinog Mutagen* 13:35–45, 1993.

233. Dawson BV, Johnson PD, Goldberg SJ, Ulreich JB. Cardiac teratogenesis of trichloroethylene and dichloroethylene in a mammalian model. *J Am Coll Cardiol* 16:1304–1309, 1990.

234. Dawson BV, Johnson PD, Goldberg SJ, et al. Cardiac teratogenesis of halogenated hydrocarbon-contaminated drinking water. *J Am Coll Cardiol* 21:1466–1472, 1993.

235. Loeber CP, Hendrix MJC, Diez de Pinos S, Goldberg SJ. Trichloroethylene: A cardiac teratogen in developing chick embryos. *Pediatr Res* 24:740–744, 1988.

236. Ishikawa S, Nozaki T, Tsunemi T, Chikaoka H. Effects of trichloroethylene on embryonic chick heart. *Jpn Teratol Soc Abst* 42:25A, 1990.

237. Isaacson LG, Taylor DH. Maternal exposures to 1,1,2-trichloroethylene affects myelin in the hippocampal formation of the developing rat. *Brain Res* 488:403–407, 1989.

238. Noland-Gerbec EA, Pfohl RJ, Taylor DH, et al. 2-Deoxyglucose uptake in the developing rat brain upon pre- and postnatal exposure to trichloroethylene. *Neurotoxicology* 7:157–164, 1986.

239. Taylor DH, Lagory KE, Zaccaro DJ, et al. Effect of trichloroethylene on the exploratory and locomotor activity of rats exposed during development. *Sci Total Environ* 47:415–420, 1985.

240. Fredriksson A, Danielsson BRG, Eriksson P. Altered behavior in adult mice orally exposed to tri- and tetrachloroethylene as neonates. *Toxicol Lett* 66:13–19, 1993.

241. Corbett TH, Cornell RG, Enders JL, Leiding K. Birth defects among children of nurse anesthetists. *Anesthesiology* 41:341–344, 1974.

242. Tola S, Vilhunen R, Jarvinen E, et al. A cohort study on workers exposed to trichloroethylene. *J Occup Med* 22:737–740, 1980.

243. Chia S-E, Goh VHH, Ong CN. Endocrine profiles of male workers with exposure to trichloroethylene. *Am J Ind Med* 32:217–222, 1997.

244. Goh VH-H, Chia S-E, Ong C-N. Effects of chronic exposure to low doses of trichloroethylene on steroid hormone and insulin levels in normal men. *Environ Health Perspect* 106:41–44, 1998.

245. Chia S-E, Ong C-N, Tsakok MF, Ho A. Semen parameters in workers exposed to trichloroethylene. *Reprod Toxicol* 10(4):295–299, 1996.

246. Durant JL, Chen J, Hemond HF, Thilly WG. Elevated incidence of childhood leukemia in Woburn, Massachusetts: NIEHS Superfund Basic Research Program Searches for Causes. *Environ Health Perspect* 103(Suppl 6):93–98, 1995.

247. Mirkova E, Zaikov C, Antov G, Mikhailova A, Khinkova L, Benchev I. Prenatal toxicity of xylene. *J Hyg Epi Micro Immun* 27:337–343, 1983.

248. NIOSH. *Pocket Guide to Chemical Hazards.* Washington, DC: U.S. Department of Health and Human Services, June 1990.

249. Marks TA, Ledoux TA, Moore JA. Teratogenicity of a commercial xylene mixture in the mouse. *J Toxicol Environ Health* 9:97–105, 1982.

250. Hudak A, Ungvary G. Embryotoxic effects of benzene and its methyl derivatives: Toluene, xylene. *Toxicology* 11:55–63, 1978.

251. Ungvary G, Tatrai E. Studies on the embryotoxic effects of ortho- meta- and para-xylene. *Toxicology* 18:61–74, 1980.

252. Ungvary G. The possible contribution of industrial chemicals (organic solvents) to the incidence of congenital defects caused by teratogenic drugs and consumer goods—an experimental study. *Prog Clin Biol Res* 163B:295–300, 1985.

253. Hass U, Lund SP, Simonsen L, Fries AS. Effects of prenatal exposure to xylene on postnatal development and behavior in rats. *Neurotoxicol Teratol* 17(3):341–349, 1995.

254. Ungvary G, Bertalan V, Horvath E, Tatrai E, Folly G. Study on the role of maternal sex steroid production and metabolism in the embryotoxicity of para-xylene. *Toxicology* 19:263–268, 1981.

255. Ungvary G. Solvent effects on reproduction: Experimental toxicity. *Prog Clin Biol Res* 220:169–177, 1986.

256. Kucera J. Exposure to fat solvents: A possible cause of sacral agenesis in man. *J Pediatr* 72:857–859, 1968.

Chapter 5

1. US EPA. Pesticides industry sales and usage: 1994 and 1995. Market estimates, 1997.

2. Weisenburger DD. Human health effects of agrichemical use. *Hum Pathol* 24(6): 571–576, 1993.

3. US EPA. *Prevention, Pesticides, and Toxic Substances, Selected Terms and Acronyms, Office of Pesticide Programs.* June 1994.

4. Whitmore RW, Immerman FW, Camann DE, et al. Nonoccupational exposures to pesticides for residents of two U.S. cities. *Arch Environ Contam Toxicol* 26: 1–13, 1993.

5. *Soc Environ Toxicol Chem News* 11(4):9, 1991.

6. Brooks P. *House of Life: Rachel Carson at Work.* Boston: Houghton Mifflin, 1972.

7. Crowcroft P. *Elton's Ecologists: A History of the Bureau of Animal Population.* Chicago: University of Chicago Press, 1991.

8. Elton C. *The Pattern of Animal Communities.* New York: Methuen, 1966.

9. Galison P, Hevely B. *Big Science: The Growth of Large-Scale Research.* Stanford: Stanford University Press, 1992.

10. Bright DA, Dushenko WT, Grundy SL, Reimer KJ. Effects of local and distant contaminant sources: Polychlorinated biphenyls and other organochlorines in bottom-dwelling animals from an Arctic estuary. *Sci Total Environ* 15:265–283, 1995.

11. Dewailly E, et al. Inuit exposure to organochlorines through the aquatic food chain in Arctic Quebec. *Environ Health Perspect* 101(7):618–20, 1993.

12. Wargo, J. *Our Children's Toxic Legacy.* New Haven, CT: Yale University Press, 1996.

13. Moses M. *Designer Poisons.* San Francisco: Pesticide Education Center, 1995.

14. Davis JR, Brownson RC, Garcia R. Family pesticide use in the home, garden, orchard, and yard. *Arch Environ Contam Toxicol* 22(3):260–266, 1992.

15. Wasserstrom R, Wiles R. *Field Duty: US Farmworkers and Pesticide Safety.* Washington DC: World Resources Institute, 1985.

16. Mobed K, Gold EB, Schenker MB. Occupational health problems among migrant and seasonal farmworkers. *West J Med* 157(3):367–373, 1992.

17. Brown P. Race, class, and environmental health: A systematization of the literature. *Environ Res* 69:15–30, 1995.

18. Committee on Pesticides in the Diets of Infants and Children. National Research Council. *Pesticides in the Diets of Infants and Children.* Washington DC: National Academy Press, 1993.

19. Simcox NJ, Fenske RA, Wolz SA, et al. Pesticides in household dust and soil: Exposure pathways for children of agricultural families. *Environ Health Perspect* 103: 1126–1134, 1995.

20. de Cock J, Westveer K, Heederik D, et al. Time to pregnancy and occupational exposure to pesticides in fruit growers in the Netherlands. *Occup Environ Med* 51: 693–699, 1994.

21. Easter EP, Nigg HN. Pesticide protective clothing. *Rev Environ Contam Toxicol* 129:1–16, 1992.

22. Fenske RA, Black KG, Elkner KP, et al. Potential exposure and health risks of infants following indoor residential pesticide applications. *Am J Public Health* 80: 689–693, 1990.

23. Guruanthan S, Robson M, Freeman N, et al. Accumulation of chlorpyrifos on residential surfaces and toys accessible to children. *Environ Health Perspect* 106: 9–16, 1998.

24. Lewis RG, Fortmann RC, Camann DE. Evaluation of methods for monitoring the potential exposure of small children to pesticides in the residential environment. *Arch Environ Contam Toxicol* 26:37–46, 1993.

25. Needham LL, Hill RH, Ashley DL, et al. The priority toxicant reference range study: Interim report. *Environ Health Perspect* 103(Suppl 3):89–94, 1995.

26. Hill RH, Head SL, Baker S, et al. Pesticide residues in urine of adults living in the United States: Reference range concentrations. *Environ Res* 71:99–108, 1995.

27. Alavanja MCR, Sandler DP, McMaster SB, et al. The agricultural health study. *Environ Health Perspect* 104:362–369, 1996.

28. Rupa DS, Reddy PP, Reddi OS. Reproductive performance in population exposed to pesticides in cotton fields in India. *Environ Res* 55:123–128, 1991.

29. Hemminki K, Niemi ML, Saloniemi I, et al. Spontaneous abortions by occupation and social class in Finland. *Int J Epidemiol* 9:149–153, 1980.

30. Lindbohm ML, Hemminki K, Kyyronen P. Parental occupational exposure and spontaneous abortions in Finland. *Am J Epidemiol* 120:370–378, 1984.

31. McDonald AD, McDonald JC, Armstrong B, et al. Occupation and pregnancy outcome. *Br J Ind Med* 44:521–526, 1987.

32. Heidam LZ. Spontaneous abortions among dental assistants, factory workers, painters, and gardening workers: A follow-up study. *J Epidemiol Community Health* 38:149–155, 1984.

33. Rita P, Reddy PP, Venkatram R. Monitoring of workers occupationally exposed to pesticides in grape gardens of Andhra Pradesh. *Environ Res* 44:1–5, 1987.

34. Restrepo M, Munoz N, Day NE, et al. Prevalence of adverse reproductive outcomes in a population occupationally exposed to pesticides in Colombia. *Scand J Work Environ Health* 16:232–238, 1990.

35. Goulet L, Theriault G. Stillbirth and chemical exposure of pregnant workers. *Scand J Work Environ Health* 17:25–31, 1991.

36. Nurminen T, Rantala K, Kurppa K, Holmberg PC. Agricultural work during pregnancy and selected structural malformations in Finland. *Epidemiology* 6:23–30, 1995.

37. Garry VF, Schreinemachers D, Harkins ME, et al. Pesticide appliers, biocides, and birth defects in rural Minnesota. *Environ Health Perspect* 104:394–399, 1996.

38. Munger R, Isacson P, Hu S, et al. Intrauterine growth retardation in Iowa communities with herbicide-contaminated drinking water supplies. *Environ Health Perspect* 105:308–314, 1997.

39. Hemminki K, Mutanen P, Luoma K, Saloniemi I. Congenital malformations by the parental occupation in Finland. *Int Arch Occup Environ Health* 46:93–98, 1980.

40. McDonald AD, McDonald JC, Armstrong B, et al. Congenital defects and work in pregnancy. *Br J Ind Med* 45:581–588, 1988.

41. Schwartz DA, Newsum LA, Markowitz-Heifetz R. Parental occupation and birth outcome in an agricultural community. *Scand J Work Environ Health* 12:51–54, 1986.

42. Schwartz DA, LoGerfo JP. Congenital limb reduction defects in the agricultural setting. *Am J Public Health* 78:654–658, 1988.

43. Bjerkedal T. Use of medical registration of birth in the study of occupational hazards to human reproduction. In: Hemminki K, Sorsa M, Vainio H (eds), *Occupational Hazards and Reproduction*. Washington DC: Hemisphere Publishing Co., 1985.

44. Restrepo M, Munoz N, Day N, et al. Birth defects among children born to a population occupationally exposed to pesticides in Colombia. *Scand J Work Environ Health* 16:239–246, 1990.

45. Brender JD, Suarez L. Paternal occupation and anencephaly. *Am J Epidemiol* 131:517–521, 1990.

46. McDonald JC, Lavoie J, Cote R, et al. Chemical exposures at work in early pregnancy and congenital defect: A case-referent study. *Br J Ind Med* 44:527–533, 1987.

47. Lin S, Marshall EG, Davidson GK. Potential parental exposure to pesticides and limb reduction defects. *Scand J Work Environ Health* 20:166–179, 1994.

48. Zhang J, Cai W, Lee DJ. Occupational hazards and pregnancy outcomes. *Am J Ind Med* 21:397–408, 1992.

49. Fenster L, Coye MJ. Birthweight of infants born to Hispanic women employed in agriculture. *Arch Environ Health* 45:46–52, 1990.

50. Robison LL, Buckley JD, Bunin G. Assessment of environmental and genetic factors in the etiology of childhood cancers: The children's cancer group epidemiology program. *Environ Health Perspect* 103(Suppl 6):111–116, 1995.

51. Daniels JL, Olshan AF, Savitz DA. Pesticides and childhood cancers. *Environ Health Perspect* 105:1068–1077, 1997.

52. Davis JR, Brownson RC, Garcia RB, et al. Family pesticide use and childhood brain cancer. *Arch Environ Contam Toxicol* 24:87–92, 1993.

53. Laval G, Tuyns AJ. Environmental factors in childhood leukaemia. *Br J Ind Med* 45:843–844, 1988.

54. Robison LL, Buckley JD, Bunin G. Assessment of environmental and genetic factors in the etiology of childhood cancers: The children's cancer group epidemiology program. *Environ Health Perspect* 103(Suppl 6):111–116, 1995.

55. Potashnik G, Porath A. Dibromochloropropane (DBCP): A 17-year reassessment of testicular function and reproductive performance. *J Occup Environ Med* 37(11):1287–1292, 1995.

56. Ratcliffe JM, Schrader SM, Steenland K, et al. Semen quality in papaya workers with long-term exposures to ethylene dibromide. *Br J Ind Med* 44:317–326, 1987.

57. Lerda D, Rizzi R. Study of reproductive function in persons occupationally exposed to 2,4-dichlorophenoxyacetic acid (2,4-D). *Mutat Res* 262:47–50, 1991.

58. Kloos H. 1,2 Dibromo-3-chloropropane (DBCP) and ethylene dibromide (EDB) in well water in the Fresno/Clovis metropolitan area, California. *Arch Environ Health* 51(4):291–299, 1996.

59. O'Toole K. Three giants battle suit against pesticide. *Oakland (CA) Tribune,* January 24, 1983, p. 7.

60. Wharton D, Milby TH, Krauss RM, Stubbs HA. Testicular function in DBCP exposed pesticide workers. *J Occup Med* 21(3):161–166, 1979.

61. Kloos H. Chemical contaminants in public drinking water wells in California. In: Majumdar SK, Brenner FJ, Miller EW, Rosenfield LM (eds), *Environmental Contaminants and Health.* Easton, PA: Pennsylvania Academy of Science, 1995.

62. Thrupp, LA. Sterilization of workers from pesticide exposure: The causes and consequences of DBCP-induced damage in Costa Rica and beyond. *Int J Health Serv* 21(4):731–757, 1991.

63. Dulout FN, Pastori MC, Olivero OA, et al. Sister-chromatid exchanges and chromosomal aberrations in a population exposed to pesticides. *Mutat Res* 143: 237–244, 1985.

64. Nehez M, Berencsi G, Paldy A, et al. Data on the chromosome examinations of workers exposed to pesticides. *Reg Toxicol Pharmacol* 1:116–122, 1981.

65. Mohammad O, Walid AA, Ghada K. Chromosomal aberrations in human lymphocytes from two groups of workers occupationally exposed to pesticides in Syria. *Environ Res* 70:24–29, 1995.

66. Gianessi LP, Anderson JE. *Pesticide Use in US Crop Production: National Summary Report.* Washington DC: National Center for Food and Agricultural Policy, February 1995.

67. Ciesielski S, Loomis DP, Mims SR, Auer A. Pesticide exposures, cholinesterase depression, and symptoms among North Carolina migrant farmworkers. *Am J Public Health* 84:446–451, 1994.

68. Chanda SM, Pope CN. Neurochemical and neurobehavioral effects of repeated gestational exposure to chlorpyrifos in maternal and developing rats. *Pharmacol Biochem Behav* 53(4):771–776, 1996.

69. Muto MA, Lobelle F, Bidanset JH, Wurpel J. Embryotoxicity and neurotoxicity in rats associated with prenatal exposure to Dursban. *Vet Hum Toxicol* 34(6): 498–501, 1992.

70. Breslin WJ, Liberacki AB, Dittenber DA, Quast JF. Evaluation of the developmental and reproductive toxicity of chlorpyrifos in the rat. *Fundam Appl Toxicol* 29(1):119–130, 1996.

71. Gupta RC, Rech RH, Lovell KL, et al. Brain cholinergic, behavioral, and morphological development in rats exposed in utero to methylparathion. *Toxicol Appl Pharmacol* 77(3):405–413, 1985.

72. Spyker JM, Avery DL. Neurobehavioral effects of prenatal exposure to the organophosphate diazinon in mice. *J Toxicol Environ Health* 3(5–6):989–1002, 1977.

73. Lauder JM. Neurotransmitters as morphogens. *Prog Brain Res* 73:365–387, 1988.

74. Ahlbom J, Fredriksson A, Eriksson P. Exposure to an organophosphate (DFP) during a defined period in neonatal life induces permanent changes in muscarinic receptors and behavior in adult mice. *Brain Res* 677:13–19, 1995.

75. Whitney KD, Seidler FJ, Slotkin TA. Developmental neurotoxicity of chlorpyrifos: Cellular mechanisms. *Toxicol Appl Pharmacol* 134:53–62, 1995.

76. Dobbing J, Sands J. Comparative aspects of the brain growth spurt. *Early Hum Dev* 3:79–83, 1979.

77. Rattner BA, Michael SD. Organophosphorous insecticide induced decrease in plasma luteinizing hormone concentration in white-footed mice. *Toxicol Lett* 24(1):65–69, 1985.

78. US EPA. IRIS database, 1986.

79. Abd el-Aziz MI, Salab AM, Abd el-Khalik M. Influence of diazinon and deltamethrin on reproductive organs and fertility of male rats. *Dtsch Tieraerztl Wochenschr* 101(6):230–232, 1994.

80. Altamirano-Lozano MA, Del Camacho-Manzanilla CM, Loyola-Alvarez R, et al. Mutagenic and teratogenic effects of diazinon. *Rev Int Contam Ambient* 5(1): 49–58, 1989.

81. Spyker JM, Avery DL. Neurobehavioral effects of prenatal exposure to the organophosphate diazinon in mice. *J Toxicol Environ Health* 3(5–6):989–1002, 1977.

82. Afifi NA, Ramadan A, Abd el-Aziz MI, et al. Influence of dimethoate on testicular and epididymal organs, testosterone plasma level and their tissue residues in rats. *Dtsch Tieraerztl Wochenschr* 98(11):419–420, 1991.

83. Prakash N, Narayana K, Murthy GS, et al. The effect of malathion, an organophosphate, on the plasma FSH, 17,beta-estradiol and progesterone concentrations and acetylcholinesterase activity and conception in dairy cattle. *Veter Hum Toxicol* 34(2):116–119, 1992.

84. US EPA. IRIS database, 1987.

85. US EPA. IRIS database, 1994.

86. Kumar KB, Devi KS. Teratogenic effects of methyl parathion in developing chick embryos. *Vet Hum Toxicol* 34(5):408–410, 1992.

87. Collins TFX, Hansen WH, Keeler HV. The effect of carbaryl (Sevin) on reproduction of the rat and gerbil. *Toxicol Appl Pharmacol* 19:202–216, 1971.

88. Strachan W, Eriksson G, Kylin H, Jensen S. Organochlorine compounds in pine needles: Methods and trends. *Environ Toxicol Chem* 13(3):443–451, 1994.

89. *EXTOXNET Pesticide Information Notebook*. Ithaca, NY: Pesticide Management Education Program, Cornell University.

90. Cooper RL, Chadwick RW, Rehnberg GL, et al. Effect of lindane on hormonal control of reproductive function in the female rat. *Toxicol Appl Pharmacol* 99(3):384–394, 1989.

91. Chadwick RW, Cooper RL, Chang J, et al. Possible antiestrogenic activity of lindane in female rats. *J Biochem Toxicol* 3:147–158, 1988.

92. Sircar S, Lahiri P. Lindane (gamma-HCH) causes reproductive failure and fetotoxicity in mice. *Toxicology* 59(2):171–177, 1989.

93. Chowdhury AR, Gautam AK, Bhatnager VK. Lindane induced changes in morphology and lipids profile of testes in rats. *Biomed Biochim Acta* 49(10):1059–1065, 1990.

94. Das SN, Paul BN, Saxena AK, Ray PK. Effect of in utero exposure to hexachlorohexane on the developing immune system of mice. *Immunopharmacol Immunotoxicol* 12(2):293–310, 1990.

95. Saxena DK, Murthy RC, Chandra SV. Embryotoxic and teratogenic effects of interaction of cadmium and lindane in rats. *Acta Pharmacol Toxicol* 59(3):175–178, 1986.

96. Fry MD, Toone KC, Speich SM, Peard JR. Sex ratio skew and breeding patterns of gulls: Demographic and toxicological considerations. *Stud Avian Biol* 10: 26–43, 1987.

97. Fry M. Reproductive effects in birds exposed to pesticides and industrial chemicals. *Environ Health Perspect* 103(Suppl 7):165–171, 1995.

98. Vom Saal FS, Nagel SC, Palanza P, et al. Estrogenic pesticides: Binding relative to estradiol in MCF-7 cells and effects of exposure during fetal life on subsequent territorial behavior in male mice. *Toxicol Lett* 77(1–3):343–350, 1995.

99. Swartz WJ, Corkern M. Effects of methoxychlor treatment of pregnant mice on female offspring of the treated and subsequent pregnancies. *Reprod Toxicol* 6(5): 431–437, 1992.

100. Soto AM, Chung KL, Sonnenschein C. The pesticides endosulfan, toxaphene, and dieldrin have estrogenic effects on human estrogen-sensitive cells. *Environ Health Perspect* 102:380–383, 1994.

101. Di Muccio A, Camoni I, Citti P, Pontecorvo D. Survey of DDT-like compounds in dicofol formulations. *Ecotoxicol Environ Saf* 16(2):129–132, 1988.

102. MacLellan KN, Bird DM, Fry DM, Cowles JL. Reproductive and morphological effects of o,p'-dicofol on two generations of captive American kestrels. *Arch Environ Toxicol* 30(3):364–372, 1996.

103. MacLellan KN, Bird DM, Shutt LJ, Fry DM. Behavior of captive American kestrels hatched from o,p'-dicofol-exposed females. *Arch Environ Contam Toxicol* 32(4):411–415, 1997.

104. Guillette LJ, Gross TS, Masson GR, et al. Developmental abnormalities of the gonad and abnormal sex hormone concentrations in juvenile alligators from contaminated and control lakes in Florida. *Environ Health Perspect* 102(8):680–688, 1995.

105. Lemonica IP, Garrido Dos Santos AM, Bernardi MM. Effect of administration of organochlorine pesticide (dicofol) during gestation on neurobehavioral development of rats. *Teratology* 46(3):25A, 1992.

106. Singh SK, Pandy RS. Effect of subchronic endosulfan exposures on plasma gonadotropins, testosterone, testicular testosterone, and enzymes of androgen biosynthesis in rats. *Indian J Exp Biol* 28(10):953–956, 1990.

107. Pandey N, Gundevia F, Prem AS, Ray PK. Studies on the genotoxicity of endosulfan, an organochlorine insecticide, in mammalian germ cells. *Mutat Res* 242(1):1–7, 1990.

108. Hastings FL, Brady UE, Jones AS. Lindane and fenitrothion reduce soil and litter mesofauna on Piedmont and Appalachian sites. *Environ Entomol* 18(2):245–250, 1989.

109. Dikshith TS, Srivastava MK, Raizada RB. Fetotoxicity of hexachlorocyclohexane in mice: Morphological, biochemical, and residue evaluations. *Vet Hum Toxicol* 32(6):524–527, 1990.

110. Lindenau A, Fischer B, Seiler P, Beier HM. Effects of persistent chlorinated hydrocarbons on reproductive tissues in female rabbits. *Hum Reprod* 9(5):772–780, 1994.

111. US EPA. IRIS database, 1990.

112. Malaviya M, Husain R, Seth PK, Husain R. Perinatal effects of two pyrethroid insecticides on brain neurotransmitter function in the neonatal rat. *Vet Hum Toxicol* 35(2):119–122, 1993.

113. Ahlbom J, Fredriksson A, Eriksson P. Neonatal exposure to a type-1 pyrethroid (bioallethrin) induces dose-response changes in brain muscarinic receptors and behavior in neonatal and adult mice. *Brain Res* 645:318–324, 1994.

114. US EPA. IRIS database, 1988.

115. Eil C, Nisula BC. The binding properties of pyrethroids to human skin fibroblast androgen receptors and to sex hormone binding globulin. *J Steroid Biochem* 35 (3/4):409–414, 1990.

116. Husain R, Malaviya M, Seth PK. Differential responses of regional brain poly-amines following an in utero exposure to synthetic pyrethroid insecticides: A pre-liminary report. *Bull Environ Contam Tox* 49:402–409, 1992.

117. US EPA. IRIS database, 1986.

118. Ecobichon DJ. Toxic effects of pesticides. In: Amdur MO, Doull J, Klaassen CD (eds), *Casarett and Doull's Toxicology,* 4th ed. New York: McGraw-Hill, 1991.

119. Houeto P, Bindoula G, Hoffman JR. Ethylenebisdithiocarbamates and ethyl-enethiourea: Possible human health hazards. *Environ Health Perspect* 103:568–573, 1995.

120. Larsson KS, Arnander C, Cekanova E, Kjellberg M. Studies of teratogenic effects of the dithiocarbamates maneb, mancozeb, and propineb. *Teratology* 14(2): 171–183, 1976.

121. Lu MH, Kennedy GL. Teratogenic evaluation of mancozeb in the rat follow-ing inhalation exposure. *Toxicol Appl Pharmacol* 84(2):355–368, 1986.

122. Beck SL. Prenatal and postnatal assessment of maneb-exposed CD-1 mice. *Reprod Toxicol* 4(4):283–290, 1990.

123. Maci R, Arias E. Teratogenic effects of the fungicide maneb on chick em-bryos. *Ecotoxicol Environ Saf* 13(2):169–173, 1987.

124. Munk R, Schulz V. Study of possible teratogenic effects of the fungicide maneb on chick embryos. *Ecotoxicol Environ Saf* 17(2):112–118, 1989.

125. Kaloyanova F, Ivanova-Chemishanska L. Dose effect relationship for some specific effects of dithiocarbamates. *J Hyg Epidem Microbiol Immunol* 33(1):11–17, 1989.

126. Kackar R, Srivastava MK, Raizada RB. Induction of gonadal toxicity to male rats after chronic exposure to mancozeb. *Ind Health* 35(1):104–111, 1997.

127. Lankas GR, Wise DL. Developmental toxicity of orally administered thiaben-dazole in Sprague-Dawley rats and New Zealand white rabbits. *Food Chem Toxicol* 31(3):199–207, 1993.

128. See also Registry of Toxic Effects of Chemical Substances (NIOSH) database for summary of reproductive animal studies.

129. Lim J, Miller MG. The role of the benomyl metabolite carbendazim in beno-myl-induced testicular toxicity. *Toxicol Appl Pharmacol* 142(2):401–410, 1997.

130. Munley SM, Hurtt ME. Developmental toxicity study of benomyl in rabbits. *Toxicologist* 30(1 pt 2):192, 1996.

131. Cummings AM, Ebron-McCoy MT, Rogers JM, et al. Exposure to carben-dazim during early pregnancy produces embryolethality and developmental defects. *Biol Reprod* 44(Suppl 1):131, 1991.

132. Hess RA, Moore B. The fungicide benomyl (methyl 1-(butylcarbamoyl)-2-benzamidol-carbamate) causes testicular dysfunction by inducing the sloughing of germ cells and occlusion of efferent ductules. *Fund Appl Toxicol* 17:733–745, 1991.

---

133. Gray LE, Ostby JS, Kelce WR. Developmental effects of an environmental antiandrogen. *Toxicol Appl Pharmacol* 129(1):46–52, 1994.

134. Janardan A, Sattur PB, Sisodia P. Teratogenicity of methyl benzimidazole carbamate in rats and rabbits. *Bull Environ Contam Toxicol* 33:257–263, 1984.

135. Nakai M, Hess RA. Morphological changes in the rat Sertoli cell induced by the microtubule poison carbendazim. *Tissue Cell* 26(6):917–927, 1994.

136. Petrova-Vergieva T, Ivanova-Tchemishanska L. Assessment of the teratogenic activity of dithiocarbamate fungicides. *Food Cosmet Toxicol* 11:239–244, 1973.

137. Stoker TE, Goldman JM, Cooper RL, et al. The dithiocarbamate fungicide thiram disrupts the hormonal control of ovulation in the female rat. *Reprod Toxicol* 7(3):211–218, 1993.

138. Mishra VK, Srivastava MK. Testicular toxicity of thiram in rat: Morphological and biochemical evaluations. *Ind Health* 31(2):59–67, 1993.

139. Calderoni P. *Herbicides: Chemical Economics Handbook.* Menlo Park, CA: SRI Consulting, 1994.

140. US EPA, IRIS database, 1993.

141. Tennant MK, Hill DS, Elderidge JC, et al. Chloro-S-triazine antagonism of estrogen action: Limited interaction with estrogen receptor binding. *J Toxicol Environ Health* 43:197–211, 1994.

142. Tennant MK, Hill DS, Eldridge JC, et al. Anti-estrogenic properties of chloro-S-triazines in rat uterus. *J Toxicol Environ Health* 43:183–186, 1994.

143. Wetzel LT, Luempert LG, Breckenridge CB, et al. Chronic effects of atrazine on estrus and mammary tumor formation in female Sprague-Dawley and Fischer 344 rats. *J Toxicol Environ Health* 43(2):169–182, 1994.

144. Connor K, Howell J, Chen I, et al. Failure of chloro-S-triazine-derived compounds to induce estrogen receptor–mediated responses in vivo and in vitro. *Fundam Appl Toxicol* 30:93–101, 1995.

145. Tran DQ, Kow KY, McLachlan JA, Arnold SF. The inhibition of estrogen receptor–mediated responses by chloro-S-triazine-derived compounds is dependent on estradiol concentration in yeast. *Biochem Biophys Res Community* 227(1):140–146, 1996.

146. Bradlow HL, Davis DL, Lin G, et al. Effects of pesticides on the ratio of 16-alpha/2-hydroxyestrone: A biologic marker of breast cancer risk. *Environ Health Perspect* 103(Suppl 7):147–150, 1995.

147. 59 FR 1788–1844.

148. Hansen WH, Quaife ML, Haberman RT, et al. Chronic toxicity of 2,4-dichlorophenoxyacetic acid in rats and dogs. *Toxicol Appl Pharmacol* 20(1):122–129, 1971.

149. Lerda D, Rizzi R. Study of reproductive function in persons occupationally exposed to 2,4-D. *Mutat Res* 262:47–50, 1991.

150. Munro IC, Carlow GL, Orr JC, et al. A comprehensive, integrated review and evaluation of the scientific evidence relating to the safety of the herbicide 2,4-D. *J Am Coll Toxicol* 11(5):559–664, 1992.

151. US EPA. IRIS database, 1984.

152. Simic B, Kniewald J, Kniewald Z. Effects of atrazine on reproductive performance in the rat. *J Appl Toxicol* 14(6):401–404, 1994.

153. Kniewald J, Osredecki V, Gojmerac T, et al. Effect of s-triazine compounds on testosterone metabolism in the rat prostate. *J Appl Toxicol* 15(3):215–218, 1995.

154. US EPA. IRIS database, 1993.

155. Savage P, Scheidt B, Brockington L. A cyanazine–birth defects link? *Chem Week* 136(20):11–12, 1985.

156. Kutz FW, Cook BT, Carter-Pokras OD, et al. Selected pesticide residues and metabolites in urine from a survey of the US general population. *J Toxicol Environ Health* 37:277–291, 1992.

157. US EPA. IRIS database, 1986.

158. Final report on the reproductive toxicity of Iowa pesticide/fertilizer mixture (IWA) in CD-1 Swiss mice: vol 1. NTIS Technical Report (NTIS/PB93-109270), 1992.

159. Meisner LF, Roloff BD, Belluck DA. In vitro effects of n-nitrosoatrazine on chromosome breakage. *Arch Environ Contam Toxicol* 24:108–112, 1993.

160. *Pesticides for Evaluation as Candidate Toxic Air Contaminants.* HEH 96-01, Sacramento, California EPA, Department of Pesticide Regulation, 1996.

161. Landrigan PJ. Ethylene oxide. In: Rom WN (ed), *Environmental and Occupational Medicine,* 2d ed. Boston: Little, Brown, 1992.

162. Frink CR, Bugbee GJ. Ethylene dibromide: Persistence in soil and uptake by plants. *Soil Sci* 148(4):303–307, 1989.

163. Amir D, Volcano R. Effects of dietary ethylene dibromide on bull semen. *Nature* 206:99–100, 1965.

164. Amir D. The sites of the spermicidal action of ethylene dibromide in bulls. *J Reprod Fertil* 35:519–525, 1973.

165. Snellings WM, Maronpot RR, Zelenak JP, et al. Teratology study in Fischer 344 rats exposed to ethylene oxide by inhalation. *Toxicol Appl Pharmacol* 64:476–481, 1982.

166. Lynch DW, Lewis TR, Moorman WJ, et al. Toxic and mutagenic effects of ethylene oxide and propylene oxide on spermatogenic functions in cynomolgus monkeys. *Toxicologist* 3:60–68, 1983.

167. Generoso WM, Cain KT, Krishan M, et al. Heritable translocation and dominant lethal mutation induction with ethylene oxide in mice. *Mutat Res* 129:89–102, 1980.

---

168. LaBorde JB, Kimmel CA. The teratogenicity of ethylene oxide administered intravenously to mice. *Toxicol Appl Pharmacol* 56:16–22, 1980.

169. Methyl Bromide Fact Sheet. San Francisco: Pesticide Action Network North America, 1996.

170. Eustis SL, Haber SB, Drew RT, et al. Toxicology and pathology of methyl bromide in F344 rats and B6C3F1 mice following repeated inhalation exposure. *Fundam Appl Toxicol* 11:594–610, 1988.

171. Kato N, Morinobu S, Ishizu S. Subacute inhalation experiment for methyl bromide in rats. *Ind Health* 24:87–103, 1986.

172. Hurtt ME, Working PK. Evaluation of spermatogenesis and sperm quality in the rat following acute inhalation exposure to methyl bromide. *Fundam Appl Toxicol* 10(3):490–498, 1988.

173. Kaneda M, Hatakenada N, Teramoto S, Maita K. A two-generation reproduction study in rats with methyl bromide-fumigated diets. *Food Chem Toxicol* 31(8):533–542, 1993.

174. Jackson RJ. California EPA. *Evaluation of the Health Risks Associated with the Metam Spill in the Upper Sacramento River.* September 1992.

175. Schneider K. EPA failed to evaluate warnings on at least 10 dangerous pesticides. *NY Times,* August 23, 1991.

176. Coye MJ. *An Investigation of Spontaneous Abortions Following a Metam Sodium Spill into the Sacramento River.* California Department of Health Services, March 1993.

177. Morrissey RE, Schwetz BA, Lamb JC, et al. Evaluation of rodent sperm, vaginal cytology, and reproductive weight data from National Toxicology Program 13-week studies. *Fund Appl Toxicol* 11:343–358, 1988.

Chapter 6

1. Dodds EC, Lawson W. Molecular structure in relation to oestrogenic activity. Compounds without a phenanthrene nucleus. *Proc Royal Soc London, B,* 125:222–232, 1938.

2. Burlington H, Lindeman VF. Effect of DDT on testes and secondary sex characters of white leghorn cockerels. *Proc Soc Exp Biol Med* 74:48–51, 1950.

3. Colborn T, Dumanoski D, Myers JP. *Our Stolen Future.* New York: Dutton, 1996.

4. McLachlan JA (ed). *Estrogens in the Environment.* Amsterdam: Elsevier Science, 1985.

5. Guillette LJ, Gross TS, Masson GR, et al. Developmental abnormalities of the gonad and abnormal sex hormone concentrations in juvenile alligators from contaminated and control lakes in Florida. *Environ Health Perspect* 102(8):680–688, 1994.

6. Tillet DE, Ankley GT, Giesy JP, et al. Polychlorinated biphenyl residues and egg mortality in double-breasted cormorants from the Great Lakes. *Environ Toxicol Chem* 11:1281–1288, 1992.

7. McMaster ME, Portt CB, Munkittrick KR, Dixon DG. Milt characteristics, reproductive performance, and larval survival and development of white sucker exposed to bleached kraft mill effluent. *Ecotoxicol Environ Saf* 23:103–117, 1992.

8. Rajpert-De-Meyts E, Skakkeboek NE. The possible role of sex hormones in the development of testicular cancer. *Eur Urol* 23:54–61, 1993.

9. Chilvers C, Pike MC, Forman D, et al. Apparent doubling of frequency of undescended testicles in England and Wales 1962–81. *Lancet* i:330–332, 1984.

10. Jackson MB, Chilvers C, Pike MC, et al. Cryptorchidism: An apparent substantial increase since 1960. *Br Med J* 293:1401–1404, 1986.

11. Paulozzi LJ, Erickson JD, Jackson RJ. Hypospadias trends in two US surveillance systems. *Pediatrics* 100:831–834, 1997.

12. Kimmel CA. Approaches to evaluating reproductive hazards and risks. *Environ Health Perspect* 101(Suppl 2):137–143, 1993.

13. US General Accounting Office. *Reproductive and Developmental Toxicants.* Washington DC: US General Accounting Office, October 1991.

14. Giusti RM, Iwamoto K, Hatch EE. Diethylstilbesterol revisited: A review of the long-term health effects. *Ann Intern Med* 122(10):778–788, 1995.

15. Herbst AL, Scully RE. Adenocarcinoma of the vagina in adolescence: A report of 7 cases including 6 clear-cell carcinomas (so-called mesonephromas). *Cancer* 25: 745–747, 1970.

16. Gill WB, Schumacher GFB, Bibbo M, et al. Association of diethylstilbesterol exposure in utero with cryptorchidism, testicular hypoplasia, and semen abnormalities. *J Urol* 122:36–39, 1979.

17. Colton T, Greenberg ER, Noller K, et al. Breast cancer in mothers prescribed diethylstilbesterol in pregnancy. Further follow-up. *JAMA* 269(16):2096–2100, 1993.

18. Tarttelin MF, Gorski RA. Postnatal influence of diethylstilbesterol on the differentiation of the sexually dimorphic nucleus in the rat is as effective as perinatal treatment. *Brain Res* 456:271–274, 1988.

19. Reinisch JM, Ziemba-Davis M, Sanders SA. Hormonal contributions to sexually dimorphic behavior in humans. *Psychoneuroendocrinology* 16(1–3):213–278, 1991.

20. Bolander F. *Molecular Endocrinology.* 2d ed. San Diego, CA: Academic Press, 1994.

21. Damassa DA, Cates JM. Sex hormone-binding globulin and male sexual development. *Neurosci Biobehav Rev* 19(2):165–175, 1995.

22. Eil C, Nisula BC. The binding properties of pyrethroids to human skin fibroblast androgen receptors and to sex hormone binding globulin. *J Steroid Biochem* 35(3–4):409–414, 1990.

23. Adlercreutz H, Hockerstedt K, Bannwart C, et al. Effect of dietary components, including lignans and phytoestrogens, on enterohepatic circulation and liver metabolism of estrogens and on sex hormone binding globulin (SHBG). *J Steroid Biochem* 27(4–6):1135–1144, 1987.

24. McKinney JD, Waller CL. PCBs as hormonally active structural analogues. *Environ Health Perspect* 102(3):290–297, 1994.

25. Goldman JM, Parrish MB, Cooper RL, McElroy WK. Blockade of ovulation in the rat by systemic and ovarian intrabursal administration of the fungicide sodium dimethyldithiocarbamate. *Reprod Toxicol* 11(2–3):185–190, 1997.

26. Mably TA, Moore RW, Peterson RE. In utero and lactational exposure of male rats to 2,3,7,8-tetrachlorodibenzo-p-dioxin: 1. Effects on androgenic status. *Toxicol Appl Pharmacol* 114:97–107, 1992.

27. DeVito MJ, Birnbaum LS, Farland WH, Gasiewicz TA. Comparisons of estimated body burdens of dioxinlike chemicals and TCDD body burdens in experimentally exposed animals. *Environ Health Perspect* 103(9):820–831, 1995.

28. Gibbs PE, Pascoe PL, Burt GR. Sex change in the female dog-whelk, Nucella lapillus, induced by tributyl tin from antifouling paints. *J Mar Bio Assoc UK* 68:715–731, 1988.

29. Purdom CE, Hardiman PA, Bye VJ, et al. Estrogenic effects of effluents from sewage treatment works. *Chem Ecol* 8:275–285, 1994.

30. Jobling S, Sheahan D, Osborne JA, et al. Inhibition of testicular growth in rainbow trout (*Oncorhyncus mykiss*) exposed to alkylphenolic chemicals. *Environ Toxicol Chem* 15:194–202, 1996.

31. Guillette LJ, Crain DA, Rooney A, Pickford DB. Organization versus activation: The role of endocrine-disrupting contaminants (EDCs) during embryonic development in wildlife. *Environ Health Perspect* 103(Suppl 7):157–164, 1995.

32. Fry M. Reproductive effects in birds exposed to pesticides and industrial chemicals. *Environ Health Perspect* 103(Suppl 7):165–171, 1995.

33. Fox GA. Epidemiological and pathobiological evidence of contaminant-induced alterations in sexual development in free-living wildlife. In: Colborn T, Clement C (eds), *Chemically-Induced Alterations in Sexual and Functional Development: The Wildlife/Human Connection*. Princeton, NJ: Princeton Scientific Publishing Co. 1992.

34. Reijnders PJH. Reproductive failure in common seals feeding on fish from polluted coastal waters. *Nature* 324:456–457, 1986.

35. Brouwer A, Reijnders PJH, Koeman JH. Polychlorinated biphenyl (PCB)-contaminated fish induces vitamin A and thyroid hormone deficiency in the common seal (*Phoca vitulina*). *Aquatic Toxicol* 15:99–105, 1989.

36. Woodward AR, Percival HF, Jennings ML, Moore CT. Low clutch viability of American alligators on Lake Apopka. *Fla Sci* 56:52–63, 1993.

37. Guillette Jr LJ, Gross TS, Masson GR, Matter JM, Percival HF, Woodward AR. Developmental abnormalities of the gonad and abnormal sex hormone concentrations in juvenile alligators from contaminated and control lakes in Florida. *Environ Health Perspect* 102:680–688, 1994.

38. Parker SL, Tong T, Bolden S, et al. Cancer statistics. *CA Cancer J Clin* 65:5–27, 1996.

39. Telang NT, Katdare M, Bradlow HL, Osborne MP. Estradiol metabolism: An endocrine biomarker for modulation of human mammary carcinogenesis. *Environ Health Perspect* 105(Suppl 3):559–564, 1997.

40. Wilson JD, Foster DW (eds). *Williams' Textbook of Endocrinology.* 8th ed. Philadelphia: WB Saunders, 1992.

41. Toniolo PG, Levitz M, Zeleniuch-Jacquotte A, et al. A prospective study of endogenous estrogens and breast cancer in post-menopausal women. *J Natl Cancer Inst* 87:190–197, 1995.

42. Adlercreutz H, Gorbach SL, Goldin BR, et al. Estrogen metabolism and excretion in oriental and caucasian women. *J Natl Cancer Inst* 86:1076–1082, 1994.

43. Wolff MS, Weston A. Breast cancer risk and environmental exposures. *Environ Health Perspect* 105(Suppl 4):891–896, 1997.

44. Labreche FP, Goldberg MS. Exposure to organic solvents and breast cancer in women: A hypothesis. *Am J Ind Med* 32:1–14, 1997.

45. Morris JJ, Seifter E. The role of aromatic hydrocarbons in the genesis of breast cancer. *Med Hypotheses* 38:177–184, 1992.

46. Mutti A, Vescovi PP, Falzoi M, et al. Neuroendocrine effects of styrene on occupationally exposed workers. *Scand J Work Environ Health* 10:225–228, 1984.

47. Alessio L, Apostoli P, Feriolo A, Lombardi S. Interference of manganese on neuroendocrinal system in exposed workers. *Biol Trace Element Res* 21:249–253, 1989.

48. Lucchi L, Govoni S, Memo M, et al. Chronic lead exposure alters dopamine mechanisms in rat pituitary. *Toxicol Lett* 32:255–260, 1986.

49. Newcomb PA, Storer BE, Longnecker MP, et al. Lactation and a reduced risk of premenopausal breast cancer. *N Engl J Med* 330:81–87, 1994.

50. Yuan JM, Yu MC, Ross RK, et al. Risk factors for breast cancer in Chinese women in Shanghai. *Cancer Res* 48:1949–1953, 1988.

51. Liehr JG. Hormone-associated cancer: Mechanistic similarities between human breast cancer and estrogen-induced kidney carcinogenesis in hamsters. *Environ Health Perspect* 105(Suppl 3):565–569, 1997.

52. Wolff MS, Toniolo PG, Lee EW, et al. Blood levels of organochlorine residues and risk of breast cancer. *J Natl Cancer Inst* 85(8):648–652, 1993.

53. Mussalo-Rauhamaa H, Hasanen E, Pyysalo H, et al. Occurrence of beta-hexachlorocyclohexane in breast cancer patients. *Cancer* 66:2124–2128, 1990.

54. Falck F, Ricci A, Wolff M, et al. Pesticides and polychlorinated biphenyl residues in human breast lipids and their relation to breast cancer. *Arch Env Health* 47(2):143–146, 1992.

55. Dewailly E, Dodin S, Verreault R, et al. High organochlorine body burden in women with estrogen-responsive breast cancer. *J Natl Cancer Inst* 86:232–234, 1994.

56. Krieger N, Wolff MS, Hiatt RA, et al. Breast cancer and serum organochlorines: A prospective study among white, black, and Asian women. *J Natl Cancer Inst* 86:589–599, 1994.

57. Unger M, Kiaer H, Blichert-Toft M, et al. Organochlorine compounds in human breast fat from deceased with and without breast cancer and in a biopsy material from newly-diagnosed patients undergoing breast surgery. *Environ Res* 34:24–28, 1984.

58. Ahlborg UG, Lipworth L, Titus-Ernstoff L, et al. Organochlorine compounds in relation to breast cancer, endometrial cancer, and endometriosis: An assessment of the biological and epidemiological evidence. *Crit Rev Toxicol* 25(6):463–531, 1995.

59. Hunter DJ, Hankinson SE, Laden F, et al. Plasma organochlorine levels and the risk of breast cancer. *N Engl J Med* 337(18):1253–1258, 1997.

60. Safe S. Is there an association between exposure to environmental estrogens and breast cancer? *Environ Health Perspect* 105(Suppl 3):675–678, 1997.

61. Garnic MB. The dilemmas of prostate cancer. *Sci Am* 270:72–81, 1994.

62. Nagel SC, vom Saal F, Thayer KA, et al. Relative binding affinity-serum modified access (RBA-SMA) assay predicts the relative in vivo bioactivity of the xenoestrogens bisphenol-A and octylphenol. *Environ Health Perspect* 105:70–76, 1997.

63. Makela SI, Pylkkanen LH, Santti RS, et al. Dietary soybean may be antiestrogenic in male mice. *J Nutr* 125:437–445, 1995.

64. Habenicht UF, Schwartz K, Schweikert HU, et al. Development of a model for the induction of estrogen-related prostate hyperplasia in the dog and its response to the aromatase inhibitor 4-hydroxy-r-androstene-3, 17-dione. *Prostate* 8:181–194, 1986.

65. Habenicht UF, El Eltreby MF. The periurethral zone of the prostate of the cynomologus monkey is the most sensitive prostate part for an estrogenic stimulus. *Prostate* 13:305–316, 1988.

66. Strinivasan G, Campbell E, Bashirelahi N. Androgen, estrogen, and progesterone receptors in normal and aging prostates. *Micros Res Tech* 30(4):293–304, 1995.

67. Ho S, Roy R. Sex hormone–induced nuclear DNA damage and lipid peroxidation in the dorsolateral prostates of Noble rats. *Cancer Lett* 84:155, 1994.

68. Malins DC, Polissar NL, Gunselman SJ. Models of DNA structure achieve almost perfect discrimination between normal prostate, benign prostatic hyperplasia (BPH), and adenocarcinoma and have a high potential for predicting BPH and prostate cancer. *Proc Natl Acad Sci* 94:259, 1997

69. Sakr WA, Haas GP, Cassin BF. The frequency of carcinoma and intraepithelial neoplasia of the prostate in young male patients. *J Urol* 150:379–385, 1993.

70. Yasuda Y, Kihara T, Tanimura T. Effect of ethinyl estradiol on the differentiation of mouse fetal testis. *Teratology* 32:113–118, 1985.

71. Prener A, Hseih C, Engholm G et al. Birth order and risk of testicular cancer. *Cancer Causes Control* 3:265–272, 1992.

72. Depue RH, Pike MC, Henderson BE. Estrogen exposure during gestation and risk of testicular cancer. *J Natl Cancer Inst* 71:1151–1155, 1983.

73. Henderson BE, Benton B, Jing J, et al. Risk factors for cancer of the testis in young men. *Int J Cancer* 23:589, 1979.

74. Moss AR, Osmond D, Bacchetti P, et al. Hormonal risk factors in testicular cancer. *Am J Epidemiol* 124:39–52, 1986.

75. Stone R. Environmental estrogens stir debate. *Science* 265:308–310, 1994.

76. Carlsen E, Giwercman A, Keiding N, Skakkebaek NE. Evidence for decreasing quality of semen during the past 50 years. *Br Med J* 305:609–613, 1992.

77. Olsen GW, Bodner KM, Ramlow JM. Have sperm counts been reduced 50 percent in 50 years? A statistical model revisited. *Fertil Steril* 63:887–893, 1995.

78. Brake A, Kraus W. Decreasing quality of semen. *Br Med J* 305:1498, 1992.

79. Swan S, Elkin EP, Fenster L. Have sperm densities declined? A reanalysis of global trend data. *Environ Health Perspect* 105:1228–1232, 1997.

80. Fisch H, Goluboff ET. Geographic variations in sperm counts: A potential cause of bias in studies of semen quality. *Fertil Steril* 65:1044–1046, 1996.

81. Auger J, Kunstmann JM, Czyglik F, et al. Decline in semen quality among fertile men in Paris during the past 20 years. *N Engl J Med* 332(5):281–285, 1995.

82. Irvine S, Cawood E, Richardson D, et al. Evidence of deteriorating semen quality in the United Kingdom: Birth cohort study in 577 men in Scotland over 11 years. *Br Med J* 312:467–471, 1996.

83. Fisch H, Goluboff ET, Olson JH, et al. Semen analyses in 1,283 men from the United States over a 25-year period: No decline in quality. *Fertil Steril* 65: 1009–1014, 1996.

84. Paulsen CA, Berman NG, Wang C. Data from men in greater Seattle area reveals no downward trend in semen quality: Further evidence that deterioration of semen quality is not geographically uniform. *Fertil Steril* 65:1015–1020, 1996.

85. Pajarinen J, Laippala P, Penttila A, et al. Incidence of disorders of spermatogenesis in middle-aged Finnish men, 1981–1991: Two necropsy series. *Br Med J* 314: 13–18, 1997.

86. Giwercman A, Carlsen E, Keiding N, Skakkebaek NE. Evidence for increasing incidence of abnormalities of the human testis: A review. *Environ Health Perspect* 101(Suppl 2):65–71, 1993.

87. Anonymous. An increasing incidence of cryptorchidism and hypospadias? *Lancet* i:1311, 1985.

88. Jacobson JL, Jacobson SW. Intellectual impairment in children exposed to polychlorinated biphenyls in utero. *N Engl J Med* 335:783–789, 1996.

89. Porterfield SP. Vulnerability of the developing brain to thyroid abnormalities: Environmental insults to the thyroid system. *Environ Health Perspect* 102(Suppl 2): 125–130, 1994.

90. Koopman-Esseboom C, Morse D, Weisglas-Kuperus N, et al. Effects of dioxins and polychlorinated biphenyls on thyroid hormone status of pregnant women and their infants. *Pediatr Res* 36:468–473, 1994.

91. Kornilovskaya IN, Gorelaya MV, Usenko VS, et al. Histological studies of atrazine toxicity on the thyroid gland in rats. *Biomed Environ Sci* 9(1):60–66, 1996.

92. Cooper RL, Goldman JM, Rehnberg GL. Pituitary function following treatment with reproductive toxins. *Environ Health Perspect* 70:177–184, 1986.

93. Van den Berg KJ, van Raaij JA, Bragt PC, Notten WR. Interactions of halogenated industrial chemicals with transthyretin and effects on thyroid hormone levels in vivo. *Arch Toxicol* 65(1):15–19, 1991.

94. Ferris CF. The rage of innocents. *Sciences* Mar–Apr: 22–26, 1996.

95. Safe S. Environmental and dietary estrogens and human health: Is there a problem? *Environ Health Perspect* 103:346–351, 1995.

96. Hajek RA, Van NT, Johnston DA, et al. Early exposure to 17-alpha estradiol is tumorigenic in mice. *Proc Am Assoc Cancer Res* 36:632, 1995.

97. Bedding ND, McIntyre AE, Perry L, Lester JN. Organic contaminants in the aquatic environment. I. Sources and occurrence. *Sci Total Environ* 25:143–167, 1982.

98. Jobling S, Reynolds T, White R, et al. A variety of environmentally persistent chemicals, including some phthalate plasticizers, are weakly estrogenic. *Environ Health Perspect* 103(6):582–587, 1995.

99. Smith AH. Infant exposure assessment for breast milk dioxins and furans derived from waste incineration emissions. *Risk Anal* 7(3):347–353, 1987.

100. Wall St J, Feb 20, 1992.

101. Birnbaum LS. The mechanism of dioxin toxicity: Relationship to risk assessment. *Environ Health Perspect* 102(Suppl 9):157–167, 1994.

102. Huff J. Dioxins and mammalian carcinogenesis. In: Schecter A (ed), *Dioxins and health*. New York: Plenum Press, 1994.

103. Olson JR, Holscher MA, Neal RA. Toxicity of 2,3,7,8-tetrachlorodibenzo-p-dioxin in the Golden Syrian hamster. *Toxicol Appl Pharmacol* 55:67–78, 1980.

---

104. Gilbertson M. Effects on fish and wildlife populations. In: Kimbrough RD, Jensen AA (eds), *Halogenated Biphenyls, Terphenyls, Naphthalenes, Dibenzodioxins, and Related Products,* 2d ed. Amsterdam: Elsevier Science Publishers, 1989.

105. Walker MK, Peterson RE. Potencies of polychlorinated dibenzo-p-dioxins, dibenzofurans, and biphenyl congeners for producing early life stage mortality in rainbow trout (*Oncorhyncus mykiss*). *Aquatic Toxicol* 21:219–238, 1991.

106. Couture LA, Abbott BD, Birnbaum LS. A critical review of the developmental toxicity and teratogenicity of 2,3,7,8-tetrachlorodibenzo-p-dioxin: Recent advances toward understanding the mechanism. *Teratology* 42:619–627, 1990.

107. Mably TA, Moore RW, Goy RW, et al. In utero and lactational exposure of male rats to 2,3,7,8-tetrachlordibenzo-p-dioxin. 2. Effects on sexual behavior and the regulation of LH secretion in adulthood. *Toxicol Appl Pharmacol* 114:108–117, 1992.

108. McKinney JD, Waller CL. PCBs as hormonally active structural analogues. *Environ Health Perspect* 102(3):290–97, 1994.

109. Murray FJ, Smith FA, Nitschke CG, et al. Three-generation reproduction study of rats given 2,3,7,8-tetrachlorodibenzo-p-dioxin (TCDD) in the diet. *Toxicol Appl Pharmacol* 50:241–252, 1979.

110. Allen JR, Barsotti DA, Lambrecht LK, et al. Reproductive effects of halogenated aromatic hydrocarbons on nonhuman primates. *Ann NY Acad Sci* 320:419–425, 1979.

111. Barsotti DA, Abrahamson LJ, Allen JR. Hormonal alterations in female rhesus monkeys fed a diet containing 2,3,7,8-TCDD. *Bull Environ Contam Toxicol* 21:463–469, 1979.

112. Rier SE, Martin DC, Bowman RE, et al. Endometriosis in rhesus monkeys (Macaca mulatta) following chronic exposure to 2,3,7,8-tetrachlorodibenzo-p-dioxin. *Fund Appl Toxicol* 21:433–441, 1993.

113. Kociba RJ, Keeler PA, Park GN, et al. 2,3,7,8-tetrachlorodibenzo-p-dioxin: Results of a 13 week oral toxicity study in rats. *Toxicol Appl Pharmacol* 35:553–574, 1976.

114. McConnell EE, Moore JA, Haseman JK, et al. The comparative toxicity of chlorinated dibenzo-p-dioxins in mice and guinea pigs. *Toxicol Appl Pharmacol* 44:335–356, 1978.

115. Chahoud I, Krowke R, Schimmel A, et al. Reproductive toxicity and pharmacokinetics of 2,3,7,8-TCDD. Effects of high doses on the fertility of male rats. *Arch Toxicol* 63:432–439, 1989.

116. Moore RW, Jefcoate CR, Peterson RE. 2,3,7,8-tetrachlorodibenzo-p-dioxin inhibits steroidogenesis in the rat testis by inhibiting the mobilization of cholesterol to cytochrome p450. *Toxicol Appl Pharmacol* 109:85–97, 1991.

117. Grieg JB, Jones G, Butler WH, et al. Toxic effects of 2,3,7,8-tetrachlorodibenzo-p-dioxin. *Food Cosmet Toxicol* 11:585–595, 1973.

118. Cheung MO, Gilbert EF, Peterson RE. Cardiovascular teratogenicity of 2,3,7,8-tetrachlorodibenzo-p-dioxin in the chick embryo. *Toxicol Appl Pharmacol* 61:197–204, 1981.

119. Weber H, Harris MW, Haseman JK, Birnbaum LS. Teratogenic potency of TCDD, TCDF, and TCDD-TCDF combinations in C57BL/6N mice. *Toxicol Lett* 26:159–167, 1985.

120. Schantz SL, Bowman RE. Learning in monkeys exposed perinatally to 2,3,7,8-tetrachlorodibenzo-p-dioxin (TCDD). *Neurotoxicol Teratol* 11:13–19, 1989.

121. Wolfe WH, Michalek JE, Miner JC, et al. Paternal serum dioxin and reproductive outcomes among veterans of Operation Ranch Hand. *Epidemiology* 6(1): 17–22, 1995.

122. Erickson JD, Mulinare J, McClain PW, et al. Vietnam veterans' risks for fathering babies with birth defects. *JAMA* 252:903–912, 1984.

123. Egeland GM, Sweeney MH, Fingerhut MA, et al. Total serum testosterone and gonadotropins in workers exposed to dioxin. *Am J Epidemiol* 139(3):272–281, 1994.

124. Mocarelli P, Brambilla P, Gerthoux PM, et al. Change in sex ratio with exposure to dioxin. *Lancet* 348:409, 1996.

125. Mastroiacovo P, Spagnolo A, Marni E, et al. Birth defects in the Seveso area after TCDD contamination. *JAMA* 259:1668–1672, 1988.

126. Birnbaum LS. Endocrine effects of prenatal exposure to PCBs, dioxins, and other xenobiotics: Implications for policy and future research. *Environ Health Perspect* 102:676–679, 1994.

127. Stockbauer JW, Hoffman RE, Schramm WF, et al. Reproductive outcomes of mothers with potential exposure to 2,3,7,8-tetrachlorodibenzo-p-dioxin. *Am J Epidemiol* 128:410–419, 1988.

128. den Ouden AL, Kok JH, Verkerk PH, et al. The relation between neonatal thyroxine levels and neurodevelopmental outcome at age 5 and 9 years in a national cohort of very preterm and/or very low birth weight infants. *Pediatr Res* 39:142–145, 1996.

129. Schmitt CJ, Zajicek JL, Ribick MA. National pesticide monitoring program. Residues of organochlorine chemicals in freshwater fish, 1980–81. *Arch Environ Contam Toxicol* 14:225–260, 1985.

130. U.S. EPA. *Environmental Transport and Transformation of PCBs*. EPA-560/5-83-05. Washington, DC: EPA, 1983.

131. Dewailly E, Ayotte P, Bruneau S, et al. Inuit exposure to organochlorines through the aquatic food chain in arctic Quebec. *Environ Health Perspect* 101:618–620, 1993.

132. Fielden MR, Chen I, Chitten B, et al. Examination of the estrogenicity of 2,4,6,2′,6′-pentachlorobiphenyl (PCB 104), its hydroxylated metabolite

2,4,2',4',6'-pentachloro-4-bephenylol (HO-PCB 104), and a further chlorinated derivative, 2,4,2',4',6'-hexachlorobiphenyl (PCB 155). *Environ Health Perspect* 105(11):1238–1248, 1997.

133. Bernhoft A, Nafstad I, Engen P, Skaare JU. Effects of pre- and postnatal exposure to 3,3',4,4',5-pentachlorobiphenyl on physical development, neurobehavior and xenobiotic metabolizing enzymes in rats. *Environ Toxicol Chem* 13(10): 1589–1597, 1994.

134. Holene E, Nafstad I, Skaare JU, et al. Behavioral effects of pre- and postnatal exposure to individual polychlorinated biphenyl congeners in rats. *Environ Toxicol Chem* 14(6):967–976, 1995.

135. Rice DC, Hayward S. Effects of postnatal exposure to a PCB mixture in monkeys on nonspatial discrimination reversal and delayed alternation performance. *Neurotoxicol* 18(2):479–494, 1997.

136. Crews D, Bergeron JM, McLachlan JA. The role of estrogen in turtle sex determination and the effect of PCBs. *Environ Health Perspect* 103(Suppl 7):73–77, 1995.

137. Bergeron JM, Crews D, McLachlan JA. PCBs as environmental estrogens: Turtle sex determination as a biomarker of environmental contamination. *Environ Health Perspect* 102:780–781, 1994.

138. Battershill JM. Review of the safety assessment of polychlorinated biphenyls (PCBs) with particular reference to reproductive toxicity. *Hum Exp Toxicol* 13: 581–597, 1994.

139. Sager DB, Shih-Schroeder W, Girard D. Effect of early postnatal exposure to polychlorinated biphenyls (PCB) on fertility in male rats. *Bull Environ Contam Toxicol* 38:946–953, 1987.

140. Barsotti DA, Marlar RJ, Allen JR. Reproductive dysfunction in rhesus monkeys exposed to low levels of polychlorinated biphenyls (Aroclor 1248). *Food Cosmet Toxicol* 14:99–103, 1976.

141. Muller WF, Hobson W, Fuller GB, et al. Endocrine effects of chlorinated hydrocarbons in rhesus monkeys. *Ecotoxicol Environ Safety* 2:161–172, 1978.

142. Jansen HT, Cooke PS, Porcelli J, et al. Estrogenic and anti-estrogenic actions of PCBs in the female rat: In-vitro and in-vivo studies. *Reprod Toxicol* 7:237–248, 1993.

143. Tilson HA, Davis GJ, McLachlan JA, Lucier GW. The effects of polychlorinated biphenyls given prenatally on the neurobehavioral development of mice. *Environ Res* 18:466–474, 1979.

144. Agrawal AK, Tilson HA, Bondy SC. 3,4,3',4'-tetrachlorobiphenyl given to mice prenatally produces long term decreases in striatal dopamine and receptor binding sites in the caudate nucleus. *Toxicol Lett* 7:417–424, 1981.

145. White RD, Allen SD, Bradshaw WS. Delay in the onset of parturition in the rat following prenatal administration of developmental toxicants. *Toxicol Lett* 18:185–192, 1983.

146. Barsotti DA, Marlar RJ, Allen JR. Reproductive dysfunction in rhesus monkeys exposed to low levels of polychlorinated biphenyls (Aroclor 1248). *Food Cosmet Toxicol* 14:99–103, 1976.

147. Collins WT, Capen CC. Fine structural lesions and hormonal alterations in thyroid glands of perinatal rats exposed in utero and by the milk to polychlorinated biphenyls. *Am J Pathol* 99:125–142, 1980.

148. Bowman RE, Heironimus MP, Barsotti DA. Locomotor hyperactivity in PCB-exposed rhesus monkeys. *Neurotoxicology* 2:251–268, 1981.

149. Rice DC, Hayward S. Effects of postnatal exposure to a PCB mixture in monkeys on nonspatial discrimination reversal and delayed alternation performance. *Neurotoxicology* 18(2):479–494, 1997.

150. Taylor PR, Stelma JM, Lawrence CE. The relation of PCBs to birth weight and gestational age in the offspring of occupationally exposed mothers. *Am J Epidemiol* 129:395–406, 1989.

151. Rogan WJ, Gladen BC, Hung KL, et al. Congenital poisoning by polychlorinated biphenyls and their contaminants in Taiwan. *Science* 241:334–338, 1988.

152. Chen YC, Guo YL, Hsu CC, Rogan WJ. Cognitive development of Yu-Cheng ("oil disease") children prenatally exposed to heat-degraded PCBs. *JAMA* 268(22):3213–3218, 1992.

153. Safe, S. Toxicology, structure-function relationship, and human and environmental health impacts of polychlorinated biphenyls: Progress and problems. *Environ Health Perspect* 100:259–268, 1992.

154. Jacobson JL, Jacobson SW. Effects of in utero exposure to PCBs and related contaminants on cognitive functioning in young children. *J Pediatr* 116(1):38–45, 1990.

155. Gladen BC, Rogan WJ. Effects of perinatal polychlorinated biphenyls and dichlorodiphenyl dichloroethene on later development. *J Pediatr* 119:58–63, 1991.

156. Lonky E, Reihman J, Darvill T, et al. Neonatal behavioral assessment scale performance in humans influenced by maternal consumption of environmentally contaminated Lake Ontario fish. *J Great Lakes Res* 22(2):198–212, 1996.

157. White R, Jobling S, Hoare SA, et al. Environmentally persistent alkylphenolic compounds are estrogenic. *Endocrinology* 135(1):175–182, 1994.

158. Clark LB, Rosen RT, Hartman TG, et al. Determination of alkylphenol ethoxylates and their acetic acid derivatives in drinking water by particle beam liquid chromatography/mass spectrometry. *Int J Environ Anal Chem* 47:167–180, 1992.

159. Junk GA, Svec HJ, Richard JJ, et al. Contamination of water by synthetic polymer tubes. *Environ Sci Technol* 8:1100–1106, 1974.

160. Soto AM, Justicia H, Wray JW, Sonnenschein C. p-Nonyl-phenol: An estrogenic xenobiotic released from "modified" polystyrene. *Environ Health Perspect* 92: 167–173, 1991.

161. Purdom CE, Hardiman PA, Bye VJ, et al. Estrogenic effects of effluent from sewage treatment works. *Chem Ecol* 8:275–285, 1994.

162. Stahl FW, Mulach R, Sakuma Y. Bisphenol-A. In: *Chemical Economics Handbook.* Menlo Park, CA: SRI Consulting, 1996.

163. Krishnan AV, Stathis P, Permuth SF, et al. Bisphenol-A: An estrogenic substance is released from polycarbonate flasks during autoclaving. *Endocrinology* 132: 2279–2286, 1993.

164. Brotons JA, Olea-Serrano MF, Villalobos M, et al. Xenoestrogens released from lacquer coatings in food cans. *Environ Health Perspect* 103:608–612, 1995.

165. Olea N, Pulgar R, Perez P, et al. Estrogenicity of resin-based composites and sealants used in dentistry. *Environ Health Perspect* 104:298–305, 1996.

166. Olea N, Pulgar R, Perez P, et al. Estrogenicity of resin-based composites and sealants used in dentistry. *Environ Health Perspect* 104:298–305, 1996.

167. Nagel SC, vom Saal F, Thayer KA, et al. Relative binding affinity-serum modified access (RBA-SMA) assay predicts the relative in vivo bioactivity of the xenoestrogens bisphenol-A and octylphenol. *Environ Health Perspect* 105:70–76, 1997.

168. Steinmetz R, Brown NG, Allen DL, et al. The environmental estrogen bisphenol-A stimulates prolactin release in vitro and in vivo. *Endocrinology* 138:1780–1786, 1997.

169. vom Saal FS, Cooke PS, Buchanan DL, et al. A physiologically based approach to the study of biphenol A and other estrogenic chemicals on the size of reproductive organs, daily sperm production, and behavior. *Toxicol Ind Health* 14(1–2):239–260, 1998.

170. Hoyle WC, Budway R. Bisphenol A in food cans: An update. *Environ Health Perspect* 105(6):570–571, 1997.

171. Welshons W, vom Saal FS, Nagel S. Response. *Environ Health Perspect* 105(6): 571–572, 1997.

172. Menzer RE. Water and soil pollutants. In: Amdur MO, Doull J, Klaassen CD (eds), *Casarett and Doull's Toxicology, The Basic Science of Poisons,* 4th ed. New York: McGraw-Hill, 1991.

173. Harris CA, Henttu P, Parker M, et al. The estrogenic activity of phthalate esters in vitro. *Environ Health Perspect* 105(8):802–811, 1997.

174. Heindel JJ, Gulati DK, Mounce RC, Russell SR, Lamb JC. Reproductive toxicity of three phthalic acid esters in a continuous breeding protocol. *Fund Appl Toxicol* 12:508–518, 1989.

175. Lloyd SC, Foster PMD. Effect of mono-(2-ethylhexyl)phthalate on follicle-stimulating hormone responsiveness of cultured rat Sertoli cells. *Toxicol Appl Pharmacol* 95:484–489, 1988.

176. Davis BJ, Maronpot RR, Heindel JJ. Di-(2-ethylhexyl) phthalate suppresses estradiol and ovulation in cycling rats. *Toxicol Appl Pharmacol* 128:216–223, 1994.

177. Ema M, Itami T, Kawasaki H. Teratogenic phase specificity of butyl benzyl phthalate in rats. *Toxicology* 79(1):11–19, 1993.

178. Ema M, Itami T, Kawasaki H. Embryolethality and teratogenicity of butyl benzyl phthalate in rats. *J Appl Toxicol* 12(3):179–183, 1992.

179. Gulati DK, Barnes LH, Chapin RE, Heindel J. *Final Report on the Reproductive Toxicity of Di(N-butyl)phthalate in Sprague-Dawley Rats.* NTIS Technical Report (NTIS/PB92-111996) September 1991.

180. Ministry of Agriculture, Fisheries, and Food Safety Directorate. Food surveillance information sheet No. 82. London, March 1996.

181. Jaeger R, Rubin R. Migration of a phthalate ester plasticizer from polyvinyl chloride blood bags into stored human blood and its localization in human tissue. *New Engl J Med* 287:1114–1118, 1972.

182. Menzer RE. Water and soil pollutants. In: Amdur MO, Doull J, Klaassen CD (eds), *Casarett and Doull's Toxicology, The Basic Science of Poisons,* 4th ed. New York: McGraw-Hill, 1991.

183. Guzelian PS. Comparative toxicology of chlordecone (kepone) in humans and experimental animals. *Annu Rev Pharmacol Toxicol* 22:89–113, 1982.

184. Kelce WR, Stone CR, Laws SC, et al. Persistent DDT metabolite p,p'-DDE is a potent androgen receptor antagonist. *Nature* 375:581–585, 1995.

185. Eroschenko VP, Cooke PS. Morphological and biochemical alterations in reproductive tracts of neonatal female mice treated with the pesticide methoxychlor. *Biol Repro* 42(3):573–583, 1990.

186. Gray LE, Ostby JS, Ferrell JM, et al. Methoxychlor induces estrogen-like alterations of behavior and the reproductive tract in the female rat and hamster: Effects on sex behavior, running wheel activity, and uterine morphology. *Toxicol Appl Pharmacol* 96(3):525–540, 1988.

187. Gray LE, Ostby JS, Ferrell JM, et al. A dose-response analysis of methoxychlor-induced alterations of reproductive development and function in the rat. *Fund Appl Toxicol* 12(1):92–108, 1989.

188. vom Saal FS, Nagel SC, Palanza P, et al. Estrogenic pesticides: Binding relative to estradiol in MCF-7 cells and effects of exposure during fetal life on subsequent territorial behavior in male mice. *Toxicol Lett* 77:343–350, 1995.

189. Soto AM, Chung KL, Sonnenschein C. The pesticides endosulfan, toxaphene, and dieldrin have estrogenic effects on human estrogen-sensitive cells. *Environ Health Perspect* 102:380–383, 1994.

190. Sircar S, Lahiri P. Lindane (gamma-HCH) causes reproductive failure and fetotoxicity in mice. *Toxicol* 59(2):171–177, 1989.

191. Cooper RL, Chadwick RW, Rehnberg GL, et al. Effect of lindane on hormonal control of reproductive function in the female rat. *Toxicol Appl Pharmacol* 99(3):384–394, 1989.

192. MacLellan KN, Bird DM, Fry DM, et al. Reproductive and morphological

effects of o,p′-dicofol on two generations of captive American kestrels. *Arch Environ Toxicol* 30(3):364–372, 1996.

193. Van den Berg KJ, van Raaij AGM, Bragt PC, Notten WRF. Interactions of halogenated industrial chemicals with transthyretin and effects on thyroid hormone levels in vivo. *Arch Toxicol* 65:15–19, 1991.

194. Hill RH, Head SL, Baker S, et al. Pesticide residues in urine of adults living in the United States: reference range concentrations. *Environ Res* 71:99–108, 1995.

195. Van den Berg KJ. Interaction of chlorinated phenols with thyroxine binding sites of human transthyretin, albumin, and thyroid binding globulin. *Chem Biol Interact* 76:63–75, 1990.

196. Jekat FW, Meisel ML, Eckard R, Winterhoff H. Effects of pentachlorophenol (PCP) on the pituitary and thyroidal hormone regulation in the rat. *Toxicol Lett* 71:9–25, 1994.

197. van Raaij JA, Frijters CM, Kong LW, et al. Reduction of thyroxine uptake into cerebrospinal fluid and rat brain by hexachlorobenzene and pentachlorophenol. *Toxicol* 94(1–3):197–208, 1994.

198. Van den Berg KJ, van Raaij AGM, Bragt PC, Notten WRF. Interactions of halogenated industrial chemicals with transthyretin and effects on thyroid hormone levels in vivo. *Arch Toxicol* 65:15–19, 1991.

199. Kelce WR, Monosson E, Gray LE. An environmental anti-androgen. *Recent Prog Horm Res* 50:449–453, 1995.

200. Gray LE, Ostby JS, Kelce WR. Developmental effects of an environmental antiandrogen: The fungicide vinclozolin alters sex differentiation of the male rat. *Toxicol Appl Pharmacol* 129(1):46–52, 1994.

201. Kelce WR, Monosson E, Gamcsik MP, et al. Environmental hormone disrupters: Evidence that vinclozolin developmental toxicity is mediated by antiandrogenic metabolites. *Toxicol Appl Pharmacol* 126(2):276–285, 1994.

202. Wong C, Kelce WR, Sar M, Wilson EM. Androgen receptor antagonist versus agonist activities of the fungicide vinclozolin relative to hydroxyflutamide. *J Biol Chem* 270(34):19998–20003, 1995.

203. Ronis MJJ, Barger TM, Gandy J, et al. Anti-androgenic effects of perinatal cypermethrin exposure in the developing rat. *Abstracts of the 13th International Neurotoxicology Conference*. Hot Springs, AK, 1995.

204. Connor K, Howell J, Chen I, et al. Failure of chloro-s-triazine-derived compounds to induce estrogen receptor-mediated responses in vivo and in vitro. *Fundam Appl Toxicol* 30:93–101, 1996.

205. Babic-Gojmerac T, Kniewald Z, Kniewald J. Testosterone metabolism in neuroendocrine organs in male rats under atrazine and deethylatrazine influence. *J Steroid Biochem* 33:141–146, 1989.

206. Davis DL, Bradlow HL. Can environmental estrogens cause breast cancer? *Sci Am* 166–172, Oct 1995.

---

207. Kniewald J, Osredecki V, Gojmerac T, et al. Effect of s-triazine compounds on testosterone metabolism in the rat prostate. *J Appl Toxicol* 15(3):215–218, 1995.

208. Cooper RL, Stoker TE, Goldman JM, et al. Atrazine disrupts hypothalamic control of pituitary-ovarian function. *Toxicologist* 30:66, 1996.

209. Cooper RL, Stoker TE, Goldman JM, et al. Effect of atrazine on ovarian function in the rat. *Reprod Toxicol* 10(4):257–264, 1996.

210. Houeto P, Bindoula G, Hoffman JR. Ethylenebisdithiocarbamates and ethylenethiourea: Possible human health hazards. *Environ Health Perspect* 103:568–573, 1995.

211. Steenland K, Cedillo L, Tucker J, et al. Thyroid hormones and cytogenetic outcomes in backpack sprayers using ethylenebis(dithiocarbamate) (EBDC) fungicides in Mexico. *Environ Health Perspect* 105:1126–1130, 1997.

212. Whitten Pl, Lewis C, Russell E, Naftolin F. Potential adverse effects of phytoestrogens. *J Nutrition* 125(3 Suppl):771S–776S, 1995.

213. Whitten PL, Lewis C, Russel E, Naftolin F. Phytoestrogen influences on the development of behavior and gonadotropin function. *Proc Soc Exp Biol Med* 208(1): 82–86, 1995.

214. Levy JR, Faber KA, Ayyash L, Hughes CL. The effect of prenatal exposure to the phytoestrogen genistein on sexual differentiation in rats. *Proc Soc Exp Biol Med* 208(1):60–66, 1995.

215. Whitten PL, Russel E, Naftolin F. Influence of phytoestrogen diets on estradiol action in the rat uterus. *Steroids* 59(7):443–449, 1994.

216. Whitten PL, Lewis C, Naftolin F. A phytoestrogen diet induces the premature anovulatory syndrome in lactationally exposed female rats. *Biol Reprod* 49(5): 1117–1121, 1993.

217. Faber KA, Hughes CL. Dose-response characteristics of neonatal exposure to genistein on pituitary responsiveness to gonadotropin releasing hormone and volume of the sexually dimorphic nucleus of the preoptic area (SDN-POA) in postpubertal castrated female rats. *Reprod Toxicol* 7(1):35–39, 1993.

218. Adlercreutz H. Phytoestrogens: Epidemiology and a possible role in cancer protection. *Environ Health Perspect* 103(Suppl 7):103–112, 1995.

Chapter 7

1. U.S. EPA, 57 Federal Register 22888–22938 (1992)

2. United States International Trade Commission. *Synthetic Organic Chemicals: United States Production and Sales, 1994.* USITC Publication 2933. November 1995.

3. Letter from Sam Gibbons and Bill Archer, Committee on Ways and Means, to Peter Watson, Chairman, U.S. International Trade Commission, October 17, 1995.

4. Gianessi LP, Anderson JE. *Pesticide Use in U.S. Crop Production: National Summary Report*. Washington, DC: National Center for Food and Agriculture Policy, February 1995.

5. U.S. EPA. *Pesticides Industry Sales and Usage: 1994 and 1995 Market Estimates*. EPA 733-R-97-002. August 1997.

6. U.S. EPA. *1995 Toxics Release Inventory: Public Data Release*. EPA 745-R-97-005. April 1997.

7. USDA. *Pesticide Data Program Progress Report*. Washington DC, February 1998.

8. U.S. EPA. *Update: Listing of Fish and Wildlife Advisories*. EPA-823-F-98-009. March 1998.

9. U.S. Department of Agriculture Food Safety and Inspection Service. *Domestic Residue Data Book: National Residue Program, 1994*.

10. U.S. EPA. *Nonoccupational Pesticide Exposure Study (NOPES) Final Report*. EPA/600/3-90/003. Office of Research and Development, January 1990.

11. Wallace LA. *The Total Exposure Assessment Methodology (TEAM) Study: Summary and Analysis: Vol. 1*. EPA/600/6-87/002a. Office of Research and Development, June 1987.

12. U.S. Department of Health and Human Services. *Plan and Operation of the Third National Health and Nutrition Examination Survey, 1988–94*. (PHS) 94-1308. July 1994.

13. Ezzati-Rice TM, Murphy RS. Issues associated with the design of a national probability sample for human exposure assessment. *Environ Health Perspect* 103(Suppl 3):55–60, 1995.

14. Holzman D. Banking on tissues. *Environ Health Perspect* 104(6):606–610, 1996.

15. Committee on National Monitoring of Human Tissues. *Monitoring Human Tissues for Toxic Substances*. Washington, DC: National Academy Press, 1991.

16. Pellizzari E, Lioy P, Whitmore R, Clayton A, et al. Population-based exposure measurements in EPA Region 5: A phase I field study in support of the National Human Exposure Assessment Survey. *J Exp Anal Environ Epidemiol* 5(3):327–358, 1995.

17. Lebowitz MD, O'Rourke MK, Gordon S, et al. Population-based exposure measurements in Arizona: A phase I field study in support of the National Human Exposure Assessment Survey. *Anal Environ Epidemiol* 5(3):297–325, 1995.

18. Schreiber JS. Predicted infant exposure to tetrachloroethylene in human breastmilk. *Risk Anal* 13:515–524, 1993.

19. Labreche FP, Goldberg MS. Exposure to organic solvents and breast cancer in women: A hypothesis. *Am J Ind Med* 32:1–14, 1997.

20. Bagnell PC, Ellenberger HA. Obstructive jaundice due to a chlorinated hydrocarbon in breast milk. *Can Med J* 5:1047–1048, 1977.

21. Schecter A, Startin J, Wright C, et al. Dioxins in U.S. food and estimated daily intake. *Chemosphere* 29:2261–2265, 1994.

22. Rogan WJ, Bagniewska A, Damstra T. Pollutants in breast milk. *New Engl J Med* 302:1450–1453, 1980.

23. Gladen BC, Rogan WJ. DDE and shortened duration of lactation in a northern Mexican town. *Am J Public Health* 85(4):504–508, 1995.

24. Newman J. How breast milk protects newborns. *Sci Am* 76–79, December 1995.

25. Aniansson G, Alm B, Andersson B, et al. A prospective cohort study on breastfeeding and otitis media in Swedish infants. *Pediatr Infect Dis J* 13:183–188, 1994.

26. Howie PW, Forsyth JS, Ogston SA, Clark A, Florey C. Protective effect of breastfeeding against infection. *Br Med J* 300:11–16, 1990.

27. Van-Coric M. Antibody responses to parenteral and oral vaccines were impaired by conventional and low-protein formulas as compared to breast feeding. *Acta Paedr Scand* 79:1137–1142, 1990.

28. Saarinen UM, Kajosaari M. Breastfeeding as prophylaxis against atopic disease: Prospective follow-up study until 17 years old. *Lancet* 346:1065–1069, 1995.

29. Koletzo S, Sherman P, Corey M, Griffiths A, Smith C. Role of infant feeding practices in development of Crohn's disease in childhood. *Br Med J* 298:1617–1618, 1989.

30. Mayer EJ, Hamman RF, Gay EC, Lezotte DC, Savitz DA, Klingensmith, GJ. Reduced risk of insulin-dependent diabetes mellitus among breastfed children. *Diabetes* 37:1625–1632, 1988.

31. Morley R, Cole TJ, Powell R, Lucas A. Mother's choice to provide breastmilk and developmental outcome. *Arch Dis Child* 63:1382–1385, 1988.

32. Lucas A, Morley R, Cole TJ, Lister G, Leeson Payne C. Breast milk and subsequent intelligence quotient in children born preterm. *Lancet* 339:261–264, 1992.

33. U.S. EPA. *National Air Quality and Emissions Trends Report,* 1995. EPA 454/R-96-005. October 1996.

34. U.S. EPA. *Mercury Study Report to Congress.* EPA-452/R-96-001c. Office of Air Quality Planning and Standards and Office of Research and Development, December 1997.

35. U.S. EPA. *Environmental Indicators of Water Quality in the United States.* EPA 841-R-96-002. June 1996.

36. Bolger PM, Yess NJ, Gunderson EL, Troxell TC, Carrington CD. Identification and reduction of dietary lead in the United States. *Food Addit Contam* 13:53–60, 1996.

37. Adams MA. FDA Total Diet Study: Dietary intake of lead and other chemicals. *Chem Spec Bioavail* 3:37–41, 1991.

38. Bolger PM, Carrington CD, Capar SG, Adams MA. Reductions in dietary lead exposure in the United States. *Chem Spec Bioavail* 3:31–36, 1991.

39. Galal-Gorchev H. Dietary intake, levels in food, and estimated intake of lead, cadmium, and mercury. *Food Addit Contam* 10:115–128, 1993.

40. Pirkle JL, Brody DJ, Gunter EW, et al. The decline in blood lead levels in the United States: The National Health and Nutrition Examination Surveys (NHANES). *JAMA* 272:284-291, 1994.

41. Update: Blood lead levels—United States, 1991–1994. *MMWR* 46(7):141–146, 1997.

42. Brody DJ, Pirkle JL, Kramer RA, et al. Blood lead levels in the U.S. population: Phase 1 of the third National Health and Nutrition Examination Survey (NHANES III, 1988–1991). *JAMA* 272:277–283, 1994.

43. Whittemore AS, DiCiccio Y, Provenzano G. Urinary cadmium and blood pressure: Results from the NHANES II survey. *Environ Health Perspect* 91:133–140, 1991.

44. Somogyi A, Beck H. Nurturing and breast-feeding: Exposure to chemicals in breast milk. *Environ Health Perspect* 101 (Suppl 2):45–52, 1993.

45. Grandjean P, Jorgensen PJ, Weihe P. Human milk as a source of methylmercury exposure in infants. *Environ Health Perspect* 102:74–77, 1994.

46. Andelman JB. Human exposures to volatile halogenated organic chemicals in indoor and outdoor air. *Environ Health Perspect* 62:313–318, 1985.

47. Wallace L, Nelson W, Ziegenfus R, et al. The Los Angeles TEAM study: Personal exposures, indoor-outdoor air concentrations, and breath concentrations of 25 volatile organic compounds. *J Expo Anal Environ Epidemiol* 1(2):157–192, 1991.

48. Wallace LA, Pellizzari ED, Hartwell TD, et al. The influence of personal activities on exposure to volatile organic compounds. *Environ Res* 50:37–55, 1989.

49. Wallace L, Buckley T, Pellizzari E, Gordon S. Breath measurements as volatile organic compound biomarkers. *Environ Health Perspect* 104(Suppl 5):861–859, 1996.

50. Thompson KM, Evans JS. Workers' breath as a source of perchloroethylene (perc) in the home. *J Expo Anal Environ Epidemiol* 3:417–430, 1993.

51. Baelum J. Toluene in alveolar air during controlled exposure to constant and to varying concentrations. *Int Arch Occup Environ Health* 62:59–64, 1990.

52. Jo WK, Weisel CP, Lioy PJ. Routes of chloroform exposure and body burden from showering with contaminated tap water. *Risk Anal* 10:575–580, 1990.

53. Aggazzotti G, Fantuzzi G, Righi E, et al. Chloroform in alveolar air of individuals attending indoor swimming pools. *Arch Environ Health* 48:250–254, 1993.

54. Levesque B, Ayotte P, LeBlanc A, et al. Evaluation of dermal and respiratory chloroform exposure in humans. *Environ Health Perspect* 102:1082–1087, 1994.

55. Ashley DL, Bonin MA, Cardinali FL, McCraw JM, Wooten JV. Blood concentrations of volatile organic compounds in a nonoccupationally exposed U.S. population and in groups with suspected exposure. *Clin Chem* 40(7):1401–1404, 1994.

56. Ashley DL, Bonin MA, Cardinali FL, McCraw JM, Wooten JV. Measurement of volatile organic compounds in human blood. *Environ Health Perspect* 104 (Suppl 5):871–877, 1996.

57. Pirkle JL, Needham LL, Sexton K. Improving exposure assessment by monitoring human tissues for toxic chemicals. *J Exp Anal Environ Epidemiol* 5:405–424, 1995.

58. Phillips LJ, Birchard GF. Regional variation in human toxics exposure in the USA: An analysis based on the National Human Adipose Tissue Survey. *Arch Environ Contam Toxicol* 21:159–168, 1991.

59. Fisher J, Mahle D, Bankston L, Greene R, Gearhart J. Lactational transfer of volatile chemicals in breast milk. *Am Ind Hyg Assoc J* 58:425–431, 1997.

60. U.S. Geological Survey. National Water Quality Assessment Pesticide National Synthesis Project. Provisional data, August 1997. http://water.wr.usgs.gov/pnsp/gwsw1.html.

61. Ohio EPA Pesticide Special Study. http://www.epa.ohio.gov/ddagw/pestspst.html.

62. Cohen BA, Wiles R. *Tough to Swallow: How Pesticide Companies Profit from Poisoning America's Tap Water*. Washington, DC: Environmental Working Group, August 1997.

63. U.S. Food and Drug Administration. Pesticide Program Residue Monitoring—1996. http://vm.cfsan.fda.gov/~dms/pes96rep.html.

64. FDA. Food and Drug Administration Pesticide Program: Residue monitoring 1993. *J Assoc Anal Chem Int* 77(5):163A–185A, 1994.

65. Simcox NJ, Fenske RA, Wolz SA, et al. Pesticides in household dust and soil: Exposure pathways for children of agricultural families. *Environ Health Perspect* 103:1126–1134, 1995.

66. Fenske RA, Black KG, Elkner KP, et al. Potential exposure and health risks of infants following indoor residential pesticide applications. *Am J Public Health* 80:689–693, 1990.

67. Stehr-Green P. Demographic and seasonal influences on human serum pesticide residue levels. *J Tox Environ Health* 27:405–421, 1989.

68. Murphy RS, Kutz FW, Strassman SC. Selected pesticide residues or metabolites in blood and urine specimens from a general population survey. *Environ Health Perspect* 48:81–86, 1983.

69. Needham LL, Hill RH, Ashley DL, Pirkle JL, Sampson EJ. The priority toxicant reference range study: Interim report. *Environ Health Perspect* 103(Suppl 3):89–94, 1995.

70. Hill RH, Head SL, Baker S, et al. Pesticide residues in urine of adults living in the United States: Reference range concentrations. *Environ Res* 71:99–108, 1995.

71. Lordo RA, Dinh KT, Schwemberger JG. Semivolatile organic compounds in adipose tissue: Estimated averages for the US population and selected subpopulations. *Am J Public Health* 86:1253–1259, 1996.

72. Rogan WJ, Gladen BC, McKinney, et al. Polychlorinated biphenyls (PCBs) and dichlorodiphenyl dichloroethene (DDE) in human milk: Effects of maternal factors and previous lactation. *Am J Public Health* 76:1172–1177, 1986.

73. Thomas VM, Spiro TG. An estimation of dioxin emissions in the United States. *Toxicol Environ Chem* 50:1–37, 1995.

74. Fensterheim RJ. Documenting temporal trends of polychlorinated biphenyls in the environment. *Regul Toxicol Pharmacol* 18:181–201, 1993.

75. Orban JE, Stanley JS, Schwemberger JG, Remmers JC. Dioxins and dibenzo-furans in adipose tissue of the general U.S. population and selected subpopulations. *Am J Public Health* 84:439–445, 1994.

76. Robinson PE, Mack GA, Remmers J, Levy R, Mohandjer L. Trends of PCB, hexachlorobenzene, and benzene hexachloride levels in the adipose tissue of the U.S. population. *Environ Res* 53:175–192, 1990.

77. Laug EP, Kunze FM, Prickett CS. Occurrence of DDT in human fat and milk. *Arch Ind Hyg* 3:245–246, 1951.

78. Rogan WJ, Gladen BC, McKinney JD, et al. Polychlorinated biphenyls (PCBs) and dichlorodiphenyl dichloroethene (DDE) in human milk: Effects on growth, morbidity, and duration of lactation. *Am J Public Health* 77(10):1294–1297, 1987.

Chapter 8

1. Merrill RA. Regulatory toxicology. In: Klaassen CD (ed), *Casarett and Doull's Toxicology*. New York: McGraw-Hill, 1996.

2. Dahl R. Can you keep a secret? *Environ Health Perspect* 103:914–916, 1995.

3. US GAO. *Toxics Substances Control Act: Legislative Changes Could Make the Act More Effective.* GAO/RCED-94-103. 1994.

4. Ashford NA, Caldart CC. *Technology, Law, and the Working Environment.* New York: Van Nostrand Reinhold, 1991.

5. Chemical Manufacturers Association v. EPA, 899 F.2d 344 (5th Cir. 1990).

6. Personal communication. Gary Timm, US EPA.

7. A microcomputer database of Massachusetts pesticide use estimates for tree fruits, small fruits, cranberries, and vegetables. University of Massachusetts Cooperative Extension Service, 1990.

8. GAO report. *Pesticides: Better Data Can Improve the Usefulness of EPA's Benefits Assessments.* GAO/RCED-92-32.

9. EPA. Proposed guidelines for ecological risk assessment. *Fed Reg* 61(175), September 9, 1996.

10. Orum P, MacLean A. *Progress Report: Community Right-to-Know.* Washington, DC: U.S. Public Interest Research Group Education Fund, July 1992.

11. Cushman J. Court backs EPA authority on disclosure of toxic agents. *New York Times,* May 2, 1996, sec A20.

12. OSHA Public Law 91-596, 91st Congress, S. 2193.

13. GAO. *Status of Regulation of Toxic Chemicals Under Selected Laws.* GAO/RCED-91-154. 1991.

14. Industrial Union Dept. v. American Petroleum Institute, 48 US 607, 1980.

15. Wargo, J. *Our Children's Toxic Legacy.* New Haven, CT: Yale University Press, 1996.

16. FDA. Indirect food additives: Polymers: Acrylonitrile/styrene copolymers. *FR* 49(183):36635-36644, September 19, 1984.

17. FD&C Act Sec. 601(a).

18. Rimkus GG, Wolf M. Polycyclic musk fragrances in human adipose tissue and human milk. *Chemosphere* 33(10):2033–2043, 1996.

19. Merrill RA. Regulatory toxicology. In: Klaassen, CD (ed), *Casarett and Doull's Toxicology.* New York: McGraw-Hill, 1996.

20. Gulf South Insulation v. CPSC, 701 F.2d 1137 (5th Cir. 1983).

21. Rawls J. *A Theory of Justice.* Cambridge, MA: Harvard University Press, 1971.

22. Paul M, Kurtz S. Analysis of reproductive health hazard information on Material Safety Data Sheets for lead and the ethylene glycol ethers. *Am J Ind Med* 25: 403–415, 1994.

23. Kolp P, Sattler B, Blayney M, Sherwood T. Comprehensibility of material safety data sheets. *Am J Ind Med* 23:135–141, 1993.

24. Environmental law—FIFRA after *Wisconsin Intervenor v. Mortrei:* What next? *J Corp L* 17(4):887, 1992.

25. Nelson L, Kenen R, Klitzman S. *Turning Things Around: A Woman's Occupational and Environmental Health Resource Guide.* Washington, DC: National Women's Health Network, 1990.

26. Walbott G. *Health Effects of Environmental Pollutants.* St. Louis: CV Mosby, 1978.

27. Agent orange update supports link with veterans' health problems. *Nation's Health* 26(4):7, 1996.

28. FOIA, it's always there. *Quill* 84(8):10, 1996.

29. Opheim T. Fire on the Cuyahoga. *EPA J* 19(2):14, 1993.

30. Taylor R. *Ahead of the Curve.* Washington, DC: Environmental Defense Fund, 1990.

31. Wild R (ed). *Earth Care Annual 1990*. Rodale Press, 1990.

32. Nixon W. Making Earth Day count. *E magazine* 6(2):30, 1995.

33. EPA. *Preserving Our Future Today*. EPA 21K-1012. October 1991.

34. Kovach K, Sullivan J, Alston T, Hamilton N. New prescriptions for a healthier OSHA. *Business Horizons* 40(1):45, 1997.

35. Econotes. *Environ Action* 28(1–2):6, 1996.

36. Kates RW, Clark WC. Expecting the unexpected. *Environment* 38(2):6, 1996.

37. EPA administrator Carol Browner news conference, May 20, 1997.

Chapter 9

1. Commonwealth of Massachusetts, Department of Environmental Protection. *Mercury in Massachusetts: an Evaluation of Sources, Emissions, Impacts and Controls.* June 1996.

2. Feinberg L. The day LBJ signed FOIA. *Quill* 84(8):13, 1996.

3. FOIA, it's always there. *Quill* 84(8):10, 1996.

4. Burns P. *Handy Products, Hidden Poisons: The Danger of Toxic Chemicals in Household Products.* Boston: MASSPIRG, 1996.

5. Dadd-Redalia D. *Sustaining the Earth: Choosing Consumer Products That Are Safe for You, Your Family and the Earth.* New York: Morrow, 1994.

6. Dickey P. *Buy Smart, Buy Safe: A Consumer Guide to Less Toxic Products.* Seattle, WA: Washington Toxics Coalition, 1994.

7. Shoemaker JM, Vitale CY. *Healthy Homes, Healthy Kids: Protecting Your Children from Everyday Environmental Hazards.* Washington, DC: Island Press, 1991.

8. Mott L, Vance F, Curtis J. *Handle with Care: Children and Environmental Carcinogens.* New York: Natural Resources Defense Council, 1994.

9. National Institutes of Health, NIEHS. *Lead and Your Health.* NIH 92-3465. Washington, DC: NIH, 1992.

10. U.S. EPA, U.S. CPSC, U.S. HUD. *Protect Your Family from Lead in Your Home.* EPA747-K-94-001. 1995.

11. Personal communication, National Lead Information Center, September 2, 1997.

12. National Lead Information Center. Information packet, 1997.

13. NRDC et al. Citizens' petition to initiate rulemaking concerning the presence of lead in certain dietary calcium supplements and antacids, January 27, 1997.

14. Neergaard L. Danger found in hair dyes with lead. *San Francisco Examiner,* February 4, 1997.

15. U.S. Food and Drug Administration statement. Lead acetate used in hair dye products. February 5, 1997.

16. Olson E. *Think Before You Drink: The Failure of the Nation's Drinking Water System to Protect Public Health.* New York: Natural Resources Defense Council, 1993.

17. U.S. General Accounting Office. *Drinking Water: Information on the Quality of Water Found at Community Water Systems and Private Wells.* GAO/RCED-97-123. June 1997.

18. Ingram C. *The Drinking Water Book: A Complete Guide to Safe Drinking Water.* Berkeley, CA: Ten Speed Press, 1991.

19. Personal communication, safe drinking water hotline, August 25, 1997.

20. U.S. General Acounting Office. *Drinking water.* GAO/RCED-92-34. Washington, DC: US Government Printing Office, 1991.

21. Raloff J. Home carpets: Shoeing in toxic pollution. *Sci News* 138:86, 1990.

22. EPA's OPPT:www.epa.gov/opptintr/labeling/readme1.htm

23. U.S. EPA, Office of Air and Radiation. *The Inside Story: A Guide to Indoor Air Quality.* EPA 402-K-93-007. April 1995.

24. Special Legislative Commission on Air Pollution, Commonwealth of Massachusetts. *Indoor Air Pollution in Massachusetts.* April 1989.

25. Ibid.

26. Roberts J. *Reducing Exposure to Lead in Older Homes.* Seattle, WA: Washington Toxics Coalition, 1990.

27. Brown P. Popular epidemiology challenges the system. *Environment* 35(8):16, 1993.

28. Hoffman AJ. An uneasy rebirth at Love Canal. *Environment* 37(2):4–16, 1995.

29. John Snow Institute and U.S. EPA. *Every Community's Right to Know.* Appendix IV, September 1977.

30. U.S. EPA. *Estimating Exposure to Dioxin-like Compounds.* Review draft. EPA/600/6-88/005Ca. June 1994.

31. U.S. EPA, Office of Solid Waste and Emergency Response. Cleaning up the nation's waste sites: Markets and technology trends. EPA 542-R-96-005A. April 1997.

32. U.S. EPA. Office of Solid Waste and Emergency Response. Sara Title III fact sheet. EPA 550-F-93-002. January 1993.

33. Adams W, Burns S, Handwerk P. *Nationwide LEPC Survey: Summary Report.* Washington, DC: George Washington University, 1994.

34. EnviroFacts Warehouse online queries. July 1997.

35. U.S. EPA, Office of Solid Waste and Emergency Response. LandView factsheet. EPA 550-F-95-003. April 1995.

36. Wallace D. *Upstairs, Downstairs: Perchloroethylene in the Air in Apartments Above N.Y. City Dry Cleaners; A Special Report from Consumers Union.* New York: Consumers Union, October 1995.

37. Cantin J. *Overview of Exposure Pathways.* Round Table Proceedings: Falls Church, VA, May 27–28, 1992.

38. Leubuscher S. *Dry Cleaning—Hidden Hazards.* Amsterdam: Greenpeace International, 1992.

39. Green Corps, Pesticide Watch Education Fund. *An Evaluation of San Francisco's Recreation and Parks Department Pesticide Control Program.* San Francisco: Pesticide Watch, January 1997.

40. Personal communication, Greg Small, Pesticide Watch. September 12, 1997.

41. MASSCOSH. Agencies and laws factsheet, 1997.

42. Paul M, Kurtz S. Analysis of reproductive hazard information on material safety data sheets for lead and ethylene glycol ethers. *Am J Ind Med* 25:403–415, 1994.

43. NIOSH hotline instructions for HHE evaluations.

Index

organic solvents and, 79, 81*t*
pesticides and, 117*t*–118*t*, 119–120
phenol and, 97
xylene and, 104
bisphenol-A, 151, 162, 180–181,
  335*t*
blood, human
  lead in, 53, 54*f*, 55, 210
  pesticides in, 224–225, 224*t*
  solvents in, 216, 302*t*
blood-brain barrier, 12
blood-testis barrier, 15
bone formation
  benzene and, 84
  lead and, 53
brain
  birth defects. *See* anencephaly; mi-
    croencephaly; neural tube defects
  cancer
    pesticides and, 120–121
    solvents and, 83
  development
    methoxychlor and, 133
    methyl-mercury and, 60
    neurotransmitters in, 129
    TCE and, 103
  learning/behavioral deficits. *See* be-
    havioral problems; learning impair-
    ments
  sexual differentiation of, 12, 13*f*
breast cancer, endocrine disruptors
  and, 159–161
breast milk
  environmental contaminants in, 160
  dioxin, 175
  organic solvents, 216
  pesticides, 225, 226*t*
  monitoring, 193*t*, 204–205
bromoxynil
  environmental characteristics, 109*t*
  release, 218*t*
  reproductive/developmental effects,
    334*t*
  in animals, 141*t*, 144
  endocrine disruption, 183, 184*t*
  uses, 139
Brownsville, Texas, 42

building/finishing products, 279–280
butyl benzyl phthalate (BBP), 182

CAA (Clean Air Act), 236, 262
cadmium
  absorbed dose, 210
  biological monitoring tests, 302*t*
  breast milk levels, 211*t*
  lindane and, 131
  production/release, 208*t*
  reproductive/developmental effects,
    51–52, 63–65, 334*t*
  routes of exposure, 62–63, 209
calcium, fetal requirements, lead and,
  53
California
  pesticide use reporting legislation, 263
  Proposition 65, 258–259, 263
California Birth Defects Monitoring
  Program (CBDMP), 40
cancer. *See also specific cancers*
  benzene and, 84
  childhood
    benzene and, 84–85
    organic solvents and, 80, 82–83
    pesticides and, 120–121
  solvents and, 80, 82–83
CAP (Compliance Audit Program),
  242
carbamates. *See also specific carbamates*
  in indoor air/house dust, 223*t*
  reproductive/developmental effects,
    126–127, 128*t*, 129
  uses, 126
carbaryl
  biological monitoring tests, 302*t*
  environmental characteristics, 109*t*
  in food, 221, 223
  in human urine, 224*t*
  in indoor air/house dust, 223*t*
  uses, 126, 217
carbendazim, 33–34
carbon tetrachloride, low birth weight
  and, 79
carcinogens. *See also specific carcinogens*
  dose-response curves for, 31, 32*f*
  endocrine disruption and, 159

in humans, 173–175
thyroid hormone function and, 166
toxic equivalents, 228–229
disinfection by-products (DBPs), 87–88
dithiocarbamates
endocrine disruption effects, 155, 184t, 187
ethylene bisdithiocarbamates, 302t
reproductive/developmental effects, 334t
uses, 136
diuron, 109t, 139, 143
dopamine, 6
dose-response curves, 31, 32f, 33–34, 43
dry cleaners, exposures from, 286–287
Dursban. See chlorpyrifos
dust, pesticides in, 113, 222–223, 223t

EBDCs (ethylene bisdithiocarbamates), 302t. See also specific ethylene bisdithiocarbamates
EDB (ethylene dibromide), 121, 145–146, 146t, 334t
EDF (Environmental Defense Fund), 285, 324
EDs. See endocrine disruptors
EDSTAC (Endocrine Disruptor Screening and Testing Advisory Committee), 153
EGEE (ethylene glycol monoethyl ether), 91, 302t
eggs (oocytes), 9, 10, 14
EGME (ethylene glycol monomethyl ether), 91
EGMEA (ethylene glycol monoethyl ether acetate), 91
Endocrine Disruptor Screening and Testing Advisory Committee (EDSTAC), 153
endocrine disruptors (EDs), 226–227, 230. See also specific endocrine disruptors
absorbed dose, 228–229
behavioral abnormalities, 165–166
biological effects, 151–152

definition of, 151
environmental concentrations, 227–228
human intake, 228
mechanism of action, 154–156
pesticides, 183, 184t, 185–187
production/release, 227, 227t
reproductive/developmental effects, 156, 168–169
in animals, 156–158
in humans, 159–166
from low-dose exposures, 166–168
sperm counts, 164–165
undescended testicles, 165
endosulfan
endocrine disruption effects, 184t
environmental characteristics, 109t
in food supply, 221
reproductive/developmental effects, 132t, 133, 334t
uses, 130, 219t
Environmental Defense Fund (EDF), 285, 324
environmental history, 295–296
environmental monitoring, 301
environmental persistence
dioxin, 170
mercury, 312–313
PCBs, 227–228
pesticides, 109–111, 109t
Environmental Protection Agency (EPA)
air quality and emissions trends, 193t, 196–197
animal tests, 19, 123–124
Compliance Audit Program, 242
creation of, 262
definition of exposure, 189
drinking water testing and, 274–275
EDSTAC, 153
fish and wildlife advisories, 193t, 199
inert ingredients and, 108
information resources, 317–318, 330
Listing of Fish and Wildlife Advisories, 199
maximum contaminant levels, 250
NOPES study, 193t, 200

Federal Insecticide, Fungicide, and Rodenticide Act (FIFRA), 236, 244–246, 261

Federal Register, 238, 239

Federal Water Pollution and Control Act (Clean Water Act), 236, 262

feedback loops
in men, 5, 7*f,* 8*f*
negative, 4, 5, 7*f*–9*f*
positive, 4, 5–6, 9*f*
in women, 5–6, 9*f*

fenvalerate
reproductive/developmental effects, 135, 135*t*
uses, 134, 219*t*

fetal development. *See* development, fetal

fetal resorption. *See* spontaneous abortion

fetal solvent syndrome, toluene and, 101

fetotoxicity
chloroform and, 86
lead and, 51
methylene chloride and, 93

FIFRA (Federal Insecticide, Fungicide, and Rodenticide Act), 236, 244–246, 261

fish
consumption, safety guidelines for, 276
hermaphroditic, 157
mercury in, 57, 61, 313
PCBs in, 228

fish advisories, 193*t,* 199, 222, 228

Florida, Lake Apopka, 157, 158

fluvalinate, 186

folate, 42

follicle-stimulating hormone (FSH)
control, 5, 9
phthalates and, 182
receptor interactions, 4, 5, 9
Sertoli cells and, 10

food
FDCA and, 253–254
pesticide residues in, 112–113, 114, 223. *See also under specific pesticides*

Pesticide Data Program and, 193*t,* 198–199

Total Diet Study and, 193*t,* 198–199

safety, guidelines for, 275–276

Food, Drug and Cosmetic Act (FDCA), 253–255, 261

Food Additives Amendment, 253

Food and Drug Administration (FDA)
FDCA and, 253–255, 261
pesticide residues in food and, 112–113
Total Diet Studies, 193*t,* 198–199, 221, 228

food chain
mercury and, 59*t,* 312–313
pesticides in, 112–114, 220–222, 220*t,* 221*t*

Food Quality Protection Act (FQPA), 113, 153, 245, 247–248, 263

formaldehyde
production, 213*t*
release, 213*t*
reproductive/developmental effects, 89–90, 333*t*
spontaneous abortion and, 77*t*

FQPA (Food Quality Protection Act), 153, 245, 247–248, 263

Freedom of Information Act (FOIA), 261–262, 268–271

FSH. *See* follicle-stimulating hormone

fumigants. *See also specific fumigants*
agricultural usage, 219*t*
release, 218*t*
reproductive/developmental effects, 145–149, 146*t*

fungicides. *See also specific fungicides*
in building/finishing products, 279
endocrine disruption effects, 185–186
environmental characteristics, 109*t*
release, 218*t*
reproductive/developmental effects, 136–138, 137*t*
uses, 136, 219*t*

furniture refinishing, chemicals used in. *See* methylene chloride; N-methyl-2-pyrrolidone

medical devices, FDCA and, 254
medical history, 295–296
medical waste, 181. *See also* dioxin; mercury
Medline, 303
memory impairments, 104
menstrual abnormalities
  benzene and, 84
  formaldehyde and, 90
  mercury and, 62
  phenol and, 97
  styrene and, 99
  toluene and, 84
  xylene and, 84
mental retardation
  lead and, 51
  PCBs and, 155, 166
mercury
  biological monitoring tests, 302*t*
  breast milk levels, 211*t*
  cycle, 59*f*
  dietary intake, 209–210
  elemental vapor, 58, 58*t*, 334*t*
  environmental concentrations, 208
  environmental exposures, 312–313
  fetal exposure, 16
  inorganic, 58*t*, 61–62, 334*t*
  organic, 57–61, 58*t*
  production/release, 208*t*
  reproductive/developmental effects, 51, 57–61
  routes of exposure, 266
  uses, 58*t*
metallothionein, 63
metals. *See also specific metals*
  absorbed dose, 210, 211*t*
  biological monitoring tests, 302*t*
  in breast milk, 211, 211*t*
  environmental concentrations, 207–208
  human exposure, 207–212, 208*t*, 211*t*
  production/release, 207, 208*t*
  reproductive/developmental effects, 51, 334*t*

metam sodium (sodium N-methyldi-thiocarbamate)
  release, 218*t*
  reproductive/developmental effects, 145, 146*t*, 148, 334*t*
  uses, 219*t*
methoxychlor
  endocrine disruption effects, 183, 184*t*
  environmental characteristics, 109*t*
  in food supply, 223
  in indoor air/house dust, 223*t*
  reproductive/developmental effects, 132*t*, 133, 335*t*
  uses, 130, 219*t*
methyl bromide
  biological monitoring tests, 302*t*
  release, 218*t*
  reproductive/developmental effects, 145, 146*t*, 147–148, 334*t*
  uses, 219*t*
methylcyclopentadienyl manganese tricarbonyl (MMT), 69–70
methylene chloride, 213*t*
  biological monitoring tests, 302*t*
  human intake, 214
  reproductive/developmental effects, 77*t*, 92–93
methylisothiocyanate (MITC), 148
methyl-mercury. *See* mercury, organic
N-methyl-2-pyrrolidone (NMP), 94–95, 213*t*, 333*t*
metolachlor, 218
metribuzin, 139, 144
microencephaly, lead and, 51
Minamata Bay, Japan, 60, 61*f*, 261
miscarriage. *See* spontaneous abortion
MITC (methylisothiocyanate), 148
MMT (methylcyclopentadienyl manganese tricarbonyl), 69–70
molinate, 109*t*, 139, 144
monitoring tests, biological. *See* bio-monitoring
Montreal Protocol, 147
MSDS. *See* Material Safety Data Sheet
Müllerian inhibiting substance (MIS), 11
Multigenerational studies, 28

environmental, 74
  human, 212, 213t, 214–217
  risks from, 73–74
  physical properties, 74–75
  production/release, 212, 213t
  reproductive/developmental effects,
    334t
    in animals, 73
    birth defects, 76, 78t, 83
    childhood cancer, 80, 82–83
    epidemiologic studies, 83
    infertility, 76, 79
    low birth weight, 79
    in men, 80, 81t
    preeclampsia, 79
    spontaneous abortion, 73, 75–76
  types of, 74
  volatile organic compounds. *See* volatile organic compounds
organochlorines
  bioaccumulation, 131
  breast cancer and, 160–161
  endocrine disruption effects, 183,
    185
  environmental persistence, 131
  in fish, 222
  in human fat, 225, 225t
  in indoor air/house dust, 223t
  release, 218t
  reproductive/developmental effects,
    130–131, 132t, 133–134
  uses, 130, 219t
organogenesis, 11–12
organophosphates
  biological monitoring tests, 302t
  in indoor air/house dust, 223t
  release, 218t
  reproductive/developmental effects,
    126–127, 128t, 129
  symptoms of intoxication/exposure,
    126–127
  uses, 126, 219t
OSHA. *See* Occupational Safety and
  Health Administration
OSHAct (Occupational Safety and
  Health Act), 251–253, 262
Our Stolen Future, 151

ovaries
  functions of, 6, 9
  structures in, 9–10
  toxic chemical effects on, 14–15
ovulation, malfunction of, 14

PAHs (polycyclic aromatic hydrocarbons), 15, 157
paint, lead-based, 272–273
paraquat, 141t, 218t
parathion
  agricultural usage, 219t
  biological monitoring tests, 302t
  environmental characteristics, 109t
  in human urine, 224t
  reproductive/developmental effects,
    127, 128t, 129, 335t
Parkinsonian syndrome, manganese
  and, 72
parks, pesticides in, 288–289
passive surveillance, 40
PCBs. *See* polychlorinated biphenyls
pentachlorophenol (PCP)
  biological monitoring tests, 302t
  endocrine disruption effects, 183,
    184t, 185
  environmental characteristics, 109t
  in human urine, 224t, 225
peptide hormones, 4
perchloroethylene (PCE)
  absorbed dose, 216
  biological monitoring tests, 302t
  production/release, 213t
  reproductive/developmental effects,
    76, 77t, 81t, 95–97, 333t
  routes of exposure, 95, 215, 286–
    287
  uses, 95
permethrin
  endocrine disruption effects, 186
  environmental characteristics, 109t
  in food supply, 221
  reproductive/developmental effects,
    135t
  uses, 134
persistence, environmental. *See* environmental persistence

Pesticide Data Program (PDP), 193*t*, 198–199

pesticides. *See also specific pesticide or class of pesticide*
absorbed dose, 223–225, 224*t*
active ingredients, 107, 108
biological monitoring tests, 302*t*
in breast milk, 225, 226*t*
classifications, 107
endocrine disruption effects, 183, 184*t*, 185–187
environmental concentrations, 217–218, 220
fate and transport, 109–112
in food supply, 112–114, 220–222, 220*t*, 221*t*
in household products, 276–279
human exposure, 112–114, 217–218, 218*t*–221*t*, 220–226
inert ingredients, 107, 108
information resources, 329–330
labeling instructions, 112
multipathway exposures, 222–223, 223*t*
persistence, environmental, 109–111, 109*t*
production/release, 217, 218*t*, 219*t*
registration and regulatory processes, 107–108, 124, 244
reproductive/developmental effects, 114–115, 334*t*
in animals, 123–124
birth defects, 117*t*–118*t*, 119–120
childhood cancer, 120–121
chromosome abnormalities, 122–123
epidemiological studies, 115, 116*t*–118*t*
fertility problems, 114–115, 115*t*, 118–119
low birth weight, 117*t*–118*t*, 119–120
spermatotoxicity, 121–122
spontaneous abortions, 114–115, 115*t*, 118–119
reregistration, 244–245
routes of exposure, 113–114

TRI list, 125
uses, 112, 217, 219*t*
warning labels, 246
water solubility, 109*t*, 110

phenol
biological monitoring tests, 302*t*
production/release, 213*t*
reproductive/developmental effects, 97–98, 333*t*
routes of exposure, 97
uses, 97

phenotype, 3

photochemical assessment monitoring stations (PAMS), 212

phthalates, 181–182, 335*t*

physician
consulting, 293–294
evaluation of occupational/environmental concern, 295–298

phytoestrogens, 167, 187–188

pink disease, 62

pituitary hormones, 4, 5, 6. *See also specific pituitary hormones*

placental toxicity, of cadmium, 64

plastics. *See* bisphenol-A; cadmium; nonylphenol; phthalates; polyvinyl-chloride; styrene; trichloroethylene

playgrounds, exposures in, 288–289

politics, scientific objectivity and, 309–310

pollution
community concerns, 283–284
information sources, 284
researching, 284–286

Pollution Prevention Act, 236, 263

polybrominated biphenyls, 229

polycarbonate plastics, 180

polychlorinated biphenyls (PCBs)
absorbed dose, 229
Ah-receptor and, 173
biological monitoring tests, 302*t*
in breast milk, 204–205, 229
endocrine disruption effects, 155, 166, 175–176
environmental concentrations, 228
epidemiological studies, 177–179
in fish, 228

resmethrin
  endocrine disruption effects, 186
  reproductive/developmental effects,
    135–136, 135t, 335t
  uses, 134
Resource Conservation and Recovery
  Act (RCRA), 236, 263
respiratory distress syndrome, cadmium and, 51–52, 63, 65
retrospective cohort studies, 36
RfD (reference dose), 61, 124
right-to-know laws, 256–260, 311–
  312, 327
risk assessment, quantitative. See quantitative risk assessment
Ronilan. See vinclozolin

schools
  exposures in, 292–293
  information resources for, 330
science
  limits of, 26–27
  objectivity of, 309–310
  soundness of, 310–311
  tools of, 45–48
scientific method
  issue of proof and, 21–23
  limits of, 26–27
scientific objectivity, 27
scoliosis, 72
second messengers, 154
segment studies, 28
seizures, toxins associated with, 62,
  79
semiconductor manufacturing. See cadmium; epichlorohydrin; glycol
  ethers; N-methyl-2-pyrrolidone
Sertoli cells, 10, 15
Seveso, Italy, 39, 174
Sevin. See carbaryl
sex hormones. see also estrogen; progesterone; prolactin; testosterone
  xylene and, 104–105
sex ratio alterations, 18, 60
sex-hormone-binding globulin
  (SHBG), 155
sexual differentiation, 11–12, 13f

Seychelles Islands, mercury exposure
  in, 60
SIDS (sudden infant death syndrome;
  respiratory distress syndrome), 51–
  52, 63, 65
Silent Spring (Carson), 110–111, 130,
  151, 261
simazine
  in drinking water, 218
  release, 218t
  reproductive/developmental effects,
    140, 141t
  uses, 139, 219t
skeletal abnormalities, manganese and,
  72
skin absorption, pesticide exposure
  and, 222–223
smelters, toxic exposures from. See arsenic; lead
smoking, cadmium exposure and, 63
sodium N-methyldithiocarbamate. See
  metam sodium
soil binding, of pesticides, 110
solvents, hydrocarbon-based. See organic solvents
species differences, in toxicological
  studies, 29
sperm
  counts, endocrine disruptors and,
    164–165
  production, 10
  quality, of perchloroethylene-
    exposed men, 80
  toxic effects on, 15
spermatotoxicity
  benzene and, 84
  cadmium and, 63
  chloroform and, 86, 88
  epichlorohydrin and, 88–89
  formaldehyde and, 89
  fumigants and, 145
  glycol ethers and, 92
  herbicides and, 139
  lead and, 55
  manganese and, 70
  methylene chloride and, 93
  pesticides and, 121–122

definition of, 74
environmental concentrations, 197, 212, 213*t*
human exposure, 212, 213*t*, 214–217
TEAM studies, 193*t*, 200–201
volatility, of pesticides, 110
vulnerability, critical windows of, 33

water
benzene in, 84
bottled, 83, 87
discharges, 285
disinfection by-products, 87–88
filters, 275
home testing of, 273–275
lead in, 208
mercury in, 208
monitoring programs, 193*t*, 197
pesticides in, 217–218, 220
Safe Drinking Water Act, 153, 236, 247, 250–251, 262, 274
Safe Drinking Water and Toxic Enforcement Act (California Proposition 65), 258–259, 263
TCE in, 103
water chlorination by-products, 85, 88–89
welding, toxic exposures from. *See* cadmium
wildlife advisories, 193*t*, 199
Woburn, Massachusetts, 82–83
women
infertility. *See* infertility, female
menstrual abnormalities. *See* menstrual abnormalities
pregnancy. *See* pregnancy
wood products, arsenic in, 66–67
workers' rights advocates, 268
Working Group on Community Right-to-Know, 327
workplace exposures. *See also* Occupational Safety and Health Administration
assessment of, 289–293
Coalition for Occupational Safety and Health, 324

employee safety guidelines for, 290–291
monitoring, 301
National Institute for Occupational Safety and Health, 251, 301
office workers safety and, 292
physician questions on, 296
right to know and, 289–291
schools, day care centers, public buildings and, 292–293
World Health Organization (WHO), 40

xylene
absorbed dose, 216
biological monitoring tests, 302*t*
production/release, 213*t*
reproductive/developmental effects, 77*t*, 104–105, 334*t*
routes of exposure, 104, 214
uses, 104

Y chromosome, 11

zineb, 137*t*, 138